**Dictionary of
Food Microbiology**

Dictionary
of Food
Microbiology

HANNS K. FRANK

TECHNOMIC
PUBLISHING CO., INC.

LANCASTER · BASEL

Dictionary of Food Microbiology

a **TECHNOMIC** publication

Published in the Western Hemisphere by
Technomic Publishing Company, Inc.
851 New Holland Avenue, Box 3535
Lancaster, Pennsylvania 17604 U.S.A.

Distributed in the Rest of the World by
Technomic Publishing AG
Missionsstrasse 44
CH-4055 Basel, Switzerland

Printed in the United States of America
10 9 8 7 6 5 4 3 2 1

Main entry under title:
  Dictionary of Food Microbiology

Library of Congress Catalog Card No. 92-62872
ISBN No. 1-56676-010-0

This Technomic Publishing Company, Inc. edition is produced under license between
Technomic Publishing Company, Inc. and B. Behr's Verlag.

Foreword

Professor Frank with his *Dictionary of Food Microbiology* has created a source for basic information for all decision makers working in a laboratory or in the field of food science and food law. This reference book with its preciseness and numerous explanations of specific terms should be a first choice when looking for information and orientation.

I hope that this book which fits well into the list of publications of the Behr's publishing company, will be received well by the readers. It may contribute by giving the interested public an understanding of food microbiology and hygiene.

Berlin. May 1989   H.-J. Sinell

The Author

From 1949 to 1953, Prof. Dr. rer. nat. Hanns K. Frank studied biology, chemistry and physics in Munich. He received his doctorate at the Botanical Institute of the University in Munich under Prof. O. Renner for whom he worked as scientific assistant. In 1966, he qualified as a university lecturer at the Technical University in Munich where he gave lectures until 1969. Subsequently he lectured in Heidelberg until 1972 and then in Karlsruhe until 1987.

From 1955 to 1960 he headed the botanical department and the cultivation department for medical plants in a pharmaceutical company. From 1960 to 1965 he worked as a scientific employee at the Southern German Board for Dairy Research in Weihenstephan and from 1965 to 1987 on the Federal Board for Nutrition in Karlsruhe (formerly named Federal Board for Preserving Food Freshness). Here he was appointed head of the Biological Institute in 1970. From 1976 to 1978 he was elected chairman of the Senate of the Federal Boards of the BML. He published 152 scientific works and worked several times in Brazil, Peru, Columbia and India on the post harvest protection of agricultural products on the order of the GTZ and other organisations for foreign aid. Prof. Frank is a bearer of the Order of the Federal Republic of Germany.

Preface

This dictionary is based on lectures on applied and food microbiology for process engineers and food technologists which the author had given for several years. For selection of the keywords, indexes of relevant educational and specific textbooks of adjacent fields such as toxicology, technology, etc. were consulted.

However, a lexicon-type reference book is not intended to substitute for a specific textbook which presents the facts with a more comprehensive view based on different didactics. It should, rather, enable non-microbiologists working with foods, as well as technologists, food chemists, nutritionists, economists, and last but not least the technical staff, to understand specialised literature. It should provide useful information for the physician in regard to food poisoning and food infections. And it should support the jurist in finding his way through the unknown terminology when confronted with expert reports.

Besides the general microbiological facts, related toxicological and medical problems, maximum values, and other regulations have been taken into consideration to help avoid difficulties and complications in companies and during distribution, and to prepare as much as possible for the EC-year 1992.

The names of the microorganisms are to a large degree cited according to the currently valid nomenclature. Often used synonyms from the applied literature are set in brackets. However, in view of the eagerness of taxonomists and the rapid development in systematics, it did not always succeed. The tables found under the keyword "synonyms" should help to locate the valid names. Often used abbreviations in the text are listed after the Preface. An alphabetical index stating names of authors is available at the end of the book for further reference. Only in exceptions are original works cited directly under the keywords.

I would like to thank Mr. Benecke who initiated this dictionary. My special thanks are due to my wife for her patience during the preparation of the manuscript and her help in proofreading. I appreciate and would like to encourage constructive criticism and information on missing keywords.

Ettlingen, June 1989   Hanns K. Frank

Preface to the English Edition

Luckily the Behr's Verlag was able to win an excellent expert for the English translation of the first edition of the *Dictionary of Food Microbiology*. Prof. Wilhelm Holzapfel grew up in South Africa where he went to school, too. He studied microbiology, dairy technology and chemistry at the Agricultural Facility in Bloemfontein. He received his promotion in microbiology in Munich at the Technical University under Prof. Kandler. Until his move to Germany he held a chair for microbiology at the University in Pretoria. Since l987 he heads the Institute for Hygiene and Toxicology at the Federal Center for Nutrition in Karlsruhe (Germany).

It is always a certain risk to translate a dictionary destined for a specific language and legislative area into another language. This does not concern the scientific facts, but rather, the legislative statements associated with foods which inevitably are different on a national level and may change from time to time. This is especially true for the EC region where a harmonisation in the field of food law is not yet completed. Thus it should be stressed that this dictionary is not intended to provide quick references for valid hygiene regulations. However, it should assist the user who might not be very familiar with food microbiology such as a food chemist, food technologist, nutritionist or the staff in a laboratory or in companies producing or processing foods.

Ettlingen, March 1992   Hanns K. Frank

Table of Contents

Foreword 5

The Author 7

Preface 9

Preface to the English Edition 11

Abbreviations 15

Dictionary of Food Microbiology A – Z 17

References 293

Abbreviations

a	year (annum)	LAB	lactic acid bacteria
A.	*Aspergillus*	*Lb.*	*Lactobacillus*
ADI	Acceptable Daily Intake in mg/kg BW	LMBG	"Lebensmittel- und Bedarfsgegenständsgesetz"
a$_w$	water activity	MAK	maximum working place concentration
B.	*Bacillus*		
BGesBl.	Bundesgesundheitsblatt	MW	molecular weight (mol mass)
BSG	Bundes-Seuchengesetz	min	minute(s)
b.p.	boiling point	MO	microorganism(s)
BW	body weight	OECD	Organisation for Economic Cooperation and Development, Paris
CAC	Codex Alimentarius Commission of the FAO/WHO		
C + D	cleaning and disinfection	*P.*	*Penicillium*
CH	carbohydrate	ppb	parts per billion; 1:10^9
Cl.	*Clostridium*	ppm	parts per million; 1: 10^6
conc.	concentration	*Ps.*	*Pseudomonas*
d	day(s)	rH	relative humidity
DM	dry matter	s	second(s)
DW	dry weight	sp.	species (singular)
EC	European Commission	spp.	species (plural)
FAO	Food and Agriculture Organisation, Rome	ssp.	subspecies
		Staph.	*Staphylococcus*
GI	gastro-intestinal (tract)	*Str.*	*Streptococcus*
GLP	Good Laboratory Practices	t	ton(nes)
GMP	Good Manufacturing Practice(s)	temp.	temperature
		TrinkwV	"Trinkwasserverordnung"; drinking water regulation
Gram +	Gram-positive		
Gram −	Gram-negative	var.	variety
h	hour(s)	VO	"Verordnung"; regulation
IAEA	International Atomic Energy Organisation, Vienna	WHO	World Health Organisation, Geneva
IU	international units	ZZulV	"Zusatzstoff-Zulassungsverordnung"; food additive regulation
K	degrees Kelvin		
kGy	kilo-Gray; measured irradiated dose; 1 kGy = 0.1 Mrad = 1 J/g		

A

α-**amylase**
(E.C. 3.2.1.1). A hydrolase (endoamylase) that cleaves the α-1,4-glycosidic bonds in oligosaccharides, i.e., of sugars with 3 or more glucose units. Dextrins are produced from starch. Trivial names: dextrinogen amylase, glycogenase, diastase. Systematic name: 1,4-α-D-glucan glucanhydrolase. Present in a wide range of plant and animal materials, bacteria and fungi, from which it may be produced commercially. – Uses: hydrolysis of starch to dextrins, e.g., in the production of glucose syrup (corn syrup) and alcohol; processing of cereals, pectin extraction, processing of fruit juices. – MO used for production: *Bacillus licheniformis, B. stearothermophilus, B. subtilis, Aspergillus niger, Asp. oryzae, Rhizopus delamar, Rh. oryzae* (GDCh 1983).

β-**amylase**
(E.C. 3.2.1.2). An exoamylase that hydrolytically cleaves oligo- or polysaccharides to produce maltose units from the non-reducing end of the chain. Trivial names: saccharogenic amylase, diastase, glycogenase. Systematic name: 1,4-α-D-glucan maltohydrolase. – Uses: production of maltose and maltose containing products from polysaccharides. – MO used for production: *Bacillus cereus, B. megaterium, B. subtilis* (GDCh 1983).

abiotic spoilage
Adverse chemical changes in foods. Expression common in food processing when reference is made to non-microbial spoilage, although microbial and product specific enzymes may be involved.
– hydrolytic changes, e.g.,→degradation of sugars, lipids or proteins
– oxidative changes, e.g., fat oxidation (rancidity)
– Maillard reactions (non-enzymatic

browning), formation of melanoids (brown polymers, containing N) in plant products, milk powder, etc.
After inactivation or destruction of microbial cells their enzymes can still be involved in these reactions. Example: breakdown of lipids by bacterial lipases at – 18 °C (Heiss and Eichner 1984).

Absidia
Genus (→ Mycota) belonging to the Zygomycetes, Mucoraceae, with asepate hyphae. Sporangia (→Sporangium) are pear-shaped. Mainly saprophytic, found in soil in association with vegetable matter and grains, and also on meat. Optimum temp. 35 – 37 °C. *A. corymbifera, A. italiana, A. lichtheimii* and *A. ramosa* may cause skin→mycosis of humans and warm-blooded animals, or allergic reactions resulting from inhalation of cereal dust, flour, etc. (Gedek 1980; Reiss 1986).

acceleration phase
Adaptation phase. Stage between the→lag phase (practically no cell divisions occur) and the→exponential phase (regular/constant cell divisions) during the→multiplication (growth) of a bacterial population. Transition period during which an increasing number of cells become adapted to environmental conditions and start dividing.

accommodation disturbances
Paralysing reactions, released after oral intake of botulinum toxin (→botulism), affecting rhythmic muscle movement.

acetaldehyde
CH_3-CHO. Acetic acid aldehyde; ethanal. Liquid (m.p. – 123 °C; b.p. 20.1 °C) with a stinging smell and irritating influence on mucous membranes. Soluble in water and ethanol. Maximum working-place con-

centration (German: "MAK"): 90 mg/m^3 of air. In the alcoholic fermentation pyruvate is decarboxylated by pyruvate decarboxylase (E.C. 4.1.1.1) to acetaldehyde, which is then reduced to ethanol by alcohol dehydrogenase (E.C. 1.1.1.1). It may be present in cell-free fermenting must; and in wine from 20 to 275 mg/l. It combines with sulfurous acid (SO_2) in winemaking where it is undesirable; → sulphuration. During sherry manufacture ethanol is oxidised by the yeasts to acetaldehyde (Dittrich 1987). By the same mechanism acetic acid is produced by → acetic acid bacteria. Some lactic acid bacteria are able to form → acetoin from acetaldehyde. Acetaldehyde is of importance, either as a desirable or undesirable component in a number of fermented products.

acetate
Salt of → acetic acid.

acetic acid
Ethanoic acid. CH_3COOH. Colourless pungent smelling liquid. "MAK" value 25 mg/m^3 of air. Metabolically formed by species of *Acetobacter* and *Gluconobacter*, by dehydration of acetaldehyde:

$$CH_3CH_2OH + O_2 \rightarrow$$
$$CH_3COOH + H_2O + 118 \text{ kcal}$$

Produced by → acetic acid bacteria from alcohol containing liquid in the → submerged culture process ("acetator"), equipped with a special aeration system. Under sufficient aeration levels, the yield may be up to 97% of the theoretical (calculated) value. – Main applications in the chemical and pharmaceutical industries. Acetic acid also becomes available as by-product of mineral oil (Rehm 1985).

acetic acid bacteria
Acetobacter spp. Ellipsoid to rod shaped cells, single, in pairs or chains; motile by

Table 1 Typical oxidants by acetic acid bacteria (modified acc. to Rehm 1985).

Substrate	Product
ethanol	acetic acid
ethandiol	glycolic acid
glycerol	dihydroxyacetone
lactic acid	acetoin
D-2,3-butanediol	D-acetoin
meso-2,3-butanediol	L-acetoin
D-mannitol	D-fructose
D-sorbitol	L-sorbose*
D-glucose	D-gluconic acid
D-gluconic acid	D-5-ketogluconic acid
D-gluconic acid	D-2-ketogluconic acid
fructose	5-ketofructose

*Basic product for vitamin C production.

peritrichous flagellation, or immotile. Obligatory aerobic; catalase +; Opt.: 30°C and pH 5.4 – 6.3. – Strains used for biotechnical production mainly subspp. or variants of the few known spp. Common feature is the oxidation of a large number of organic substances (see Table 1), and especially ethanol to acetic acid. Tendency of several strains to "over-oxidation", i.e., acetate and even lactate may be end-oxidised to CO_2 and H_2O; such strains are undesirable for commercial acetic acid production. Some strains produce mucilagous substances, e.g., cellulose by *Ac. xylinum,* that may be responsible for a "cum" (→ zoogloea) on the surface of static liquids. – Occurrence on fruit and vegetables as surface microbes. – *Gluconobacter* (formerly *Acetomonas*): Gram –, catalase –, strictly aerobic rods or cocci, single in pairs or chains, commonly producing slime; typically motile by polar flagellation. Opt.: 25 – 30°C; pH 5.5 – 6. Glucose is oxidised to → gluconic acid, and ethanol to acetic acid. Over-oxidation uncommon. – Found on plants and in soil. May spoil beer or wine.

Acetobacter
Genus representing Gram-variable strictly aerobic rod-shaped bacteria ($\rightarrow$acetic acid bacteria) that are able to oxidise $\rightarrow$ethanol to$\rightarrow$ acetic acid, or even to CO_2 + H_2. The species *Ac. aceti* and *Ac. aceti* ssp. *xylinum*, and especially *A. acetigenum* and *A. schuezenbachii* are used for the commercial production of$\rightarrow$vinegar.

acetoin
CH_3-CO-HCOH-CH_3. 3-hydroxy-2-butanone; acetyl methyl carbinol. Metabolic product formed by some bacteria by reduction of$\rightarrow$diacetyl. Typical for *Enterobacter aerogenes*, but is absent in *E. coli;* $\rightarrow$coliforms. Determination:$\rightarrow$Voges-Proskauer test ($\rightarrow$IMVIC test). Some$\rightarrow$lactic acid bacteria (LAB), especially *Pediococcus* spp., produce acetoin, which may be oxidised to$\rightarrow$diacetyl (butter aroma) by acetoin decarboxylase (E.C. 1.1.1.5), and may cause off-odours in wine and beer (Dittrich 1987; Priest and Campbell 1987).

Acetomonas
Former name of those$\rightarrow$acetic acid bacteria that now belong to the genus$\rightarrow$*Gluconobacter* (Bergey 1986).

acetone
CH_3-CO-CH_3. Propanone. Metabolic product formed via acetate-aceto-acetate by some *Clostridium* spp. during aerobic degradation of carbohydrates;$\rightarrow$Figure 19, see page 235. Only traces (<0.3 mg/l) detectable in wines (Dittrich 1987).

Achromobacter
An obsolete bacterial genus in which several Gram-negative water associated spoilage bacteria were formerly grouped (Kunz 1988). Some representatives have now been designated to either$\rightarrow$*Acinetobacter* or$\rightarrow$*Alcaligenes*.

acid production
Incomplete oxidation of intermediate glycolysis products, etc. of MO, with the production of organic$\rightarrow$acids. Desired effect in many processes; enhanced by reducing the oxygen tension. Basis of practically all traditional biotechnical processes for shelf life increase of foods (fermented milk products, cheese, sauerkraut, etc.).

acid tolerance
Several MO are inhibited or even killed by acids ($\rightarrow$disinfectants;$\rightarrow$lactic acid). They are not only sensitive to free H ions ($\rightarrow$pH value) of the dissociated acids, but also to undissociated acids that may penetrate into the cell and inhibit important metabolic processes. The tolerance is dependent on the type of acid (Wallhäusser 1988) and the type of organism. The pH limits of some MO are summarised exemplary in Table 2, irrespective of the type of acid.

acidophilic bacteria
Bacteria that are able to proliferate well at pH values around 2 to 4, and may cause$\rightarrow$"flat sour" spoilage without gas production of high acid foods. Of special importance are the aerobic or microaerophilic species *Bacillus acidocaldarus* and *B. coagulans,* although some lactobacilli may also be involved. – Anaerobic acidophiles (*Thiobacillus* spp.) are of importance in corrosion and leaching.

Acidophilus
Abbreviated form of *Lb. acidophilus* ($\rightarrow$LAB), often used in product promotions. Produces DL-lactic acid. Distribution: mouth (caries), gastro-intestinal tract, vagina ("Döderlein's bacillus"). Included in some$\rightarrow$starter cultures for fermented dairy products.

acids (organic)
Most aerobic bacteria oxidise organic nutrients to CO_2 + H_2O, provided that sufficient O_2 is available ("complete

Table 2   Minimum and maximum pH values for growth of MO (acc. to Krämer 1987).

Microorganisms	Minimum pH	Maximum pH	Acid Tolerance
Micrococcus spp.	5.6	8.1	slight acid
Pseudomonas aeruginosa	5.1	8.0	tolerance
Bacillus stearothermophilus	5.2	9.2	$pH_{min} > 5.0$
Clostridium sporogenes	5.0	9.0	moderate acid
Bacillus cereus	4.9	9.3	tolerance
Vibrio parahaemolyticus	4.8	11.0	pH_{min} 5.0 – 4.0
Clostridium botulinum Type A, B	4.5	8.5	
Staphylococcus aureus	4.0	9.8	
Salmonella	4.0 – 4.5	8.0 – 9.6	
Escherichia coli	4.4	9.0	
Proteus vulgaris	4.4	9.2	
Streptococcus lactis	4.3 – 4.8	9.2	
Lactic acid bacteria			high acid
Lactobacillus spp.	3.8 – 4.4	7.2	tolerance
Acetic acid bacteria			$pH_{min} < 4.0$
Acetobacter acidophilus	2.6	4.3	
Yeasts			
Saccharomyces cerevisiae	2.3	8.6	
Moulds			
Penicillium italicum	1.9	9.3	
Aspergillus oryzae	1.6	9.3	

oxidation"). This respiration process provides the highest energy yield to the cell. Products of incomplete oxidation are organic compounds (acetic acid, fumaric acid, gluconic acid, keto-acids, citric acid, lactic acid, etc.) that are excreted by the cell (→oxygen demand). During→fermentation similar or even identical compounds are formed, however, under different metabolic conditions and from different MO, e.g., propionic acid, butyric acid, succinic acid, lactic acid, etc.,→sensorics. Fungi are aerobic with a strong glycolytic metabolism; under rich nutrient availability as may be found in foods (and in the laboratory or fermentor) blocking or "bottle-necks" may occur in intermediary metabolism. The "excessive" products, often org. acids, may be excreted in a modified form, e.g., ithaconic acid, citric acid, oxalic acid, oxalo-acetic acid, etc. Trace elements play an important role, either when present in minimal or excessive concentrations (Zn, Fe, Mn, Cu), a phenomenon man has made use of in biotechnology (Schlegel 1985). Under anaerobic conditions little growth takes place; however, ethanol (yeast fermentation), lactic acid (*Rhizopus* spp.) or fumaric acid (*Cunninghamella,* etc.) are formed. – Organic acids may also be released in foods by→lipases of microbial origin or intrinsic to a food. Fatty acids are thus hydrolytically released from fats (lipids). The number of released fatty acids depends on the number of different acids present in the product (fat) (e.g., more than 60 different types in butter). These fatty acids are mostly undesired, and deleteriously influence the sensory quality of the product. In the ripening of cheese some fatty acids are desirable.

Acinetobacter
Gram-negative, rod-shaped, psychrotrophic bacteria. Min. a_w for growth: 0.96. Distribution: soil, water, raw milk, surface of chilled meat and fish. *Ac. calcoaceticus* is an example of an → opportunist, and can grow in presence of 6.5% NaCl. Heat resistance: D_{65} 4.5 – 6.7 min (Krämer 1987; Kunz 1988).

Acremonium
(*Cephalosporium*). Genus of the fungi belonging to the Deuteromycetes. Teleomorphic examples: → *Nectria* and *Emericellopsis*. Saprophytes and plant parasites. During growth on foods cephalosporins (antibiotics) may be produced; this is undesired in view of the therapeutic use of Cephalosporin C (selection of resistant strains). The excretion of → trichothecenes has also been reported. – Cultivation on malt extract agar or oatmeal agar.

acrolein
CH_2CH-CHO. Acrylaldehyde; 2-propenal. Component of garlic with antibiotic properties. Irritating to skin and mucosa. Maximal working place concentration ("MAK"): 0.1 ml/m^3 air (alt. 0.25 ml/m^3 air) (DFG 1985; Roth 1989).

Actinomucor
Genus of the fungi belonging to the Zygomycetes. In addition to other fungi, *Ac. elegans* is used for the production of East Asian specialities, e.g., Sufu or Meitanza (Beuchat 1987; Krämer 1987).

Actinomycetales
Actinomycetes. Gram-positive, typically aerobic bacterial order, sometimes pigmented, immotile, often showing → mycelial growth with typical branching. Formic, acetic and lactic acid produced from glucose. Some spp. of → *Mycobacterium* and *Nocardia* are pathogens. Antibiotics produced by many representatives (e.g., tetracyclins, streptomycin), and some strains are implemented in large-scale production processes. → *Bifidobacterium* produces L(+)-lactic acid, and may be included in some dairy → starter cultures.

adaptation
Changes in some phenotypic features by means of which the growth or survival capacity of a MO is improved under changed environmental conditions, such as temp., pH, a_w, salt conc., available nutrients, etc. By contrast, genetic adaptation involves → mutation and selection within a given population. Physiological adaptation is an important factor related to growth and spoilage during processing and handling/distribution of food. This explains the importance of keeping the duration of different production steps as short as possible, even at marginal growth conditions for bacteria and fungi. Several fungi are "masters" in adaptation as a result of → heterokaryosis. Adaptation is a vital factor that may enable some "specialists" to settle in extreme habitats, such as → refrigerators, ripening rooms, → CA storage rooms → brine, etc. – Genotypic adaptation can be induced by sublethal treatments, e.g., by using toxic substances, irradiation, etc., so as to impose desired properties for purposes such as fermentation. This may also be the method of choice for the development of new → starter cultures, and also by making use of recombinant DNA technology for strain improvement: → genetically modified MO.

adaptation phase → acceleration phase of microbial growth.

additives
Colour additives and dyes, → preservatives, → stabilisers, antioxidants, vitamins, etc. According to §2 LMBG these are "substances added to foods with the aim

at influencing their condition or to bring about specific characteristics or effects''. Their use is subjected to the regulations on additives of 10.7.84 which also include hygiene specifications. The regulations on additives in the version of 20.12.84 cover their quantitative and qualitative use. The→technological necessity and absence of any→health risk (unobjectionability) is to be proved for every additive prior to its approval (Hahn and Muermann 1987).

adenosine triphosphate

(ATP). An important component in cell metabolism responsible for the supply of phosphate and the transport of energy stored in two high-energy phosphate bonds. Adenosine diphosphate (ADP) and adenosine monophosphate (AMP) provide substantially less energy (Alberts et al. 1986; Schlegle 1985).

adhesins

Cell surface adhesion components of some bacterial spp. of which the composition and structure are not completely known yet; some may be thin filamentous protein structures, projecting from the cell surface. They are important in the colonisation of the intestinal mucous membranes, e.g., by→enteropathogenic *E. coli* [Bockemühl, J.: *BGesBl.* 26 (1983) 316 – 322], as well as of the mucous membrane of the mouth cavity (→nitrosamines) and the teeth (→caries). These components also facilitate the adhesion to wet inanimate surfaces such as glass, ceramics and synthetics (→utensils), as well as for the successful colonisation of fixed-bed bioreactors (→solid-state fermentation); (→surface growth).

ADI value

Acceptable Daily Intake. By definition it is the amount of a substance that, according to existing scientific knowledge, can be taken in by a human adult on a daily basis through his whole life, without any detrimental effects to his health. The ADI value is calculated from the amount determined in an animal experiment that caused no harmful effects (No-Adverse-Effect-Level, NOEL). Based on worldwide experience in modern→toxicology, this value is normally divided by 100. In this way uncertainties in the extrapolation of data from an animal experiment to humans are taken into consideration; at the same time it should serve as protective measure towards sensitive consumer groups such as children, and sick and aged persons. As shown in Figure 1, the maximum theoretical uptake and the probable daily uptake are at least one order of magnitude below the ADI value. – An ADI value is determined after critically evaluating and assessing data of investigations by scientific institutions such as the Ministry of Health, the German Research Foundation, the FDA and USDA, etc.; internationally by JECFA (Joint Expert Committee on Food Additives), the toxicological committee of the WHO/FAO, or the relevant expert committee of the EC.

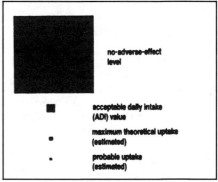

Figure 1 Relationship between ADI value and non-harmful level in animal experiments – on the one hand – and actual daily uptake – on the other hand (source: BLL, Bonn).

aerial mycelium

A cottonwool-like fungal→mycelium projecting above the level of the substrate and usually partly bearing the reproductive structures (conidia, sporangia, etc.) in contrast to the→substrate mycelium responsible for nutrition (Müller and Loeffler 1982).

Aerobacter

Obsolete (rejected) bacterial genus, most representatives are now incorporated in the genus *Enterobacter.* "*Ae. aerogenes*" strains growing at >37°C are sometimes classified as→*Klebsiella pneumoniae.* Because of gas production from lactose in presence of bile or other inhibitory substances, it is included with *E. coli* in the→coliform group,→indicator organisms, →coli titre. Its detection plays an important part in the→official methods for the microbiological examination of foods, and of drinking and household water.

aerobe

A MO requiring oxygen for growth. Opposite:→anaerobe. Includes most spoilage bacteria and fungi. Obligately aerobic MO only grow in presence of molecular oxygen, e.g.,→acetic acid bacteria. Facultative anaerobes refer to aerobes that can also grow under anaerobic conditions, and possess haemines (cytochromes and catalase); they may generate energy by respiration (aerobic) or fermentation (anaerobic). Typical examples are the *Enterobacteriaceae.* – Facultative aerobes may grow in presence of low oxygen pressure. – Original definitions suggested by Louis Pasteur.

Aerococcus

Gram-positive bacterial genus of the family Streptococcaceae (Bergey 1986). Nonmotile, producing L(+)-lactic acid homofermentatively from glucose. Significance for food processing not clear.

Aeromonas

Genus of Gram-negative facultatively anaerobic bacteria. Morphology: straight rods to coccobacilli, arranged singly, in pairs and short chains. Represented by non-motile psychrotrophs (e.g., *A. salmonicida*) and motile mesophiles (e.g., *A. hydrophila*). Mainly associated with surface waters, sewage and aquatic animals. *A. hydrophila* may be transferred by contaminated water and may cause→opportunistic acute forms of diarrhoea in man. An→enterotoxin, similar to that of *Vibrio cholerae* has been isolated (Sinell 1985). Selective medium for isolation: e.g., *Aeromonas* nutrient medium base (Ryan) + Ampicillin (Oxoid).

aeroplankton

Collective designation for solid particles of biological origin dispersed in the atmosphere: e.g., bacteria, fungus hyphae and spores, algae, pollen, fragments of plants or insects, hair of animals, mites, viruses, etc. A source of→contamination and cause of→allergies (Schachta and Jorde 1986).

aerosol

Minute drops or solid particles dispersed in air. May contain pathogenic MO or fungus spores that may cause infection diseases or allergies when inhaled. In processing plants aerosols may be caused especially by high-pressure or steam cleaning procedures (→sanitation).

aerotolerance

Tolerance of anaerobic bacteria towards reduced oxygen concentrations in the atmosphere or substrate (<1%). This property may have hazardous consequences if associated with clostridia.

aflatoxins

Mycotoxins of→*A. flavus* and *A. parasiticus*, discovered in 1960. Especially the aflatoxins B_1 and G_2, and to a lesser extent

B_2 and G_2, are formed during growth of these fungi on food and feeds. *A*. are resistant to heat and acids, but may be rapidly inactivated by alkalis and strong oxidatives. Aflatoxin B_1 is the strongest hepatocarcinogen known, acting orally (10 μg/kg rat). About 0.3% of aflatoxin B taken in by cattle is excreted as aflatoxin M in milk ($\rightarrow$carry-over). Feed-legislative tolerance level (EC 1986) is 10 μg/kg of feed for lactating cows. – In most industrialised countries the following regulations exist for maximum amounts tolerable in food: 5 μg of B_1/kg, and 10 μg of total aflatoxins/kg whereby B_1 should not exceed 5 μg. It is expected that a maximum level of 50 ng/kg for M_1 will become mandatory regulatory for milk and dairy products in Germany, and 10 ng/kg for baby foods. – Distribution: feed products from tropical regions, especially oil seed press cake; peanuts, pistachio, maize (corn), cereals, etc.; as well as dried figs and ground nutmeg. Rarely found in commodities produced in Europe (Frank 1974; Reiss 1981, 1986). Detection: e.g., by thin-layer chromatography; B and G up to >0.1 μg/kg [BGA: *BGesBl.* 18 (1975) 230 – 233]; M_1 >3 ng/kg ($\rightarrow$official methods). Quantitative and rapid semi-quantitative tests are commercially available.

agar (Malay: agar-agar)
Sulfur-containing polysaccharide (galactan) produced by several marine rhodophycean algae and commercially obtained from the genera *Gelidium, Euchema, Gracilaria,* etc. Used since Robert Koch as gelatinisation base for solid and semi-solid nutrient media in microbiology. By contrast to$\rightarrow$gelatine, its degradation (liquefaction) is rare, and is practically confined to a few marine MO. Normally 1.5 to 2.5% of agar are added to nutrient broths. It liquifies at ca. 95°C (which is reached early during$\rightarrow$autoclaving), and solidifies around 43°C. Agar may be an ingredient of commercial dehydrated culture media. Highly purified agar may be practically nitrogen-free (e.g., Merck cat. no. 1613), and is used for specific purposes, e.g., metabolic studies.

agar block method
A microcultural method applied in the$\rightarrow$identification of moulds. A square (6 × 6 mm, ca. 2 mm thick) of nutrient agar is placed aseptically on a microscope slide, and inoculated on the 4 sides. The block is covered with a cover-slip, placed onto two glass rods and incubated in a moist petri dish. The cover-slip is removed from time to time and the formation of typical organs (conidiophores, etc.) examined microscopically and measured.

Agaricus$\rightarrow$mushroom.

agaritines$\rightarrow$mushroom.

agglomeration
Clumping or heaping together of objects, e.g., after division of MO. May be an important mistake factor in the determination of colony-forming units ($\rightarrow$cfu) by plate count methods, e.g., an agglomerate of 100 cells or a single cell would each give rise to only one colony on an agar plate. Bacteria with a strong tendency towards agglomeration: staphylococci, *Sarcina,* and chain-forming strains of *Streptococcus* and$\rightarrow$*Lactococcus*.

agglutination
Clumping; formation of insoluble aggregates of viable cells in liquid media, e.g., as the result of antibodies reacting with cells (antigens). Serological reaction commonly used for the identification of pathogenic bacteria, e.g., latex particle test used for group D streptococci (*Enterococcus* spp.) (Streptex, Wellcome Co., Slide-Strepto-Kit, bioMerieux) (Baumgart 1990).

AIDS
Acquired immuno deficiency syndrome.

When first recognised in 1981 no reliable diagnostic procedure was available; observed in Europe since 1982. Transmissible, fatal human disease that affects the immune system, and is caused by an exogenous retrovirus (HIV virus). Enables infections by→ opportunists (viruses, bacteria, chlamydia, protozoa). Several of these opportunists, but not HIV, may be transmitted by foods [Brede, D. and Keller, H.: "HIV-Skriptum". *Forum Mikrobiologie* 11 (1988) 5, etc.].

air examination
Air should be tested regularly in areas where there is a constant "rain" of MO during activities of personnel. A disadvantage of all the methods is that the results are only available 1 – 2 days after the sample has been taken following culturing. – In the simple sedimentation method, plates with Standard I culture medium are placed uncovered for a period of 2 – 10 min and then incubated at 25 °C. A disadvantage is the considerable air movement so that no reliable estimation of the bacterial numbers is possible. – In the impact method a special device with adjustable volume is used to draw in a given volume of air which is blown onto a medium. A number of MO adhere to the medium. It is available at e.g., Zinsser-analytic, Frankfurt a.M.→ Table 26, see page 171, under mould counts in households (Baumgart 1990). The highest counts are to be found in flour processing industries and hairdressing salons. – The examination of the fungal counts and specific species present are of importance in the search for inhalative allergens (Schachta and Jorde 1959).

air filter
→ Bacterial filters for retaining MO from aeration gases for→ fermentation reactors, operating theatres or areas in food processing plants that are prone to aerial contamination. → Biofilters are used for vitiated air.

air supply
Air exchange by which fresh air is introduced from outside. It is cleaned and dried or humidified. If a specific contamination risk exists in a particular industry, e.g., cheese manufacturers or meat plants, special care should be taken that no unfiltered air from the environment (e.g., where MO may be emitted from sources such as breweries or rubbish heaps) enters the system. → Air filter.

airlift fermenter (loop fermenter)
Continuously operating submerse reactor in which constant circulation of the culture and substrate is effected by air rising from a draft tube. Applied (a.o.) for the production of→ single cell protein (Kunz 1988).

airline food
On-board catering. Specific part of community care, referring to the provision of meals on board during a flight. – Interruption of the cold chain, uncontrolled warm keeping at >60 °C, different locations of preparation, → excretors among the staff, insufficient cleaning of→ utensils, etc., may be responsible for outbreaks of→ food-borne infections (→ food intoxications) after a flight. Multiple causes may be responsible (see Table 16, page 117). The main causative agents include→ *Salmonella* spp., → *Staph. aureus*, and→ *Vibrio* spp. Foods most often implicated are egg salad, fish savouries, poultry, meat pastries and sweet dishes. – From 1961 to 1978 a total of 102 civil flights were involved in several thousand registered cases of food poisoning and 9 deaths (Sinell 1985).

Alcaligenes
A genus (*Incertae sedis*) of Gram-negative, mainly rod-shaped, motile bacteria; obligatory aerobic, cytochrome-oxidase

and catalase positive. NO_3^- may serve as hydrogen acceptor for some spp. Ubiquitary, e.g., in soil, water, gastro-intestinal tract and milk. Often survive pasteurisation and involved in spoilage of a number of foods. Alcalinisation of protein-rich substrates. *Al. viscolactis* grows well between 10 and 32°C, and may cause ropiness in milk. *Al. faecalis* (type species) is a typical inhabitant of the gastro-intestinal tract, reduces nitrite but not nitrate, and has an optimal growth temperature between 25 and 37°C.

alcohol-free beverages → non-alcoholic beverages.

alcohols
Derivatives of hydrocarbons with one or more hydroxyl groups. Anti-microbial action, denaturation of protein in vegetative cells. → Endospores not damaged. Concentration in alcoholic beverages too low to inhibit yeasts or bacteria. – Ethanol used for the preservation of plant extracts and fruit. Most effective concentration for rapid disinfection in laboratory is 70%; lower effectivity of propanol, isopropanol, butanol and amylalcohol, however, with optimum concentration also around 70%. Methanol not suitable for decontamination (Wallhäusser 1987).

aldehydes
Important group of → disinfectants; preservatives for non-food commodities. Formaldehyde and its two polymers paraformaldehyde and trioxane show highest effectivity. Glutaraldehyde has a more limited application spectrum; its dialdehyde has been verified as stabilising and amplifying additive to formulations containing other preservatives. → Formaldehyde (formalin) rapidly kills all MO during disinfection of surfaces; activity increases with rising temperature (Wallhäusser 1987). – Caution! Severe irritation of mucous membranes. "MAK"

value 1 ppm. Tolerance limits for irritation and odour offensiveness: 2 – 5 ppm. Carcinogenic action has not been proven yet (Classen et al. 1987). – Aldehydes are formed in many fermentations, either as detrimental or beneficial aroma and taste components (Dittrich 1987; Rehm 1985).

algae
Unicellular and multicellular, → eukaryotic organisms resembling higher plants by evolving oxygen during photosynthesis. Mostly aquatic, found in fresh, salt and brackish waters; also on moist surfaces such as bark, tree trunks, glass windows, cool towers, etc. Multicellular marine algae are harvested commercially for production of thickeners (→ agar, carrageenan and other alginates). Unicellular algae are cultivated commercially for the production of protein (→ Scenedesmus). A few algae produce toxins that may be transmitted via shell-fish to cause food-poisoning (→ shell-fish poisoning). → Blue-green algae (cyanobacteria) are prokaryotes and not considered as algae.

alkalis
Concentrated solutions of alkaline components such as NaOH and KOH, with strong sanitising and disinfecting action. Formerly used regularly, but, because of corrosive properties, they have been replaced by other substances and formulations (Kunz 1988).

alkaloids
Natural plant components containing heterocyclic nitrogen atoms, and occurring most typically in plants. Often toxic to vertebrates. Typical examples: → mycotoxins, especially → ergot (Reiss 1981, 1986).

allantiasis → botulism.

allergens
Most typically a "foreign" protein, peptide or proteinaceous component that may act

as antigen and induce hypersensitivity (a pathological immune reaction). This reaction may also be the result of previous sensitisation with the antigen. Often caused by ubiquitous moulds such as *Penicillium, Cladosporium, Aspergillus, Mucor, Alternaria, Botrytis, Fusarium, Trichoderma, Phoma, Rhizopus, Aureobasidium, Stemphylium, Helminthosporium, Epicoccum* and *Curvularia*. Allergen extracts of these moulds are commercially available for use in diagnosis of allergies in the skin test, e.g., from Pharmacia GmbH, Freiburg (Schachta and Jorde 1989). Spores of the → edible mould species *Pleurotus ostreatus* (oyster fungus) may cause allergic reactions in some people when inhaled.

allergy
A reaction condition provoked by antigen (→ allergen) contact with antibodies, being manifested as hypersensitivity. Symptoms: widening of capillaries (reddening of skin or mucous membranes), increase in permeability (local swelling) and secretion (tears, sputum, rhinitis), itching, general uneasiness. In sensitised persons allergic reactions may be provoked by the oral, dermal and especially inhalatory assimilation of (e.g.) conidia and/or fragments of hyphae, enzymes, pollen, soya dust, etc. Bronchitic allergies (asthma) may be associated with different types of food dusts, possibly as a result of moulds present, e.g., "Baker's asthma" (Schachta and Jorde 1989). – The incidence of severe allergies due to *Paxillus involutus* has increased in recent years. This species has therefore been omitted from the guidelines for moulds and mould products in the Deutsches Lebensmittelbuch (Bötticher 1974; Roth et al. 1988, 1990).

Alternaria
A mould genus represented by 40 species and belonging to the → Deuteromycetes

(Joly 1964). Typical of the darkly pigmented fungi, the hyphae and conidiophores are darkly coloured by melanins; the typically black conidia are multicellular with cross and sometimes longitudinal walls. Distribution: soil, vegetable matter; many are host specific plant parasites. Mycotoxin producers; agents of food spoilage and material destruction (Cerny and Hoffmann 1986).

Alternaria alternata
(*A. tenuis*); cosmopolitic; attacks (growth on): packaging materials, cereals (especially in moist years), etc. Growth: min. $0 - 6\,°C$; opt. $20 - 30\,°C$; max. $35\,°C$; pH $2.7 - 8$; min. O_2 0.25% (CA storage). Heat destruction: in apple juice at $63\,°C$ in 25 min. Degrades ("attacks") practically all types of organic matter, including kerosene. Produces tenuazonic acid (insecticide), atenuen, tentoxin (phytopatho genic), alternariol, alternariol-monomethylether and other mycotoxins (Dombsch et al. 1980). This species is often identified as → allergen.

Alternaria brassicae
Causes growth as black spots on different cabbage types, starting on the field, and enlarges during storage.

Alternaria citri
Causes black moist rot of grapefruit; → cross-contamination of oranges possible during transport and storage.

Alternaria radicina
Causes black rot of carrots and celery during storage.

Alternaria solani
Causes early blight (small dark spots of necrotic tissue) of potato leaves. Infection of tubers during harvest may result in dry, corky rot.

Alternaria tenuissima

(*Helminthosporium tenuissimum*). Mould species commonly found in soil and on most plants. Weakening parasite on vegetables and fruit during storage. Conidia survive in water >6 a (Müller, G. 1983a).

Alternaria toxins

Metabolites of some *Alternaria* spp., toxic to vertebrates. May be produced on different substrates, e.g., rice, fruit, vegetables. Alternariol and alternariol-monomethyl-ether cause weakly positive reactions in the→AMES test, but induce foetotoxic and teratogenic effects in mice. Altertoxins I to III show strong acute effects, and induce carcinogenic reactions in cell cultures. Relatively little information is available about the occurrence of these toxins in foods (Reiss 1981, 1986). Altenuen and isoaltenuen are produced on tomatoes by *A. alternata;* they cause leaf necroses and are antimycotic [Visconti et al.: *Mycotoxin Res.* 5 (1990) 69 – 76].

Alteromonas

Genus (*Incertae cedis*) of Gram-negative, aerobic, psychrotrophic, rod-shaped bacteria. Occur in coastal and marine environments; as the pseudomonads, they are typically associated with the surface of fish; spoilage potential comparable to that of *Pseudomonas* spp. Fish and meat may be spoiled by *A. nigrifaciens* and→ *Shewanella putrefaciens* (formerly *A. putrefaciens*).

AMES test

"Mutatest". Relatively simple test procedure for determining mutagenicity (and possible carcinogenicity) of substances. Histidine-auxotrophic mutants of *Salmonella typhimurium* (i.e., mutants requiring histidine for growth), are treated with the specific substance in different concentrations, and then plated onto a histidine-free growth medium. Mutagenic action induces back mutations to prototrophic wild types (revertants) which are able to grow selectively on histidine-free minimal medium (Classen 1987).

amines→biogenic amines.

amino acids

Aminocarboxylic acids; organic acids carrying one or two amino groups. Especially the α-amino acids are important as components of peptides and proteins; free amino acids also present in fruit, vegetables and raw materials of animal origin. Essential amino acids cannot be synthesized in adequate quantities by mammals, and must be supplied via the food. A number of MO are able to synthesize certain amino acids (prototrophy), and some are implemented for the commercial production of amino acids (Table 3). Supplementation of foods with amino acids requires formal approval in Germany. Commercially produced L-phenylalanine (4000 t/a in 1985/86) is used for the production of the sweetener Aspartame. – Amino acids play an important role in the microbiological spoilage of foods; they are either→decarboxylated and/or→deaminated enzymatically by several bacterial spp., with the production of→biogenic amines, or→ammonia (in addition to CO_2 and fatty acids).

ammonia

NH_3. Colourless gas with a pungent smell, and irritating to the skin and mucous membranes. Shows antibiotic activity. Very soluble in cold water giving NH_4OH. MAK-value 50 ml/m^3 of air (Roth 1989). Nitrogen source for many MO. – Produced during degradation of proteins by→deamination of→amino acids, especially by Gram-negative bacteria. Detrimental to sensory quality;→putrefaction. Alkaline ammonia smell typical of overripe→soft cheese.

amoebiasis→*Entamoeba.*

Table 3   Application of MO for the production of amino acids (modified acc. to DFG (1987).

Essential Amino Acids	Microorganisms
L(+)-Isoleucine	*Serratia marcescens, Brevibacterium flavum*
L(–)-Leucine	*Serratia marcescens, Brevibacterium flavum, Corynebacterium glutamicum*
L(+)-Lysine	*Corynebacterium glycinophilum, Nocardia* sp.
L-Methionine	*Corynebacterium glutamicum*
L(–)-Phenylalanine	*Corynebacterium glutamicum, Brevibacterium flavum, E. coli*
L(–)-Threonine	*E. coli*
L-Tryptophan	*Hansenula anomala, Candida utilis, B. subtilis, Proteus rettgeri, Corynebacterium glutamicum, Brevibacterium* sp.
L(+)-Valine	*Enterobacter* sp., *E.* sp.

non-essential amino acids	
L(+)-Glutamic acid (EC no. 620)	*Arthrobacter paraffineus, Brevibacterium flavum, Corynebacterium glutamicum*
L(+)-Cystein	*Enterobacter aerogenes*

amphoteric components
Surface active agents→disinfectants, e.g., "Tego" tensides used in the food and beverage industries. Ineffective against endospores. Optimum pH range: 7 – 9. If used in recommended concentration non-toxic; sustained activity through residues on food-contact surfaces. Non-irritating to skin and mucous membranes, non-smelling. Non-mutagenic [Kästner: *Arch. Lebensm. Hyg.* 32 (1981) 97-104].

amyloglucosidase
(E.C. 3.1.2.3). An endoglucosidase that hydrolyses polysaccharides to produce glucose from the non-reducing end of the chain. Trivial names: glucoamylase, α-1, 4-amyloglucosidase, τ-amylase, acid maltase. Systematic name: 1,4-α-D-glucan glucohydrolase. Uses: saccharification of polysaccharides, e.g., for the alcohol (spirits) industry, breweries, and the production of glucose syrups. – MO for production: *Aspergillus* spp., *Rhizopus*

spp., *Endomycopsis bispora* (GDCh 1983).

anabolism
Biosynthesis. Metabolic reactions in the cell by which macromolecules (e.g., amino acids, sugars, nucleotides and their polymers), are synthesised from organic and inorganic compounds; energy is provided in organotrophs by catabolism of exogenous organic compounds. Products of biosynthesis may also include antibiotics, toxins, etc. (Schlegel 1985).

anaerobes
MO that grow/multiply only in absence of molecular O_2. Oxygen is toxic but often not lethal to obligatory anaerobes. Facultative aerobes may also grow under aerobic conditions (aerotolerance), e.g.,→lactic acid bacteria (LAB). Opposite:→aerobes.

anamorph
The "imperfect" or asexual stage of some

fungi→Deuteromycetes. A separate nomenclature (genera and spp.) is used solely for anamorphous fungi, even when fructification bodies of the sexual stage (→teleomorph) is known (Moeller and Loeffler 1982).

anaphylaxis
Anaphylactic shock (type I reaction). Immediate hypersensitivity, after sensibilisation, may be caused by→biogenic amines, penicillin, etc. The manifestation of an antigen-antibody reaction resulting in a "sudden"→allergy; in severe cases as anaphylactic shock which may even result in death.

anastomoses
Cross connection formed from the hyphae of one fungus individual (thallus) to another. By this means nuclei and→cytoplasm may be exchanged. Exchange of nuclei leads to heterokaryotic "systems" that are the basis of the extreme→adaptation potential, especially of the Deuteromycetes. Example: a mesophilic strain receives nuclei from a psychrotrophic donor strain. During refrigeration the cold tolerant nuclei multiply faster, resulting in continued growth of the mould during cold storage, after a transition period. The formation and consequences of the formation of anastomoses are also referred to as parasexuality (Müller and Loeffler 1982).

anchovies
A biologically ripened fish commodity with a strong flavour, and used for sauces and as relish. Produced from herring and salted types of fish, using sugar, salt and spices; NaNO$_3$ and even→starter cultures may also be added, although the typically associated bacteria are favoured by the conditions and contribute, together with the product enzymes, to the ripening process. – In addition to traditional anchovies, related products may include "Appetitsild", "Kräuterhering" (herring with herbs), "Gabelbissen", and "Matjesfilet" Swedish or Norwegian style, and may be ripened by spontaneous fermentation and without spices (Sinell 1985).

antagonism
Opposition or counteracting of two entities of similar kind. Negative effect of MO against each other, generally resulting from their metabolic products, e.g.,→antibiotics,→organic acids,→ammonia,→bacteriocins, etc. Traditional fermentations involving the production of→cheese, →sauerkraut and→silage are based on this principle.

anthracnose
Plant diseases caused by fungi – mainly of the Melanconiales – characterised by discrete dark-coloured lesions on the leaves or fruit. Especially tropical fruit may become infected, e.g., *Colletotrichum gloeosporioides* on citrus, avocados, mango and papaya, but also on vegetables (Beuchat 1987; Müller 1983a). Mycotoxins could not be detected yet (Heinze 1983).

antibiogram
Sensitivity spectrum of an organism towards a range of antibiotics.

antibiosis
Inhibition of growth or multiplication of MO, or even lethal effect, caused by metabolic products of other MO,→antagonism,→antibiotics.

antibiotics
Originally defined as secondary metabolites of MO which, in small quantities, will inhibit or lethally harm other MO. Difference from→mycotoxins is unclear by this definition, since several mycotoxins are active against MO, plants and animals (e.g.,→patulin or→citrinin). – In the modern sense an antibiotic refers to

natural, semi-synthetic or synthetic antimicrobial compounds effective in low concentrations. Some antibiotics are used as chemotherapeutic agents in human or veterinary medicine, some, however, are too toxic or ineffective, whilst others may be used as feed additives for growth promotion. A specific group, termed→ bacteriocins, differs from the "typical" antibiotics in some respects, and shows potential as food preservatives, e.g.,→ nisin. – Penicillin is the oldest and best known antibiotic and was discovered by Fleming in 1928. Uncontrolled use of penicillin during the 50's for medical and food production purposes in Europe and North America favoured the selection of resistant *Staph. aureus* populations, with the result of severe hospital-associated cross infections of patients. Against this background the use of antibiotics in food is not permitted in any industrialised country. In this way it is hoped to minimise possible selective proliferation of resistant populations. → Natamycin can be singled out as the only exception, and is used for the surface treatment of some cheeses and sausage types, for the prevention of→ mycotoxin formation. – The treatment of domestic animals with antibiotics and their addition to feed requires careful control by authorities and the food industry; detectable amounts of antibiotics may be transferred in this way via milk, eggs, meat and offal to the consumer. In addition, starter cultures for→ yoghurt and→ cheese production may be inhibited; waiting periods (regulation on compounds with pharmacological action, of 25.09.84) (Hahn and Muermann 1987; Krämer 1987; Sinell 1985).

antigen-antibody reaction
Interaction between antigen and antibody of compatible determinant groups. This type of reaction is involved in→ allergies caused by MO,→ anaphylaxis, and is the basis of serological reactions for the→ differentiation, e.g., of food-borne pathogens and toxinogens;→ agglutination;→ precipitation;→ immunofluorescence.

antimycotics
→ Antibiotics (secondary metabolic products of MO) that act specifically as antifungal agents. Growth and/or outgrowth of spores or conidia are inhibited (fungistatic), or the mould may be killed (fungicidal). Examples: griseofulvin (*P. griseofulvum*), nikkomycin or polyoxin (*Streptomyceta-ceae*), pimaricin/natamycin (*Streptomyces natalensis*) (Müller and Loeffler 1982; Raab 1974).

apical
At the *apex* (lat.) = end, e.g., of a cell, group of cells or hypha.

apiculate yeasts
Trivial name for a group of spp. with cells that are pointed at one or both ends. They represent different families, and are undesired in industrial processes because of weak fermentation potential and for causing sensory defects. They are considered as contaminants in the beer brewing industry.

apple scab
Caused by *Venturia inaequalis*, an→ Ascomycete, that infects the leaves and fruit of the apple tree. Cool, moist conditions during the autumn favour the infection of the unripe fruit. The mould may penetrate the undamaged skin, and dark spots and fissures are formed on the surface that allow other spoilage MO to enter the fruit. The symptoms develop further during storage and may cause considerable losses. Early treatment with fungicides may largely prevent damage. – Some consumers consider apple scab as quality criterion, indicating the fruit to be free of fungicides. – Nothing known yet about mycotoxin formation.

arachnoid
(Greek) = like a spinning web. Atypical growth of mould hyphae on surfaces that contain non-physiologically high concentration of salt or → nitrite. May be observed occasionally on meat products.

arizona
Gram-negative, rod-shaped Enterobacteriaceae; genus relationship not clarified (*Paracolobactrum arizonae; Salmonella arizona*). Generally present in the intestinal tract of reptiles. May be transmitted by poultry to cause epidemic-type of foodborne diseases that show some relationship to → salmonelloses.

aroma
Flavour property, specific of a product, that the consumer may associate with quality and acceptability; → sensory. Several aromas are the result of metabolic activities by microorganisms, e.g., in cheese, sour cream butter, fermented sausages, wine (bouquet), etc. (Fricker 1984).

aroma compounds
Volatile metabolic products of MO that can be observed by their flavour (sensorics). They are desired in many products, e.g., → diacetyl in sour cream butter; and detrimental, e.g., diacetyl in beer or wine; → off-odour. Aroma compounds such as diacetyl are often not formed directly, but the primary product may be oxidised by oxygen as in the case of → acetoin. Sometimes a series of metabolic steps may be necessary, e.g., in the formation of methylketone from milk fat by *Penicillium roqueforti*. Also, metabolites from different MO may form aroma compounds by secondary reactions, e.g., → ethanol from yeasts and → acetic acid produced by some bacteria, with the formation of ethylacetate (acetic acid ethyl esther) (e.g.) in wine. – Growth of *Pseudomonas fragi* in milk may produce aroma compounds with a fruity flavour, reminding of strawberries (Dittrich 1987; Kunz 1988). Moulds may produce species-specific aroma compounds.

aroma formation → grouping of MO.

Arthrobacter
Genus of catalase-positive, obligatory aerobic, Gram-positive bacteria that typically grow as irregular, pleomorphic rods, although ovoid or spherical forms are associated with static cultures. Distribution: soil, activated sludge; some spp. are salt-tolerant. *Brevibacterium* may be related to this genus.

arthropods
Large group of jointed animals with an outer skeleton, e.g., spiders, insects, etc.; important vectors of MO. Their control is important at all stages of food production and processing.

Ascomycetes
Largest fungal group; ca. 2000 genera represented by diverse examples, e.g., yeasts, cup fungi and edible types such as the morels and truffles. Growth in vegetative phase (→ biomass) as haploid multikaryotic, septate hyphae or budding cells, and finally producing different conjugation organs. Dikaryotic, ascogenous hyphae are formed as a result of plasmogamy; karyogamy may follow either directly after plasmogamy, or may be delayed for some cell generations. A diploid ascus (cup) is produced and after meiosis and one or two mitotic divisions, haploid ascospores are formed in typical fruiting bodies. – Most fungi of interest here are included in this class. → Anamorphous types are grouped under the → Deuteromycetes, and have their own nomenclature for historical reasons. Many Ascomycetes are either beneficial or detrimental, and thus have great economic importance: e.g., yeasts used

in commercial production of bread, beer and alcohol (biotechnical products); edible fungi; plant diseases (apple scab, blackspot, peach curl leaf); and some animal diseases (Müller and Loeffler 1982).

ascorbic acid
EC no. for L-ascorbic acid: E 300 (vitamin C). No approval necessary in Germany for use as anti-oxidant or vitamin. – Industrial production: glucose is reduced catalytically (H_2/Ni) to sorbitol. A solution containing 10 – 15% sorbitol and 0.5% corn steep liquor (nutrients) is oxidised to sorbose by *Gluconobacter suboxydans* (→ acetic acid bacteria) during strong aeration and cooling to 30 – 35°C in a bioreactor. Sorbose serves as precursor for the chemical synthesis of vitamin C. A yield of 90% of the initial sorbitol concentration is achieved by this→ biotransformation, and has led to the reduction in the price of ascorbic acid.

ascus
Sac-like microscopic structure typical of the subdivison Ascomycota; formed by coalescense of two haploid nuclei (karyogamy). The diploid nucleus (zygote) may undergo meiotic division immediately, producing two haploid nuclei, followed by one or two mitotic divisions by which 4 or 8 nuclei are formed. The plasma of the ascus synthesises different layers around the nuclei leading to the formation of 4 (*Sacch. cerevisiae*) or 8 (*Schizosaccharomyces octosporus*) ascospores; this process is referred to as free cell formation. The ascospores are generally more heat resistant than other reproductive organs of fungi (→*Byssochlamys*). Groups of asci may be surrounded by→ hyphae, or they may develop inside a special structure, called an ascocarp, represented by the typical forms: apothecium, ascostroma, cleistothecium and perithecium) (Müller and Loeffler 1982).

aseptic installation
A special device for filling and packaging of a sterile product in such a way that recontamination is excluded (Reuter 1987).→ Packaging material sterilisation.

asexual state→ anamorph.

aspergilloma
Result of→ aspergillosis of the lung. Infection mainly by inhalation of conidia of some *Aspergillus* spp. (usually *A. fumigatus*), growth/colonisation resulting in the formation of thickened, often pea-sized, compact mycelial masses within the lung tissue that are called aspergillomas (Gedek 1980).

aspergillosis
A disease of man or animals that may result from the inhalation of large numbers of conidia of *Aspergillus* spp., e.g., "farmer's lung", etc. Labourers working without a dust mask at mills and malteries, and involved in filling and emptying of grain silos, are especially at risk. Mainly *A. fumigatus,* but also *A. flavus, A. niger, A. nidulans, A. versicolor* and *A. terreus* are involved (Gedek 1980). Especially *A. fumigatus* has invasive and infective properties, and is commonly found in soil of pot plants; other spp. can be considered as→ opportunists.

Aspergillus
A mould genus of the class→ Deuteromycetes (Hyphomycetes), consisting of ca. 150 spp. Several→ teleomorphs are known, e.g., *A. glaucus* (teleomorph: *Eurotium*), *A. nidulans* (teleomorph: *Emericella*), *A. fumigatus* (teleomorph: *Sartorya*). – Inhabitants of the soil, especially in warm and moist, but also in moderate, regions.→ Xerotolerant spp. include important food spoilage organisms and are part of the *A. glaucus* group (Raper and Fennell 1965). Several produce→ mycotoxins (see Table 4). *A.*

Table 4 Some mycotoxins of the aspergilli (acc. to Milczewski in Reiss 1981; Reiss 1986).

Mycotoxin	*Aspergillus*
Aflatoxins Hepatotoxin, Carcinogen	*A. flavus; A. parasiticus*
Citrinin LD_{50} 110 mg/kg KG Mouse p.o. Nephrotoxin, Carcinogen	*A. candidus; A. carnosus; A. niveus; A. terreus*
Cyclopiazon LD_{50} mg/kg KG Mouse p.o.	*A. oryzae; A. versicolor*
Kojic acid Epileptic-like symptoms, Mutagen	*A. albus; A. alliaceus; A. awamori; A. candidus; A. clavatus;* *A. flavus; A. fumigatus; A. giganteus; A. glaucus; A. nidulans;* *A. oryzae; A. parasiticus; A. tamarii; A. terricola; A. ustus; A. wenti*
β-Nitropropanic acid LD_{50} 100 mg/kg KG Mouse i.v. Blood pressure reduction	*A. flavus; A. oryzae*
Ochratoxin A LD_{50} 21 mg/kg KG Rat p.o. Nephrotoxin, Carcinogen	*A. alliaceus; A. melleus; A. ochraceus; A. ostianus; A. petrakii;* *A. sclerotiorum; A. sulfureus*
Patulin LD_{50} 35 mg/kg KG Mouse p.o. General cell toxin	*A. clavatus; A. giganteus; A. terreus*
Penicillic acid LD_{50} 35 mg/kg KG Mouse p.o. Carcinogen	*A. alliaceus; A. auricomus; A. melleus; A. ochraceus; A. ostianus;* *A. petrakii; A. sclerotiorum; A. sulfureus*
Sterigmatocystin LD_{50} 150 mg/kg KG Mouse p.o. Carcinogen	*A. amstelodami; A. chevalieri; A. flavus; A. nidulans; A. parasiticus;* *A. ruber; A. rugulosus; A. unguis; A. versicolor*

oryzae is used for the production of → east asiatic fermented foods; some are used for the production of → enzymes and organic → acids. The septate mycelium is normally colourless; aerial hyphae (aerial mycelium) are sometimes produced. An enlarged foot cell in the hypha branches vertically to form a young conidiophore, the terminal part of which swells into a vesicle that may either bear metulae

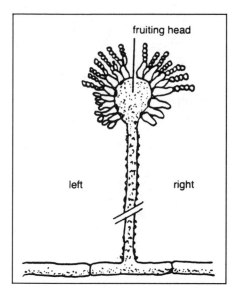

fruiting head

left   right

Figure 2 Conidiophore of *Aspergillus niger* (left) and *A. parasiticus* (right) with foot cell and stem.

(primary phialides) and phialides (secondary phialides) or only phialides on which the conidiophores are formed (Figure 2). The conidia are black, brown or green, and may be yellow in the early stage. Sometimes→sclerotia are produced. – Control: drying, →preservatives. Destruction: pasteurisation (Reiss 1986). – All species are non-fastidious; growth requirements satisfied typically on mineral salts media with nitrate and sucrose or glucose (e.g., Czapek-Dox Agar: Merck cat. no. 5460, Oxoid CM 97, Difco 0339).

Aspergillus flavus and A. parasiticus

These spp. belong to the *A. flavus-oryzae* group according to Raper and Fennell (1965). In *A. flavus* the phialides are borne on metulae (primary phialides), and in *A. parasiticus* directly on the vesicle (conidiophore). For both spp. the conidia are yellow in the early stages, and turn yellow-green and later brown-green. *A. flavus*

produces red-brown sclerotia more often than *A. parasiticus*. Both spp. can synthesise→aflatoxins,→kojic acid and→sterigmatocystin. As early as 1910 it was known that *A. flavus* is "poisonous" [Kühl, H.: *Zentralh. für Deutschland* 51 (1910) 106 – 107]; this fact, however, has been forgotten till the discovery of aflatoxin in 1960. – Distribution: soil, ubiquitary with highest incidence in warm regions. The ratio of toxinogenic isolates may range from 5 to 94%, increasing towards warmer climates. Potentially every food commodity that has not been heat treated may be contaminated with these moulds. Temperature for growth: min. 3 – 4 °C, opt. 35 °C, max. 50 °C; for toxin production: min. 9 °C, opt. 20 °C, max. 42 °C. NaCl: growth in < 15%; for toxin production: opt. 1%, max. 14%. a_w: growth to 0.80, opt .0.95; conidium production: 0.85; outgrowth of conidia: 0.80. Opt. pH 7.5, min. 2.5, max. >10.5 (Frank 1974; Reiss 1986). – Selective medium: Oxoid CM 731 with chloramphenicol Oxoid SR 78. On this medium the colonies of both spp. are coloured orange on the bottom; they may be confused with *A. oryzae* and possibly *A. niger,* which, however, turns black when conidia are formed. Toxinogenic strains cannot be identified reliably by cultural methods.

Aspergillus glaucus group

Acc. to Raper and Fennell (1965), of importance in the spoilage of products with reduced a_w: *A. amstelodami* to 0.75; *A. chevalieri* to 0.65; *A. echinulatus* to 0.62; *A. repens* to 0.65; *A. ruber* to 0.71. Growth from −8 °C (physiological drought/reduced a_w through frozen water) to 43 °C; opt. 30 °C; pH: 2 – 8.5. – May typically spoil juice concentrates, jams and marmalades, dried fruit, oats, confectioneries and marzipan ("water stains"). For→Mycotoxins see Table 2 (Reiss 1986). Cultivation on typical media for moulds with 40% sucrose added.

Aspergillus niger
Main representative of a group of *Aspergillus* spp. with black coloured conidia. Ubiquitary and an important coloniser and spoilage organism of foods. Best known *Aspergillus* sp. because of striking appearance, multiple applications and occasional pathogenicity (→ aspergillosis). – Temperature requirements: opt. 35 – 37 °C, min. 6 – 8 °C, max. 45 – 47 °C. a_w: opt. 0.96 – 0.98, min. 0.88; for conidium formation: 0.84. pH 1.5 – 9.8. – Distribution: soil, cosmopolitic. May spoil practically any type of food. Colonises and/or destroys colours, leather, synthetics, optical glass under tropical conditions, paper, packaging material, sandstone (Reiss 1986). – Selected strains of *A. niger* are used for the production of several components for the food industry, e.g., citric acid, gluconic acid, different amylases, proteases, pectinases, cellulases, invertase, lipases, → glucose oxidase, lactase, catalase, naringinase and gluconases (DFG 1987; Underkofler 1977).

Aspergillus oryzae
Close correspondence with *A. flavus* as far as habitats and cultural properties are concerned, however, does not produce → aflatoxins. Traditionally used as → starter culture for the production of different types of → East Asiatic specialities, some of which are becoming established on the market in Europe and North America. Since this organism may produce → kojic acid that gives a positive reaction in the → AMES test, attempts are being made in Japan to supply kojic acid negative strains to industry and for distribution as starter cultures (DFG 1987).

aspic products
Clear acidic jellies containing different types of meat or fish products; high spoilage potential because of high → water activity in the gelatine; refrigeration < 4 °C necessary. Limited shelf life extension achieved by acidification, e.g., with vinegar, to pH < 4.5. This is especially important when the products are sold by street vendors and in markets; → preservatives.

assimilation
Incorporation of nutrients (e.g., C-, N-, P-, S-sources), in an acceptable form into cell substances by metabolism; → anabolism. Differentiation between → autotrophic assimilation (requiring only inorganic nutrients such as plants and some bacteria) and → heterotrophic assimilation, requiring organic compounds, such as by → saprophytes, represented by most of the organisms discussed here.

ATP
Adenosine 5′-triphosphate. Most important substance involved in the energy transfer in every living cell. Chemical free energy released by respiration or fermentation is stored in the two high-energy phosphate bonds in ATP. By the cleavage of ATP to ADP (adenosine diphosphate) and ADP to AMP (adenosine monophosphate) ortophosphate groups are released, and energy supplied on sites of demand in the cell, e.g., for assimilatory processes.

Aureobasidium pullulans
(*Pullularia pullulans*). Imperfect (anamorph) mould of the class Hyphomycetes (subdivision Deuteromycotina). Threadline growth and formation of mycelium under favourable conditions. A deteriorating ecological situation induces the formation of budding (yeastlike) cells that are distributed by wind when dry. Melanines are deposited in the cell walls and conidia (black yeasts, black moulds). – Universally found in soils, on plant parts and grains, cellar walls, consumables and wall covering (Domsch et al. 1980). – Growth: 2 – 35 °C, opt. 25 °C. – Used for the commercial production of pullulan, that has

application as gelating agent for food dressings (not approved in Germany). – Involved in the deterioration of materials. May cause asthma bronchiale when the conidia are being inhaled by sensitised persons, e.g., with cereal dust or flour (Reiss 1986).

autoclave
A steam pressure chamber used for the sterilisation of laboratory utensils, nutrient media, etc., and, on industrial scale, for the heat processing of canned foods. The bench-type represents the simplest autoclave and resembles the domestic pressure cooker; it is commonly used in laboratories as are the larger type vertical autoclaves. Large autoclaves find typical application in hospitals, industry and biotechnical research. The lethal effect is achieved by a combination of moist heat (as saturated steam) and contact time, and usually amounts to 121.1 °C (15 lb/inch^2 or 103 kPa) for 15 to 30 min, and is sufficient to destroy all microorganisms typically involved in foods and their endospores, and to inactivate all enzymes. – For heat sterilisation of canned foods either static or rotation autoclaves are used; the latter allows more rapid heat transfer to the core of the product. Flexible pouches and other fragile containers are sterilised in "retort" autoclaves in which the depressurisation is regulated carefully during cooling. – The sterilising value of a heat process is expressed in terms of the → F value (F$_o$) which indicates the effectivity of physical destruction of microorganisms (ICMSF 1980).

autolysis
Degradation of structural cell components, typically the peptidoglycan of bacterial cell walls, by endogenous enzymes called autolysins. These enzymes are responsible for the lysis of dead cells, resulting in the release of intracellular enzymes. This process may play an important role in the ripening of cheese, and may influence the product specific properties. – In the propagation of pure cultures of yeasts for industrial fermentations, autolysis can be responsible for inactive cultures, or for sensory defects (by autolytic products) in wine and beer.

autotrophy
Organisms which use CO_2 or other forms of inorganic C as sole carbon source. This property may either refer to chemolitotrophs or photolitotrophs. The green plants are a typical example of the latter. Energy requirements for → assimilation are supplied either by light (photosynthesis) or by the oxidation of inorganic material (chemosynthesis). MO responsible for the spoilage of foods are → heterotrophs, as are all animals.

avocado
Persea americana and *P. gratissima* may be damaged by → anthracnose after harvesting, when grown in a moist climate. Spores of the mould *Colletotrichum gloeosporioides* grow out on the tree, and the mycelia penetrate the wax layer of the undamaged fruit. Growth extends through the epidermis during ripening. Storage and transport at 6 – 8 °C prevent anthracnose and also the development of *Diplodia natalensis* (Sommer, N. F.: in Kader 1985).

a$_w$ limits → water activity.

a$_w$ value
Symbol for → water activity, referring to the amount of water available for growth and multiplication of MO. It is defined as the ratio of water vapour pressure over a food to the saturation pressure of pure distilled water at a given temperature (Heiss and Eichner 1984). Water activity corresponds to the relative humidity (rH) above a food in a closed space after equilibration, e.g., 70% rH divided by 100 = 0.70 (a$_w$). Food

Table 5   Water activity (a_w) values of salt (NaCl), sucrose and invert sugar solutions at 25 °C (acc. to Krämer 1987, modified).

a_w Value	NaCl (weight %)	Sucrose (weight %)	Invert Sugar (weight %)
0.98	3.4	26.0	8.2
0.96	6.5	39.6	16.4
0.94	9.3	48.2	24.6
0.92	11.9	54.3	32.8
0.90	14.8	58.4	41.0
0.88	16.2	62.7	49.3
0.86	18.1	65.6*	57.5
0.85	–		61.6*
0.75	26.5*		

*Saturation.

ingredients such as protein, starch, salt, sugar, etc., are termed solutes and act to reduce the a_w, i.e., less water is available for MO (→ osmotic pressure). At increasing concentrations of a solute the water vapour pressure is reduced (and thus the a_w) (see Table 5); this principle is used for the preservation or shelf life extension of some foods, especially in combination with other factors; → hurdle concept; → water requirements.

bacilli

(Sing.: bacillus). (1) Trivial name for bacteria, or, more specifically for rod-shaped bacteria (i.e., with length 2 – 3 times that of the diameter) that may be straight or curved. (2) Refers to a member of the genus→ *Bacillus*.

Bacillus

Genus (Section 13: *Bergey's Manual of Determinative Bacteriology, Vol. 2*, 1986) of Gram-positive, anaerobic rod-shaped endospore-forming bacteria. Respiratory metabolism: aerobic to facultatively anaerobic; nitrate respiration by some strains. Nitrate reductase- and catalase-positive. Often peritrichously flagellated; produces acid but rarely gas from CH. Protein degradation accompanied by ammonia production. Important spoilage organisms; enzymes involved: proteases, pectinases, lipases, amylases. At least 40 spp. presently acknowledged; probably a heterogeneous group, as reflected a.o. by mol% G + C ranging from 30 to 70. Type species: *Bacillus subtilis*. – Detection of endospores by heating at 80 °C for 10 min in water bath, or treatment with 70% ethanol for 1 h at 37 °C to destroy vegetative cells; after serial dilution (→ dilution series) plating onto→ Standard I Agar or Plate Count Agar. Colonies appearing after incubation at 30° or 55 °C for 48 to 72 h are representative of endospore numbers. – Psychrotrophic spp. include: *B. badius, B. cereus, B. circulans, B. globisporus, B. laterosporus, B. macerans, B. megaterium, B. polymyxa, B. pumilus, B. subtilis*. Thermophilic spp.: *B. acidocaldarius, B. coagulans, B. stearothermophilus*. Alcalophiles: *B. alcalophilus, B. firmus*. – Some spp. are used for the biotechnical production of enzymes. Anthrax ("splenic fever" in herbivores) caused by *B. anthracis. B. thuringiensis* is an important insect pathogen and is used as an agent for biological control.

Bacillus cereus

Motile (peritrichously flagellated) rods, producing ellipsoidal terminal endospores; hemolytic; aerobic, although anaerobic growth in presence of glucose or nitrate. A causative agent of food poisoning; also known as opportunistic pathogen. Growth conditions: min. temp. 10 – 15 °C, opt. 22 °C, max. 35° – 45 °C; pH 4.9 – 9.3. Min. temp. for toxin production: 10 – 15 °C. – Distribution: ubiquitary, most often in soil, milk, cereals and spices. Low numbers in raw milk, although spoilage of pasteurised milk may result from surviving endospores growing out and proliferating under anaerobic conditions at a higher rate than other spp. Coagulation of milk and cream by proteases producing "bitter peptides". Rapid growth in CH rich foods; no sensoric changes by up to 10^7/g. – Destruction: vegetative cells by pasteurisation; endospores by moist heat: e.g., D_{120} 2.3 min, D_{100} 8 min, D_{90} 71 min, D_{85} 220 min; dry heat: D_{125} 8 min (Mitscherlich and Marth 1984). Inactivation of toxins: emetic toxin 120 °C >90 min; diarrhoea toxin is heat labile. Food poisoning has been associated with rice, puddings, sauces, mashed potatoes, oats, porridge, ready-to-eat foods prepared with meat and spices, containing 10^7 to 10^8 vegetative cells/g. Short→ incubation time followed by vomiting (emetic syndrome toxin). May be mistakenly identified as *Staph. aureus* food poisoning. Diarrhoea after 8 – 10 h, no fever; may be mistaken for *Cl. perfringens* food poisoning. No complications; symptoms resolve within 12 – 24 h (Doyle 1989; Sinell 1985).

Bacillus stearothermophilus

Thermophilic *Bacillus* sp., with growth

minimum around 30 °C, opt. 63 °C, max. 70 °C. May proliferate in cool towers, and endospores may reach surface water via droplets. Growth at pH $5.2 - 9.2$; a_w > 0.95. Causative agent of → "flat sour" spoilage in canned vegetables. Highly resistant endospores: D_{100} 50 h, D_{120} 5 min. For heat preservation $F_o > 5$; for tropically conserved food F_o 20 is recommended. – Used as bio-indicator for assessing effectivity of sterilisation: e.g., as paper strips containing standard concentrations of spores, e.g., Oxoid no. BR 29, or as ampules containing spores and a colour indicator, e.g., Merck no. 10285. – The organism is sensitive to antibiotics; advantages for use in the detection of antibiotics, since incubation at 68 °C permits unsterile (non-axenic) procedures.

Bacillus subtilis

Originally called the "hay bacillus". Gram-positive, slender, motile rods; no chains. Centrally arranged; ellipsoid to cylindric in shape. Starch hydrolysis, gelatinase + , nitrate-reductase + , lecithinase – , citrate fermentation. Growth: aerobic; opt. $28 - 40$°C, max. $0.40 - 55$°C; min. a_w $0.93 - 0.95$; pH $4.6 - 9.2$. Some strains produce antibiotics. *B. subtilis* var. *aterrimus* and *B. subtilis* var. *niger* produce black pigments on CH-rich media. Distribution: soil, plant material, excretions of flies and cockroaches, ubiquitary. – Destruction: vegetative cells by pasteurisation; endospores by moist heat: B_{100} $80 - 100$ min, D_{120} 7 min; dry heat: D_{95} 450 min, D_{120} 26 min, D_{180} 3.2 s. Survival of endospores: 95% ethanol > 3 a; on dry silk > 70 a; on soap > 30 min.

bacteria

The → prokaryotes are represented by the classes Bacteria and Archaea. The Bacteria constitutes the larger group and includes 12 divisions as well as several genera of unaffiliated bacteria (*The Prokaryotes, 2nd Ed.*: Balows, Trüper,

Dworkin, Harder and Schleifer 1992). Originally called the "Schizomycetes" the bacteria are unicellular microscopically small organisms that typically multiply ("grow") by binary transversal fission. Sexual reproduction is rare. The daughter cells either separate after division or may remain attached to each other, resulting in the formation of pairs, chains or irregular arrangements or aggregates (→ agglomeration). Depending on bacterial type and factors existent in the → ecosystem, the time for one cell division (doubling time or generation time) may range from 20 min to several h. Growth occurs at a maximum rate during the exponential or logarithmic phase, and it is expressed in terms of the number of cell divisions per hour. Thread-like or hyphal growth, typical for moulds, exceptional (→ Actinomycetes). Characteristics that may be species specific: presence or absence of adhering outer layers, e.g., capsules, flagella or slime production for movement, morphological shape (cocci, rods, spiral-shaped, comma-shaped; see Figure 10, page 172, for morphological forms of bacteria). Av. diameter ca. 1 μm, length 10 μm. Some types pigmented (coloured), more often colourless. Refractive index approximating that of water; examination under normal light microscope difficult, unless → staining procedures applied, or phase contrast microscopy used. Highly resistant → endospores may be produced by some genera (e.g., *Clostridium, Bacillus*). – Under ideal conditions, i.e. with constant removal of metabolic products, and excluding the influence of any external factors, bacteria may appear "immortal"; it may be extremely difficult to determine in the laboratory whether a bacterial cell is alive or dead. In practice cells not able to divide any more, i.e., not producing → colonies on appropriate media during determination of the viable cell count (in terms of colony-forming units or "cfu's"), are considered dead. However, this may lead to false-negative results, e.g., for sublethally damaged cells→resuscitation procedures may therefore be necessary. The number of dead cells in or on a food sample is given by the difference between the results obtained by the microscopical and the cultural method. – Bacteria of importance in foods are heterotrophic, i.e., dependent on organic nutrients. As the moulds, they are responsible for the mineralisation of organic residues from plant or animal origin in the ecosystem; important role in the carbon cycle and in keeping the earth habitable. → Assimilation only of small molecules by the cells; several spp. – spoilage organisms – produce extracellular→enzymes by which macromolecules such as cellulose, starch, protein, etc., and water-insoluble substances such as lipids, are hydrolysed. Metabolic products such as CO_2, CH_4, H_2, H_2S, → organic acids, → toxins, etc., are excreted and may cause spoilage of a food commodity; these products may become toxic to the culture in a closed system. – Bacteria are adapted to the most adverse conditions and ecosystems, ranging from the arctic to thermal waters. They can be considered as "permanent competitors" of moulds and animals for foods. Several spp. are facultative and seldom obligatory parasites; commensalism often found (e.g., intestinal bacteria); some are pathogenic to plants or animals;→food poisoning;→ zoonoses. – Composition and functional structure of most bacteria very similar: no mitochondria, → cell wall composition and structure different from that of all other organisms. Dry weight ca. 15 – 30% that of wet weight; specific gravity 1.1, C ca. 50%, N 10 – 15%, P 2 – 6%, S ca. 1%, as well as trace elements. – → Systematics (taxonomy) comprehensively treated in *Bergey's Manual of Determinative Bacteriology* (*Vol. 1* and *2*, 1984 and 1986) and *The Prokaryotes, 2nd Edition* (*Vol. I* to *IV*) (Balows, Trüper,

Dworkin, Harder and Schleifer 1992). Phenotypic→differentiation on the basis of physiological characteristics and morphology.

bacterial colony
A grouping or collection of bacterial cells, either visible with the naked eye or with the aid of a magnifying glass, that may result from the multiplication of a single cell on nutrient agar; basis of→viable count determination. Not produced in liquid media. The colour (pigment) (transparent, opalescent, white, cream, etc.), surface (glossy, rough or uneven), margin (entire, lobate, undulate, curved, erose, filamentous), profile (flat, convex, umbonate, raised, navel shape) are species specific; under standardised culture conditions even the diameter may be typical (Figure 3). Extremely small barely visible colonies are referred to as "pin point" in size. – Slimy transparent colonies are formed by some strains; motile strains of *Proteus vulgaris* produce "moving" colonies that may have covered a part of the agar surface after 24 h and may complicate counting.

bacterial diarrhoea→ diarrhoea.

bacterial filters
For removing bacteria from liquids or gases; the cells are retained both as a result of the microscopic size of the pores and adsorption. Made from materials such as diatomaceous earth (kieselgur, e.g., Berkefeld filter), porcelain (e.g., Chamberland filter), sintered glass, asbestos, or from polymers such as cellulose acetate and polycarbonate (membrane filters). Some are used on an industrial scale. For laboratory purposes disposable membrane filters are mainly used; commercial sterilisation by gamma irradiation; special holder for pressure or vacuum filtration (e.g., Nucleopore, Schleicher & Schüll, Sartorius, Millipore).

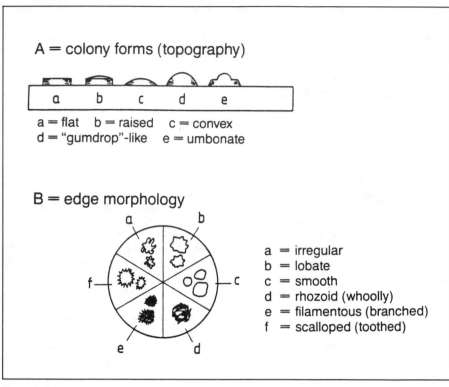

A = colony forms (topography)

a = flat   b = raised   c = convex
d = "gumdrop"-like   e = umbonate

B = edge morphology

a = irregular
b = lobate
c = smooth
d = rhozoid (whoolly)
e = filamentous (branched)
f = scalloped (toothed)

Figure 3   Morphology of bacterial colonies: A in cross section (topography) and B, edge morphology (source: Baumgart, J.: *Mikrobiologische Untersuchungen von Lebensmitteln*, Hamburg: Behr's Verlag, 1990).

bacterial flagella→flagella.

bacterial excretors→excretors.

bacterial numbers
The inner parts of healthy fruit, vegetables, potatoes, meat and eggs are practically free from microorganisms. Depending on the handling and pretreatment procedures, their surfaces, however, are contaminated to some degree with microbes (Table 6) and may cause→cross contamination in the kitchen or canteens that may result in→food poisoning.

bacterial staining→staining.

bacterial toxins
Poisons or toxins produced by bacteria that may cause illness of man and animals;→food-poisoning when transmitted by foods. The→enterotoxins may cause different forms of→gastroenteritis as a result of the oral intake of the toxin or the producer organism.→ Exotoxins or ectotoxins are excreted by the cells (*Cl. botulinum, Staph. aureus*);→endotoxins are released after digestion or lysis of the cells in the intestinal tract (e.g., from *Cl. perfringens, Salmonella*).

bacterial viruses→bacteriophages.

Table 6 Bacterial populations in the kitchen (average values).

Product	→ cfu/10 cm²
Lettuce (unwashed)	10,000 to 1,000,000
Lettuce (washed)	1.000 to 100,000
Fresh strawberries	1.000 to 1,000,000
Pork meat (fresh)	ca. 100,000
Pork meat (matured)	ca. 100,000,000
Kitchen table	>300
Kitchen cutlery	10 to > 250
Hand palm (washed)	10 to > 250

	cfu per g or ml
Tartare (raw minced meat)	100,000 to
with egg and spices	30,000,000
Italian salad (home-made)	ca. 3,000,000
Onions (minced)	ca. 20,000
Pepper (ground, natural)	ca. 30,000,000
Pepper (ground) irradiated	< 10,000
Drinking milk (pasteurised)	< 10,000
Salad mixes (shredded,	100,000 to
in pouch)	1,000,000

Berg et al. 1978.

bactericide
Chemical agents that kill bacteria, but not necessarily their endospores, at the appropriate concentration and temperature and after sufficient contact time. Examples: formaldehyde, per-acetic acid, hydrogen peroxide, activated carbon, and, with some limitations, some broad-spectrum antibiotics (Wallhäusser 1987). → Destructive disinfectants.

bacteriocins
Antimicrobial proteinaceous products of some bacteria with inhibitory or lethal effects against closely related strains or spp. Typically heat stable (e.g., 100 °C/15 min) but inactivated by proteolytic enzymes such as trypsin. Have been isolated from

E. coli, Ps. aeruginosa, B. megaterium and different lactic acid bacteria, including spp. of Lactobacillus, Lactococcus, Leuconostoc, Pediococcus and Carnobacterium. → Nisin is perhaps the most well-known example; it is produced by Lactococcus lactis and has practical importance in cheese production. Recent concepts for "biological preservation" of foods also consider the application of bacteriocinogenic strains of LAB as safety factor against Listeria monocytogenes or to counteract heterofermentative LAB causing spoilage of (e.g.) delicatessen foods or fermented cucumbers.

bacteriolysis → autolysis.

bacteriophages
Phages. → Viruses that are pathogenic to bacteria and cause type or species specific infections of "host" cells, followed by intracellular multiplication and their release by lysis of the cells. Of special importance in the dairy industry, causing inactivation of → starter cultures and reduced or slow acid production in cheese factories. – Distribution: raw milk, commercial starters, bulk tanks, cheese baths, air, fresh and dried whey, different types of cheese and other milk products. – Lower risk of phage related problems with multiple strain cultures as compared to single strain cultures. Precautions: strict axenic procedures by maintenance of mother and intermediate cultures in sterilised milk. For large scale starter production the milk is heat treated at 90 °C for ca. 30 min; all phages are not destroyed by pasteurisation. Thorough cleaning and sanitation procedures (no whey residues!); peracetic acid or chlorination will inactivate phages quickly and completely in containers, tanks, pipelines, utensils and on surfaces. Rotation of starter cultures when reduced acid production is observed (Teuber 1987).

bacteriostasis

Reversible state of a bacterium during which growth is prevented by a→bacteriostatic agent. Sublethal damage. The cells are not destroyed or killed by such inhibitory substances. In→selective media this phenomenon is used to prevent the growth of undesired microorganisms, e.g., violet-red-bile-glucose-agar for the detection of Enterobacteriaceae (→coliforms). – As soon as the inhibitory agent is diluted sufficiently growth will proceed. This phenomenon is observed, e.g., when the concentration of a disinfectant is diluted by rinsing water.

bacteriostatic agent

→Disinfection (Figure 5, see page 97);→inhibitory disinfection methods. Inhibitory agents with a bacteriostatic effect, or→bactericides applied at reduced levels. Includes several disinfectants, some antibiotics, some plant components, spices, etc.

Bacterium prodigiosum→*Serratia marcescens.*

Bacterium pyocyaneum→*Pseudomonas aeruginosa.*

Bacteroides

Bacteria division F (*The Prokaryotes, 2nd Ed.*: Balows, Trüper, Dworkin, Harder and Schleifer 1992). Genus of Gram-negative, obligatory anaerobic immotile rods. Typical habitat: lower intestinal tract; →gut microbes. Cultivation difficult; special basal agar media (e.g., Columbia Agar Base: BBL 11124, Difco 0792, Merck 10455) to be fortified with blood or serum.

bactofuge

Special centrifuges used to reduce the number of bacterial endospores (by 90 to 99%) and vegetative microorganisms (by 60 to 85%) in milk. Application in the production of milk low in spore numbers, e.g., for milk powder and for cheese production to prevent late-blowing of hard type cheeses.

baker's yeast

The yeast→*Saccharomyces cerevisiae* is the most important agent responsible for the leavening of dough during production of bread and pastries; the CO_2 produced during the→fermentation causes the "rising" of the dough. Specially selected strains of *S. cerevisiae* are used for the commercial production of baker's yeast on either sugar beet or sugar cane→molasses as substrate; the molasses is clarified by an alkaline or acid process, the concentration and pH adjusted, and (if necessary) nutrients added, followed by heat pasteurisation. Strong aeration of the→batch or→continuous culture is important. Maintenance of the pH between 4 and 5 ensures some control of bacterial contaminants. The 8- to 12-hour batch fermentation is concluded by a reduced aeration rate to allow maturation of the cells. The cells are harvested by centrifugation, washed in water to remove impurities, and concentrated to 67 – 71% water content by separation on a filter press. It is distributed commercially as "compressed" yeast after cutting the blocks into portions of suitable size and weight. "Active dry yeast" or "instant yeast" is prepared by, e.g., spray-drying to a water content of 8 – 12%, and retains activity for several months. – Spoilage:→ *Geotrichum candidum* may develop on the surface of insufficiently packaged blocks of compressed yeast. Generally surface yeasts (*Mycoderma* spp.) and *Candida utilis* are typical contaminants. "Stains" on compressed yeasts are caused by *Penicillium, Aspergillus,* and *Fusarium* spp., and also by *Serratia marcescens.* Baker's yeast with these defects cannot be used (Rehm 1980; Glaubitz et al. 1983).

bananas

Ripe bananas may become infested by moulds, e.g., *Colletotricum gloeosporioides* (causing anthracnosis), *Thievaliopsis paradoxa* or *Botryodiplodia theobromae*. Harvesting when green, and transport at 12 – 14 °C. Ripening accelerated by→ ethylene (Sommer: in Kader et al. 1985).

barn hygiene

Forms the basis for the production of→ raw milk of acceptable microbiological quality. Concerns the hygiene of the udder, milking machines, pipelines and the employees.

Basidiomycetes

Basidiomycotina. A large subgroup of the fungi (division Eumycotina) represented by ca. 15000 spp., with septate hyphae and cell walls containing→ chitin. Sexual reproduction. Basidia (sing. basidium) (homologous to the→ ascus of the Ascomycetes), are produced in the hymenium, as part of the binucleate reproductive body, the basidiocarp. Basidiospores are borne on the outside of the basidium. The haploid basidiospore germinates, giving rise to a haploid monokaryotic mycelium that later becomes septate with uninucleate cells. Karyogamy of two haploid nuclei is followed by meiosis; mitosis gives rise to 4 haploid nuclei which migrate through the sterigmata, at the tips of which the basidiospores are formed. The spores germinate to produce the primary mycelium, eventually becoming a diploid and dikaryotic clamp mycelium.→ Anamorphic forms are grouped under the→ Deuteromycetes with their own nomenclature for historical reasons. – Practically all→ edible mushrooms (→ cultivated mushrooms) as well as poisonous mushrooms (toadstools) belong to the Basidiomycetes. Saprophytic, also as mycorrhiza on tree roots. Some spp. are phytopathogens of economic importance (rust fungi, blight fungi,→ ear fungi), or wood decay-

ing moulds (house moulds). The spores of some spp. are→ allergens (Bötticher 1974; Moore-Landecker 1972; Müller and Loeffler 1982; Roth et al. 1988).

batch process

Discontinuous fermentation process; cultivation of microorganisms (yeasts, bacteria or moulds) (biocatalysts) on solid or in liquid media (→ bioreactor), for the production of metabolites, e.g., organic acids, vitamins,→ enzymes,→ biomass, or fermented foods. A process is terminated when an essential nutrient is depleted, or when the desired product concentration is reached, or by "self-toxication" resulting from metabolic products (Demain and Solomon 1988; Kunz 1988).

Baumé

(Degree Bé). Measure for the concentration of→ brines for which a calibrated Baumé-aerometer is used. The specific mass (weight) of 1 ml of liquid at 15 °C is given in grams. The salt concentration is adjusted according to the specific type of product.

bean sprouts

Grain seedlings (wheat, watercress, mung beans, soya, etc.) used for human nutrition. Germination takes place at room temperature and a wide range of utensils for domestic use exists. The bacterial load on the washed grain ranges between $10^4 - 10^6/g$. The bacterial count increases up to $10^9/g$→ aerobes in the first 48 h of the germination period with a rH of 100%. It is recommended that the grain is rinsed in hot water at 90 °C for a period of 5 minutes before germination is initiated. The reduction of the bacterial count to $10^3/g$ constitutes an insignificant health risk [Bomar, M. T.: *Ernährungs-Umschau* 34 (1987) 226 – 228].

beer

Probably the oldest fermented beverage.

Cereals (mostly barley and/or wheat) are soaked in water (steeped) and allowed to germinate under controlled conditions. → Amylases, excreted from the scutellum, hydrolyse the starch to maltose, raffinose, glucose and other sugars, including maltodextrins. The germinated grain (malt) is dried and shredded. The drying temp. ranges from 60 to 90 °C for light beers and up to 105 °C for dark beers. By the mashing process the "green" malt is mixed with warm water at 50 – 60°C and is hydrolysed to fermentable sugars. The resulting wort is rich in nutrients and serves as substrate for fermentation. The required sugar concentration is adjusted, in some countries also by using specific saccharolytic enzymes in combination with other starchy adjuncts. Insoluble residues are removed by filtration in the lauter tun, and boiled with hops (pellets or extracts of the female flower of the hop plant *Humulus lupulus*) which provides the characteristic flavour and has some preservation effect. After cooling to 5 – 10 °C the wort is inoculated with a pure culture of brewer's yeast – specially selected strains of → *Saccharomyces cerevisiae*. The fermentation continues at ca. 10 °C, and, depending on the type of beer, may take 5 – 10 days. The dissolved oxygen is utilised in the initial stages, followed by fermentation via the Emden-Meyerhof-Parnas metabolic path. In 12 enzyme catalytic steps the sugars are fermented in an exothermal reaction to → ethanol: $C_6H_{12}O_6 \rightarrow 2C_2H_5OH + 2CO_2 +$ 27 kcal. Generated heat has to be removed by cooling. The type of beer, the ethanol concentration and quality are determined by the yeast strain, the wort concentration, the amount of hops and the temperature in the fermentation vessel (see Table 7). – Bottom fermenting yeast strains (e.g., for Lager beers) sediment during → fermentation (5 – 7 °C), whilst top fermenting yeast strains (at 15 °C or higher; for Ale beers) remain in suspension and may rise with gas (CO_2) to the foam. The produced alcohol is toxic to the

Table 7 Beer types manufactured under different processing conditions.

Type	Alcohol Content	Comment
Bottom fermented beers (lager beers)		
Pilsener types	2.9 – 4.7%	light
Export beer	3.6 – 4.7%	light or dark
Strong beers ("Bock" beer)	4.6 – 5.8%	golden yellow or dark
Double strength "Bock"	5.9 – 7.0%	light or dark
"Märzen" Beer	3.9 – 4.7%	light or dark
Diet Beer	3.5 – 4.2%	light, low in carbohydrates
Top-fermented beers (ale beers)		
Altbier	2.7 – 4.2%	dark
Kölsch	3.0 – 4.4%	light
Wheat beer ("Weizen")	3.4 – 4.4%	light or dark with yeast: "Hefeweizen" filtered: "Kristallweizen"
Berlin "White" beer	2.0 – 3.0%	with yeasts and lactobacilli

yeasts, and, depending on the strain, 12 – 18% may be lethal (Priest and Campbell 1987). – As alternative to the malting process unmalted grains may be saccharified with industrial amylases. A recent approach is to apply genetically recombinated yeasts, that are able to produce amylases in contrast to normal yeasts. Approval for these processes is still pending for a major number of countries.

beer sarcinas
Traditional designation used in breweries for beer spoiling pediococci (*Pediococcus damnosus, Pediococcus cerevisiae*). Production of → diacetyl causing unpleasant sensoric changes.

beer spoiling organisms
Contamination, usually after fermentation or maturing, caused by transfer and filling procedures. Mainly *Lb. brevis, Lb. casei, Lb. plantarum, Pediococcus damnosus, Leuconostoc mesenteroides, Megasphaera*, etc. Sensoric defects in bottled (filled) beer: off-flavours and off-tastes, turbidity and/or sediment formation. – Potential spoilage organisms are represented by *Leuconostoc* spp., enterococci, *Gluconobacter, Acetobacter* and *Micrococcus* spp. and some Enterobacteriaceae. Growth promoted by composition failures and wrong storing practices. → *Zymomonas* of importance only for → raw fruit beers. – Cold tolerant *Enterobacter* and *Hafnia* spp. may develop in the → wort, also *Micrococcus kristinae* that may cause flavour defects in the end product, e.g., by production of dimethyl sulphide (Baumgart 1986; Krämer 1987; Priest and Campbell 1987).

beer wort → wort.

beer yeasts
Numerous strains of *Saccharomyces uvarum* and → *S. cerevisiae* that are used for beer brewing. Top-fermenting and bottom-fermenting strains are distinguished, each group represented by a large number of strains The bottom-fermenters have formerly been designated *Saccharomyces carlsbergensis*, but are presently classified as *S. uvarum*. The optimum fermentation temperature is 5 – 7 °C; flocculation (→ agglutination) is characteristic, resulting in sedimentation of the yeast cells during the fermentation process. Top-fermenting strains (*S. cerevisiae*) remain suspended as single cells and do not sediment; their opt. fermentation temp. is 10 – 22 °C (see Table 7). Other representatives of this group are the → bakery and → distillery yeasts (Gaubitz and Koch 1983; Priest and Campbell 1987).

benzoic acid
C_6H_5-COOH. Phenyl-formic acid. Colourless, odourless crystals. Present naturally in many resins, essential oils, in cranberries and in carrots. → Preservative. EC no. E 210; sodium benzoate, E 211; potassium benzoate, E 212; calcium benzoate, E 213. → ADI value (FAO/WHO): 0 – 5 mg/kg BW. LD_{50} rat (p.o.): 1.7 – 4 g/kg BW. Effectiveness increases with reduced pH, especially below 5.0 as a result of the increasing % of undissociated acid: at pH 4, 60.7%; pH 7, 0.15%. More effective against bacteria and yeasts than moulds (Chicester and Tanner 1977; Wallhäusser 1987). Removed by the FDA from the "GRAS" list; use under food additive regulation.

benzylalcohol
Phenylmethanol. $C_6H_5CH_2OH$. Preservative for pharmaceutical preparations. Effectivity against Gram-pos.: good; against Gram-neg. and moulds: moderate. Partly inactivated by Tween 80. – LD_{50} rat p.o.: 1.2 g/kg BW. Not approved for use in foods.

berry fruit

Fruit types such as strawberrries, raspberries, bramble berries, goose-berries, blueberries, black currants, cranberries and grapes with limited shelf life and a high spoilage potential by moulds. Most important are: *Botrytis cinerea, Rhizopus* spp., *Alternaria* sp., *P. expansum, P. patulum* and *Cladosporium* sp. Preventative measures: harvesting during dry weather conditions, after disappearance of the dew; prevention of damage during harvesting and transport, sorting, refrigeration as soon as possible after harvesting, and maintenance of cold chain. → Preservation: freezing, cooking, canning, covering with rum or brandy and sugar; grapes may be (sun-)dried (e.g., raisins, sultanas, corinths).

berry wine → fruit wine.

Betabacterium

Conventional "subgenus" designation for heterofermentative (CO_2 producing) lactobacilli; today defined as obligately heterofermentative representatives of the genus *Lactobacillus* (Bergey 1986).

Betacoccus → *Leuconostoc*.

beverages → alcohol free beverages.

Bifidobacterium bifidum

(*Lactobacillus bifidus*). Important representative of the genus *Bifidobacterium* ("bifidus"). Irregular to pleomorphic rod-shaped, Gram-positive, anaerobic bacteria producing L(+) lactic acid and acetic acid from glucose. Grouped under the → Actinomycetales. Inhabitant of the gastro-intestinal tract; special importance as "pioneer" organisms for breast-fed children, but also for bottle-fed infants and adults. May be incorporated in some → starter cultures, e.g., Biogarde, used for the manufacture of some fermented milk products.

bifidus factor

A disaccharide, β-galactosido-fructose, representing ca. 0.4% of the oligosaccharides in mother's milk. Beneficial to *Bifidobacterium* population of GI tract in adults and children acc. to Petuely (Vienna); favourable effect of L(+) lactic acid produced. Nutritional physiological value not substantially proven.

bile

Excrete of the liver; emulsifies fat in the intestines. Main components bile acids (steroid carboxylic acids)(salts: cholates); primary bile acids synthesised in the liver; secondary bile acids produced from the primary acids by bacterial action in the intestine. Inhibitory to a large number of bacteria, especially Gram+, but not to that extent to → coliforms. Dehydrated ox bile is therefore used as constituent in some → selective and → differential media. Examples: 5 g/l in crystal-violet-bile-lactose broth (VRB) acc. to Kessler-Swenarton; or 20 g/l in brilliant-green-bile-lactose broth (BRILA broth) for the selective enrichment and quantitative detection of *E. coli* and other fecal coliforms in water and food.

bio-alcohol

→ Ethanol (alcohol, spirits) produced by (anaerobic) fermentation from CH. Especially refers to alcohol of lower purity produced from "renewable raw materials", e.g., sugar cane, cassava and other starchy materials, wood waste products, etc. *Zymomonas mobilis* and other bacteria employed with increased frequency (Gottschalk et al. 1986).

bio-availability

That part of our food that can be absorbed during pasage through the gastro-intestinal (GI) tract. Only those nutrients are absorbed that are hydrolysed by enzymes present in the GI tract. Some macromolecules that reach the lower part of the intestines, including the appendix and rectum, may be partly degraded by the autochthonous microbes with the production of → biomass, whilst a smaller fraction may still be absorbed.

biocoenosis
The community (of MO), comprising the biotic components of an ecosystem (→ biotope). It is in → biological balance at a given point and time, but may constantly change under the influence of chemical and physical factors governing the biotope. In or on most foods the biocoenoses are rarely constant, and represent → successions that are (beneficially) affected by metabolites (e.g., organic acids) of one population.

biofilters
→ Contaminated air from food plants, factories, canteens and large kitchens contains aromas and droplet size components (water, lipids), in addition to MO and dust. Biofilters may be used to purify such air without leaving any residues, and constitute either soil filters or areal filters with waste compost as carrier material and a natural microbial population, or single- or multistage closed → bioreactors with natural microbes and specially isolated spp., or "bio-washing" installations acc. to the same principle as the trickle filters or activated sludge vessels. More recently, membrane reactors and parallel stream columns filled with carrier materials, have been developed [Beck et al.: *Biotechforum* 6 (1989) 94 – 99]. The molecules or particles (nutrients) in the air are being metabolised by immobilised MO in the filter. Environmentally "friendly" and economic.

Biogarde
Registered name for a fermented milk product for which a mixed strain starter culture, consisting of *Streptococcus thermophilus*, *Lactobacillus acidophilus* and *Bifidobacterium bifidum* is used. (Krämer 1987).

biogas
A mixture of CH_4 (50 – 60%), CO_2 (30 – 40%), H_2 (5 – 10%) and minor amounts of H_2S and nitrogen, resulting from the anaerobic digestion of animal and plant wastes. These gas mixtures may be produced in wastewater polluted by organic material (e.g., from breweries, dairy plants, sugar refineries, etc.) or in sludge from the activated sludge tank in a biological → sewage treatment plant. The anaerobic microbes of the anaerobic digestor convert ca. 95% of the organic materials (activated sludge, etc.) into biogas. Only ca. 4% sludge remains which is to be disposed of, e.g., by burning. Biogas can be used as energy source, e.g., for heating or generation of electricity. – In developing countries (e.g., India) small sewage treatment plants, made from oil drums, are used for supplying kitchens with gas. This has the beneficial effect that dried animal dung is used as a fuel, and that N and P are not wasted any more by the burning process, but function as fertilisers when the digested sludge is utilised in agriculture.

biogenic amines
→ Decarboxylated amino acids. Synthesis by decarboxylation and hydroxylation of amino acids affected by microbial and product specific enzymes, e.g., histamine (from histidine), tyramine (from tyrosine), tryptamine, serotonin and melatonin (from tryptophane) and 4-amino butyric acid (from glutamic acid). May be found in raw milk (ca. 1 mg/kg) or different types of cheeses (up to 2300 mg/kg). Detrimental to health in concentrations exceeding 1 g/kg of food. Example: "histamine poisoning" caused by mackerel or tuna after unrefrigerated storage. Biogenic amines are not destroyed by cooking or grilling. Some "harmless" bacteria may produce

biogenic amines leading to what has earlier been called "unspecific" food poisoning; typical toxin producers were considered non-detectable in foods or faeces. – In addition to causing allergic reactions that are amplified by some pharmaceuticals, biogenic amines may react in the acid environment of the stomach with→ nitrite from the sputum to form→ nitrosamines (Classen et al. 198.'; Sinell 1985).

Bioghurt
Registered (commercial) name for a milk product fermented by the LAB *Lb. acidophilus* (DL lactic acid) and *Lactococcus lactic* var. *taette* [L(+) lactic acid]. The latter produces slime that improves the consistency and reduces whey syneresis (Teuber 1987).

biological balance
Stability or instability of a microbial population consisting of different spp. Symbiotic or mutualistic MO contribute to stabilisation of the mixed population (e.g., yoghurt); antagonists are responsible for a shift towards one sp. or group and influences other group(s) unfavourably (e.g., in→ sourdough). By specific modification of a→ habitat (ecosystem), e.g., by addition of salt, the balance can be influenced towards domination by one or more groups in→ succession, e.g., with the production of→ sauerkraut. Temperature changes may also exert a pronounced effect on the population balance within a (food) ecosystem.

biological degradation
Refers mainly to enzymatic hydrolysis of substances not typically found in nature, e.g., pesticides, by MO. An important process for the elimination of such materials from the→ ecosystem.

biological oxygen demand
BOD (also: "biochemical oxygen demand"). Refers to the amount of dissolved oxygen required for the biological (microbiological) oxidation of organic material in an aquatic environment. BOD_5 refers to the amount of O_2 (in mg) utilised in 1 litre of (e.g.) sewage by the microbes present in 5 days. The oxygen concentration of the water at 20°C is determined iodometrically before and after this time period, and the difference expressed as mg oxygen/l. An amount of 60 g reflects an inhabitant relative consumption value of 140 l water per person per day. This is also the basis for calculating the cost for treatment – and judgement – of industrial sewage. – Presently BOD is only determined occasionally, since the→ chemical oxygen demand (COD) represents a more effective and exact method, that is not influenced by (e.g.) toxic substances present (Gottschalk et al. 1986).

biological sewage plant
Sewage from communities and food industries is treated to reduce the biological oxygen demand (BOD) of the liquid waste. The primary treatment involves the mechanical removal of solid particles (representing on average ca. 1 – 2% of the volume) by passage through a series of screens and sedimentation basins. In the secondary treatment the organic matter is degraded aerobically by microbial action, with spp. of *Pseudomonas, Alcaligenes, Flavobacterium, Acinetobacter*, etc., predominating. In the activated sludge treatment the liquid is aerated vigorously by the use of specific devices (e.g., surface aerator), and ca. 50% of the organic matter is oxidised to CO_2 and the rest assimilated into→ biomass. The greater part of the sludge is removed to the digestor where it is turned into→ biogas and "treated sludge" by anaerobic bacteria; a small part is removed as inoculum (Rehm 1985).

bioluminescence
Generation of visible light by (e.g.) some

MO's, fireflies and crustaceans. A redox reaction catalysed by the enzyme luciferase in the presence of O_2. Photobacterium spp. are found in association with deepsea fish; some halophilic MO are associated with harvested sea fish during storage at $4 - 6 °C$ ($\rightarrow$ luminescent bacteria).

biomass

Estimation of organisms/cells (usually MO), e.g., by dry weight, in a given habitat. May be used as basis for determination of growth kinetics of a culture, e.g., cells are harvested from a liquid culture medium by filtration or centrifugation. In bacteria biomass increase continues to some extent beyond the exponential growth phase, whilst in moulds it practically ends with the initiation of the $\rightarrow$ idiophase. Determination of the (viable) cell numbers ($\rightarrow$ cfu) by cultural methods only reflects the number of actively dividing cells. – Estimation of biomass by direct methods: wet or dry weight, nitrogen content, total C-content, DNA (Obst 1988) and protein content; indirect methods: nephelometric, oxygen consumption, CO_2 production, acid production, catalase activity and other parameters that can rapidly be determined, and compared in a calibration curve with direct values.

bioreactors

Open or closed fermenters (tanks) for the cultivation of MO with the aim of producing $\rightarrow$ biomass or metabolic products (organic acids, enzymes, etc.) through bioconversion. Food raw materials may be modified sensorically and their shelf life improved by the action of microbial cells and/or enzymes (biotransformation). Bioreactors/fermenters may be distinguished according to their construction and applications.
(1) Batch (discontinuous) fermenters (stirred tank reactors)

- submerged culture processes, e.g., for ripening of cream, cheese processing, fermentation tanks in breweries, etc.
- multistage submerged processes, e.g., for $\rightarrow$ single cell protein, food/feed yeast from cheese whey, etc.
- surface culture (flat pan) processes, e.g., as was used initially for penicillin production
(2) Continuous process bioreactors (chemostat, turbidostat)
- cultivation of algae in open canal systems for protein production
- generator process for vinegar production (bacteria adhering to e.g., wood chips) or biological filter (trickle filter or percolating filter) process for wastewater treatment
- chemostat for production of citric acid
(3) Solid state fermentation/fixed bed reactor with immobilised MO in water or air flow stream, e.g., for ethanol production (Chmiel et al. 1987; Kunz 1988).

biosensors

Originally developed as filter paper strips, impregnated with enzyme solutions containing reagents; further improved by L. Clark in 1962 by using an indicator electrode in which an enzyme is enclosed in a semipermeable membrane. Modern biosensors are referred to as biochips, and, in addition to enzymes, MO may also be used. A biosensor may be constructed by combining an immobilised biological system (enzyme, antibody, organelle, cell) with a transducer (thermistor, potentiometric or amperometric electrode, field effect transistor, piezoelectric sensor, optical or optoelectronical device). The complementary counterpart (substrate, antigen) is recognised by the biological device (H.-L. Schmidt and I. Saschewag: in Chmiel et al. 1987). In food analysis biosensors may be used for the quantitative determination of glucose, galactose, sucrose, ethanol, ureum, choline, L-lysine and D(+) lactate (Scheller and Schubert: *BioEngineering* 1/87, 30 – 39).

biotechnological disposal

The disposal/treatment of heavily polluted wastewater (e.g., from breweries, dairy plants) is becoming a matter of increasing concern to food processing plants. The tightening of regulatory standards for industrial effluents increases pressure on industries for disposal measurements in their own interest. Two processes may be used: (1) aerobic treatment which is a rapid and effective method, although energy costs for aeration are relatively high and, in addition, the disposal of sludge may constitute an additional problem; (2) anaerobic treatment (digestion), which is successfully used for the treatment of organically heavily polluted effluents, and requires low energy input; in addition, smaller amounts of sludge remain to be disposed of, whilst→ biogas represents a beneficial by-product. – → Biofilters and "bio-washing" units may be used for the low-cost and highly effective treatment of factory air. – Solid waste materials, e.g., from vegetable processing, may be composted.

biotechnology

The technical application of biological systems (MO, cell cultures, plants) for the production of natural substances (biogas, antibiotics, enzymes, organic acids, etc.) or for the controlled modification/change of a substrate (water purification, food fermentations, etc.). Some of these processes have been known empirically for many generations (wine, beer, sauerkraut, cheese production), but their microbiology is being understood only since the beginning of this century, and the biochemistry only since ca. 50 years. This knowledge, however, is presently being transferred and applied to new areas. One of the main aims of biotechnical processes is the application of continuous operating processes, e.g., for denitrification, removal of iron and manganese from drinking water, water purification, etc. The use of biotechnically synthesised products is limited by financial factors, e.g., the price of ethanol or biogas (as fuels) as compared to the world market price of mineral oil.

biotope

Refers to a specific environment (habitat) that is occupied by micro- and/or macro-organisms. Typically, a (natural) environment is characterised by specific intrinsic and extrinsic parameters, that provide favourable conditions for the growth and development of specific group(s) of MO: e.g., different cheese types, sauerkraut, gastro-intestinal tract, rumen, biological purification plant, anaerobic digester.

biotransformation

Conversion of a substrate by MO or enzymes to a specific product of particular value: e.g., oxidation of sorbitol to L-sorbose by *Gluconobacter suboxydans* as first step of vitamin C synthesis; production of gluconic acid from glucose by *Asperigillus niger*. Normally only some enzymes, and no synthetic chains, of the microbial cells are utilised. – Production costs of many pharmaceutics are drastically reduced in this way, e.g., cortisone and steroid hormones.

bitter peptides→ peptides; → rennet.

bitter rot

Also called→ *Gloeosporium* rot by which a wide variety of fruits are affected, e.g., apples, pears, stone fruit. Also caused by *Sclerotinia* spp. Infection by conidia of the bark-associated mould typically before harvest. Hyphae penetrate the lenticels of stone fruit and may even grow at 0 °C. Bitter taste, even away from infected areas.

black moulds

Imperfect fungi of the family Dematiaceae, of which the mycelium and/or conidia are coloured dark on account of melanine deposits. In the food area, especially the genera *Alternaria, Cladosporium, Helminthosporium* and *Stemphylium* are of interest (Ellis 1971). Some of these are destructive→field fungi of cereals, acc. to Christensen (Beuchat 1987). Other representatives colonise→filling materials of wall cracks, or→wall moulds in storage rooms and cellars.→*Alternaria* may produce→mycotoxins.

black olives→ olives.

black rot

→*Proteus* and other Enterobacteriaceae may penetrate through the shell into the→egg and produce H_2S that causes black colouration of egg white and yolk (Sinell 1985).

black spot

Type of spoilage of frozen (to −15°C) red meats and poultry. Black spots caused by surface growth of moulds: *Cladosporium,* →*Aureobasidium* and different→*Penicillium* spp. (Sinell 1985).

black yeasts

Common name for black coloured yeasts containing melanine deposits. Mainly→ *Aureobasidium.*

blanching

Short-time heating of fruit or vegetables that have been washed, cleaned and even peeled and sliced. Typical treatment prior, e.g., to filling and sterilisation, or packaging and freezing for→deep freezing. Product specific enzymes are partly inactivated and the microbial load reduced by this treatment. Temperature-time combination for treatment is dependent on the product type and varies between 70 and 95°C, and 1 to 10 min (Kunz 1988).

blight moulds

Refers to a wide range of plant pathogenic moulds, belonging to the Basidiomycetes. Produce darkly coloured spores on infected plants.→Ustilaginales are of economic importance, especially for grains and maze;→ear molds.

blood agar

Medium for isolation and cultivation of fastidious, mostly pathogenic, bacteria, and for the detection of→hemolysis. Blood agar base (e.g. Merck no. 10886; Difco 0045; BBI 11037) is cooled to 45−50°C after autoclaving and adjustment of the pH to 6.8; 4−8% defibrinated blood is mixed into the base and the medium poured. −Blood glucose agar is used in the examination of drinking water for the detection of sulfate reducing anaerobes.

blood plasma

Blood of slaughtered animals, especially pigs, out of which blood cells have been separated. The use of plasma as additive (either as liquid or dried) is approved for the processing of certain meat products, e.g.,→cooked or mildly heated cured sausages.→Ochratoxin A may be transmitted (→carry over) by feed of the animal, and is found in the plasma of 15−20% of the pigs in Germany. Ochratoxin concentrations in sausage products range between 10 and 150 μg/kg. In the plasma of herbivores detectable amounts have not been found, since ochratoxin is degraded by the rumen microbes.

blue-green algae→cyanobacteria.

blue mould

May generally include any blue-spored spp. of *Penicillium.* Specifically refers to *Peronospora tabacina* (Peronosporales, Oomycota), a pathogen of tobacco plants,

of considerable economic importance. –
Sometimes also refers to *Penicillium italicum*, causing blue-green stains on citrus
fruit.

blue-veined cheese
Soft cheese type, ripened with *Penicillium
roqueforti*. Conidia are either inoculated
together with the→starter culture during
rennetting in the cheese bath, or after
completion of the process by pin-inocula-
tion. Relativly slow development of the
mould during ripening. Examples: Roquefort, Gorgonzola, Bavaria blue, etc.

BOD→biological oxygen demand.

bone taint
Deep spoilage, especially near the bones,
of large portions of meat not sufficiently
chilled throughout, e.g., ham. *Clostridium,
Proteus, Streptococcus* and *B*. species
are responsible for the spoilage and produce a foul smell (Krämer 1987; Sinell
1985).

Botrytis
Grey mould. Genus of the Deuteromycetes belonging to the→Ascomy-
cetes (class Hyphomycetes). Leaf mould.
Noble rot (*B. cinerea*) of grapes. Cosmopolitic, including the onion mould, and
found commonly on dying leaves and
other plant tissues. Excreted mycosporin
2 promotes the growth of other MO and
thus secondary infections. – Sclerotia
produced.

Botrytis allii
Syn.: *B. aclada*. Causing onion rot during
storage. Grey mycelium, producing
masses of conidia, grows underneath the
skin, and is often recognised too late. The
infected tissue softens and obtains a
brown discolouration. Black, pea-sized
sclerotia are formed (Müller 1983a).

Botrytis cinerea
Teleomorph *Sclerotinia fuckeliana*. Common cause of post-harvest (psychrotolerant) spoilage of strawberries, other berry
fruit, lettuce, carrots, celery, cabbage and
stone fruit. Growth inhibition at < 2%
oxygen in atmosphere (→CA storage). Important disease of tobacco, especially in
moist years (Müller 1983a). – May cause
severe damage to vines, especially young
shoots; control by treatment with 8-oxy-
chinoline. – Noble rot (German: "Edelfäule") refers to growth of the mould on
ripe or over-ripe grapes, leading to loss of
water and a concentration of the sugar
and other components in the must. Part of
the organic acid is decomposed in the
wine berries with the production of (a.o.)
glycerin; the concentration of the latter
may reach up to 14 g/l. High quality, relatively sweet, wines with a "noble
bouquet" produced.

bottle fermentation – sparkling
wine.

bottom fermenting yeasts
→Yeasts able to flocculate and sediment
to the bottom at the end of the main fermentation procedure.→Beer.

botulism
Disease of man and animals with high
lethality rate. Caused by botulinum toxins
produced by toxinogenic strains of *Cl. botulinum* types A, B, E, F or G (for humans)
and D or E (for animals). These toxins are
proteins with high molecular weight
(200,000 to 900,000), and 0.1 to 1 μg per
adult may be lethal. Symptoms appear
within 12 – 36 h after consumption of toxin
containing food: nausea and vomiting,
diarrhoea and a number of neurological
symptoms, including visual impairments,
loss of normal mouth and throat functions,
fatigue and general muscle weakness,
and finally, respiratory impairment.
– Early recognition of the illness,

respiratory support and treatment with polyvalent antiserum may improve the chances of survival (Hauschild 1989). → Infant botulism.

bread
Staple food made from grain flour(s), water, salt and possibly some additives such as spices, enzymes and baking aids. The ingredients are mixed and kneaded into a paste; the dough is then fermented to produce CO_2, either by→ sourdough or→ baker's yeast, causing leavening and promoting a looser texture of the final product. During the baking process (ca. 35 min/215 °C) practically all MO are destroyed. → Ergot that may be present is detoxified to >50% during baking. – During the production of prepacked sliced bread, cut surfaces are contaminated (mainly by mould conidia from the surface). Thus, additional precautions, including possible heat treatment, may be necessary. The "red bread mould" (*Neurospora sitophila*) that caused losses earlier, is relatively heat sensitive, and is killed at 72 °C within 5 min. Toast bread may be packaged with CO_2 as protective gas, with O_2 content <1%. Up to 3% propionic acid or its salts (E 236 – 238) have been used as preservatives earlier, but approval has been withdrawn for several countries recently. – Cutting out of mould growth from bread may eliminate possible mycotoxins, which diffuse only very slowly into the bread crumbs away from the mycelium.

bread mould
Trivial name for the majority of the green moulds of the genera *Penicillium* or *Aspergillus*. More seldomly: → *Geotrichum candidum*, → *Neurospora sitophila* and *Trichosporon variabile;* → bread.

Brevibacterium linens
Obligately aerobic, catalase-positive, asporogenous, strongly proteolytic but non-lipolytic rod-shaped bacteria associated with yellow to orange-red colouration on cheese surfaces. Salt tolerant (15% NaCl), nitrite + . Growth: opt. 21 °C; min. 8 °C; max. 37 °C; pH: 6.0 – 9.8. – Suspensions in 5 – 7% saline are used for surface inoculation ("smearing") of different types of cheese during ripening (e.g., → sour milk cheese, Romadour, Limburger, Münster, etc.). Bacterial proteases diffuse into the cheese curd and contribute to the ripening process. Dense population of the surface, as well as regular rinsing with saline, prevent mold growth. – Starter cultures are commercially available as suspensions or in lyophilised form (Teuber 1987).

Brie cheese → Camembert.

BRILA broth → brilliant green bile broth.

brilliant green bile broth
Brilliant green bile lactose broth for the selective enrichment of → "coliforms", or → *E. coli*, using → Durham tubes for the detection of gas (CO_2) production from lactose. Commercially from (e.g.): Merck (no. 5454), Difco (no. 0007), BBL (no. 11080).

brine
Solution of table salt in water, also with addition of nitrate, nitrite, sugar or organic acids in order to extend the shelf life of foodstuffs to some extent. The product remains either temporarily in this liquid, e.g., → pickled products and cheese to remove part of the soluble substances, or until it is used, e.g., → fermented cucumbers, → mixed pickles and olives. The combination of brine ingredients, the period of contact with the brine and the temperature will be determined by the product. In addition to the effect on the taste, the aim is also to destroy or inhibit the undesired MO and stimulate the

beneficial MO in view of the ripening process. In some instances starter cultures are added to the brine, e.g., mixed pickles, fermented cucumbers and "sauerkraut". The salt concentration is measured in→ Baumé (Bé) and must be held constant. Heat treatment to destroy undesired MO is required from time to time (Kunz 1988).

brine injection
Rapid method of brining, by which the brine is injected directly into the muscle or veins. Has special advantage for larger meat portions, enabling the rapid penetration of→ nitrite into the deeper layers and thus safeguarding the product against *Cl.* spp.

brine spoilage
Turbidity, slime and foam formation in pickle brines commonly caused by halotolerant species of the genera *Alcaligenes, B., Ps.,* or *Vibrio.* The meat develops off-odours, a glass like appearance or sticky surface and remains raw inside.→ Surface yeasts are often responsible for the spoilage of brine used for fermentation or preservation of plant products.

broad-spectrum antibiotics
Natural metabolites of a number of MO that cause inhibition or death of other MO, mainly bacteria. Typically, Gram-positive as well as Gram-negative bacterial types may be eliminated. Examples: tetracyclines, chloramphenicol, etc. Applications: medicine, veterinary medicine and as ingredients of→ selective growth media. For the isolation of yeasts or moulds from a food sample, a "harmless" broad-spectrum antibiotic is added that will only affect (inhibit) bacterial growth; → mould count.

brown rot
→ brown (bitter) rot. General term for diseases of stored fruit and vegetables, characterised by brown soft lesions. Fruit (general): *Sclerotinia* sp., *P. expansum, P. patulum* (patulin production); carrots: *Phoma* sp.; potatoes: *Phytophthora infestans* (Müller 1983).→ Bitter rot.

Brucella
Bacterial genus under the "Alpha Subclass" (Subdivision C) of the bacteria (*The Prokaryotes, 2nd Edition*), or Section 4 (*Bergey's Manual of Determinative Bacteriology* 1986). Gram-negative, aerobic, immotile rods. Occur typically as intracellular parasites or pathogens of man and animals. Infection of pregnant animals by (e.g.) *Br. abortus* (cattle), *Br. melitensis* (goat, sheep), and *Br. suis* (pig), may result in placentitis and abortion. Persons in contact with infected animals may also become infected via mucous membranes or small skin lesions;→ zoonoses. These pathogens are excreted via the milk. – Killed by pasteurisation. Foodborne infections possible when infected raw milk or raw milk cheese (especially sheep and goat milk cheese from the Mediterranean) are consumed. *Br. abortus* causes Bang's disease (brucellosis), and *Br. melitensis*, Malta fever with symptoms reminding of typhoid fever. – Cattle in the northern industry states are practically free from brucellae; however, they may be associated with sheep and goats in Southern European countries (Krämer 1987; Sinell 1985).

budding
Asexual reproductive process of some yeasts and bacteria. An out-growth or protrusion (bud) develops from the cell wall of the mother cell into a daughter cell; the latter is composed of protoplasma and a nucleus. The daughter cell enlarges and forms a separate cell that may remain attached to the mother cell under certain conditions. Groups or bunches of cells resembling mycelia may thus be formed

(→ pseudomycelium) (Müller and Loeffler 1982).

bulbular vegetables→ potatoes.

butter
Sweet cream butter is made from cream with 30 – 35% fat content, after heating at 102 – 105 °C for a few min for elimination of recontaminating organisms. In a butter churn the butter fat is mechanically agitated at temperatures ranging from 12 to 23 °C, depending on the type of process. Butter clumps form that separate from the→ buttermilk (sweet cream whey), and after straining it is washed with drinking water for removal of protein residues and contaminating organisms. After kneading for removal of excess water, the butter is formed in standard sizes and packed. – For sour cream butter the fresh cream is pasteurised and on cooling to 20 °C inoculated with 2 – 5% of a→ "butter" culture; souring to pH 5.0 – 4.8. → Acetoin and→ diacetyl produced by *Lactococcus lactis* ssp. *diacetilactis* and *Leuconostoc mesenteroides* ssp. *cremoris*. The→ NIZO process may also be used for the production of sour cream butter. – The shelf life of the product depends to a large extent on the microbiological quality of the washing water, and the hygienic conditions in general, but specifically of the sanitation and disinfection procedures used for cleaning utensils, containers and the packaging plant.→ Coliforms (Enterobacteriaceae),→ *Pseudomonas* spp. and → moulds may constitute part of the spoilage association (Kunz 1988; Teuber 1987). Butter contains at least 80% fat, and not more than 16% of water; up to 2% salt is added in some regions. Cultured butters are better known in Europe than the USA, and have a better flavour than sweet cream butter; however, they undergo chemical deterioration faster than sweet cream butter.

butter aroma
→ Diacetyl is formed from→ acetoin under aerobic conditions. The latter is produced by *Lactococcus lactis* ssp. *diacetilactis, Leuconostoc* spp., *Pediococcus* spp. and other→ LAB in presence of citric acid. A desired flavour component in butter but undesired (defect) in→ beer and→ wine.

butter cheese
Semi hard cheese with 48% dry matter content. Starter culture and rennet added to milk at 38 – 40 °C. Processing at temperatures lower than 35 °C. Ripening at 2 – 5 °C.

butter culture
Starter culture for the controlled fermentation of cream that is used for the production of sour cream butter (ripened cream butter) and sour cream ("créme fraiche"). Typically constitutes a mixture of *Lactococcus lactis* and *Leuconostoc mesenteroides* ssp. *cremoris* strains. Formation of→ acetoin-diacetyl initiates at pH 5.5 – 4.5 and 20 – 25 °C. Strains of *Lactococcus lactis* ssp. *lactis, Lactococcus lactis* ssp. *cremoris*, *Leuconostoc lactis* and other *Leuconostoc* spp. may also be present.

buttermilk
By-product from sweet cream butter manufacture, that may be soured ("cultured") with the aid of a→ butter culture to pH of 4.6. Depending on refrigeration the product may keep well for several days; shelf life may be increased by heating, although the addition of alginates (E 401 – 405) may be necessary for stabilisation.

butyric acid
Butanoic acid. Salts: butyrates. CH_3-$(CH_2)_2$-COOH. Simplest fatty acid; colourless liquid with an unpleasant "sweaty" smell upon dilution. Naturally found as glycerin ester in butter (2.8% of

the saturated fatty acids) and some plant fats. Also in the "fruit" of the ginkgo tree (*Ginkgo biloba*) and in spoiled silage. – Butyric acid formed by clostridia in must may cause wine defects. The offensive odour of rancid butter is at least in part caused by butyric acid. – Industrial production either synthetic or with *Clostridium butyricum*.

butyric acid bacteria
→ *Clostridium*. Trivial name for bacteria that ferment CH under strictly anaerobic conditions to butyric acid, butanol, acetone, etc. (→ sensory evaluation), in addition to CO_2 and H_2. Typical spp. include *Cl. butyricum, Cl. pasteurianum, Cl. pectinovorum, Cl. butylicum, Cl. acetobutylicum, Cl. propionicum,* → *Cl. tyrobutyricum; Eubacterium limosum; Butyrovibrio* sp., etc. – May be involved in the spoilage of sauerkraut, silage and different types of (hard) cheeses.

byssochlaminic acid → *Byssochlamys.*

Byssochlamys
A genus of fungi (order Eurotiales, Ascomycetes). Anamorphs: *Paecilomyces.* Most important spp.: *By. fulva* (yellow-brown), *By. nivea* (white), *Paecilomyces variotii* (whitish) (Cerny and Hoffmann 1987; Domsch et al. 1980; Reiss 1986). – Growth generally at high temperatures (up to 45 °C), under reduced O_2 pressure (1%), at high CO_2 pressure (40%) and at NaCl concentrations of up to 7 – 10%. Ascospores and conidia are → heat resistant and the spores also resistant against → disinfectants to an extent comparable to that of bacterial endospores. Survival of pasteurisation. Ascospores preferably formed at pH 2 – 3. – Found in soil, and as typical contaminants of fruit, fruit juices and concentrates and several other food commodities. May cause rot of mechanically damaged fruit during storage. *Pae. variotii* is often isolated from cane sugar. – The synthesis of the mycotoxins → patulin and the more toxic → byssochlaminic acid have been described for all spp. (Beuchat 1987; Müller and Loeffler 1982). Special importance in storage tanks for apple juice and containers for transport of apple juice concentrates and processed fruit.

C

C + D
→ Cleaning and → disinfection. The combination of these measures is the basis of processing plant hygiene in industry. → Process control.

CA storage
Controlled atmosphere (refrigerated) storage of fruit and vegetables. Increase of the CO_2 concentration and reduction of the O_2 in air tight storage rooms result in a lowering of the metabolic activity of the (living) plant tissue, with a concomitant reduction of energy production (→ respiration energy) and inhibition of growth of aerobic MO. Several types of rot spoilage prevented, e.g., → *Gloeosporium*, or inhibited, e.g., *Penicillium expansum*. Total losses (in terms of moisture, or microbial spoilage) may be reduced from 25% to 3% for cold storage of apples; simultaneously the storage time can be extended without losing out on quality. CA storage also results in improved distribution possibilities due to increased resistance of the products against → opportunistic parasites. (Henze and Hansen 1988). – → Ethylene is produced by the stored products and MO present in the storage rooms; it enhances the ripening (aging process), and has to be disposed of with gas atmosphere exchange during storage. The type of CA conditions chosen is dependent on the type and cultivar (see Table 8). – CO_2 concentrations of up to 25% (produced by dry ice in an air tight truck) have proved beneficial for the transport of deciduous fruit sensitive to → *Botrytis cinerea* spoilage (Nicolaisen-Scupin 1985). – Packaging of → salad mixes in CO_2 tight bags (e.g., polypropylene) that allow some O_2 exchange results in comparable beneficial effects with reference to product respiration and MO. – Special CA containers have been

developed for the transport of tropical fruit and asparagus by ship freight.

Camembert cheese
→ Soft cheese with surface mould growth by *Penicillium caseicolum*. Lactic starter cultures used: *Lactococcus lactis* ssp. *cremoris* (95 – 98%) and *Lactococcus lactis* ssp. *lactis* (2 – 5%), in addition to the mould culture. After inoculation and "ripening" of pasteurized milk with 1 – 2% of a starter culture, rennet as well as a conidium suspension are added at 30 – 35°C. The curd is cut in slices of ca. 2 cm after 40 – 50 min and, allowing another 25 min, transferred into cheese moulds. Forming at 20 – 24°C, brine bath (18 – 20% NaCl, 16 – 20°C) for 40 – 200 min depending on size; drying at 17 – 19°C for 2 – 3 d; ripening at 14 – 16°C and 85 – 95% rH until desired mould growth followed by packaging. – Some cheese plants have introduced continuous processes. Main problems are related to recontamination with → coliforms (Enterobacteriaceae) and possible growth of → *Listeria* on the surface. – Brie cheese is produced under similar conditions, with the main difference being the size of the cheese loaves (1 – 3 kg); this results in a different ratio between surface and volume and thus altered partial O_2 pressure in the inner part of the cheese during ripening (Teuber 1987).

Campylobacter jejuni
Gram-negative, slender rod-shaped, low a_w sensitive bacterium, with opt. growth temp. 37°C. – Distribution: in the reproductive and intestinal tracts of man and animal; raw milk, raw poultry meat, chicken liver, raw minced meat. Survival in: faeces and soil at 4°C for 18 d, and at 25°C for 9 d; poultry meat at – 20°C for > 21 d; milk at 4°C for 22 d and at 22°C

Table 8 CA storage conditions compared to refrigerated storage of vegetables and fruit (acc. to Nicolaisen-Scupin 1985, modified). The difference in atmospheric gases is made up of nitrogen, unless otherwise stated.

Type	Refrigerated Storage			CA Storage				
	°C	r.H. %	Period	°C	r.F. %	CO_2 %	O_2 %	Period
Cauliflower	0	92 – 95	2 – 3 W	1	95	5	3	6 W
Lettuce	0 – 1	95	1 – 2 W	1	95	2	1 – 2	2 – 3 W
Green peppers	10 – 12	90 – 95	<3 W	10 – 12	95	3 – 5	2 – 3	<6 W
Brussels sprouts	– 1	<95	1 M	0 – 1	95	5	2	2 M
Asparagus	1 – 2	<95	2 W	1 – 2	95	< 15	air	6 W
Sweetcorn	0	90 – 95	1 W	0	90 – 95	20	(1 – 2)	3 W
Apples								
Cox	3 – 4	90 – 92	3 M	4	93 – 95	2	1 – 2	5 M
Golden delicious	1	93 – 95	4 M	1 – 2	95	3 – 5	1 – 2	6 – 8 M
Morgenduft	1 (– 1)	90	4 – 6 M	2 – 3	93 – 95	2 – 3	1 – 2	6 – 7 M
Pears								
Conference	– 1 – 0	95	4 – 5 M	– 1 – 0	95	2	2 – 3	7 M
Passa crassane	0 – 2	90	5 – 6 M	0 – 1	95	5	3	6 – 8 M
Strawberries	0 – 5	90 – 95	<5 D	2 – 10	90 – 93	20 – 25	air	10 D
Raspberries	– 1 – 0	90 – 95	<3 D	0 – 2	90 – 95	30	air	up to 10 D
Cherries (sour)	– 1	90 – 95	1 – 3 W	0 – 2	95	to 30	air	up to 7 W
Bananas	13	90	10 d	13 – 14	90	5 – 8	4 – 5	<2 M
Nuts/almonds	– 3 – 0	65 – 70	12 M	0	65 – 70	20	<1	24 M

D = day; W = week; M = month.

for <3 d; surface water at 4°C for 33 d, and at 25°C for 2 – 5 d. Destroyed by→pasteurisation (Mitscherlich and Marth 1984). – Pathogenic spp. generally causing campylobacteriosis; e.g., abortion in sheep and (less common) in cattle by *C. fetus* ssp. *fetus* (and *C. jejuni*) that can also be responsible for disseminated disease in man and meningitis in neonates. Main reservoirs of campylobacters include pets (dogs, cats), sea-gulls, ducks, crows, doves and surface waters. *C. jejuni* major agent of food-borne disease in Britain and other European countries. Transfer mainly by food of animal origin. Infection dose at least several hundred cells; after consumption of contaminated food the first symptoms appear after 2 – 11 d, and are charac-

terised by fever, headaches and malaise, followed by nausea, vomiting, abdominal pain and eventually watery diarrhoea, sometimes with mucus and blood. Recovery usually within 3 – 5 d. Complications in infants and immunocompromised may include meningitis, septicaemia and colitis (Sinell 1985). – Detection, e.g., by *Campylobacter* selective agar (Merck no. 2248).

Campylobacter pylori→*Heliobacter pylori.*

Candida

Large genus of yeast-like imperfect fungi, class Hyphomycetes; heterogeneous. Include pathogens of man and animals (candidiasis), commensals and several spp. of economic importance; distribution very wide, and can be isolated from soil, plants, food, etc. Candidiasis caused by *C. albicans* and *C. tropicalis*, the former of which is a commensal of the mouth as well as genital and GI tracts of man, but also a typical opportunistic pathogen. It has the ability of switching between at least seven different phenotypes. – *C. lipolytica* used for the production of→single cell protein from n-paraffins and other mineral oil products. *Saccharomycopsis* sp. and *C. utilis* are cultivated together on starch containing effluents of the food industry; *Candida* utilises the sugars that are formed by enzymatic action of *Saccharomycopsis* immediately, and eventually constitutes 90% of the biomass (symbiosis). Whey and whey products are used as substrates for the large scale production of *C. utilis* and *C. krusei*. The Waldhof process is the oldest industrial process for the production of fodder yeast (with *C. utilis*), and was originally based on sulfite waste liquor as substrate (Beuchat 1978; Reiss 1986).

capronic acids
n-Capronic acid; syn.: hexanoic acid (C_6); CH_3-$(CH_2)_4$-COOH. – n-Caprylic acid; octanoic acid (C_8); CH_3-$(CH_2)_6$-COOH. – n-Caproic acid; decanoic acid (C_{10}); CH_3-$(CH_2)_8$-COOH. Saturated fatty acids that constitute part (2.3%, 1.2% or 3.0%) of milk fat, but also of some plant fats, e.g., coconut oil; in goat's milk up to 5%. Unpleasant "sweaty" smell; released by the action of microbial→lipases; may also be formed during bacterial degradation of→lactic acid or→glycerol. Taste and flavour defects in products with a high fat content.

capsule
Several bacteria are enclosed in a gel-like layer of polysaccharides (*Str.*, etc.) or polypeptide (*B.* spp). The capsule composition is not a species specific characteristic as strains without capsules also exit. Those with capsules are termed S-forms due to their smooth and shiny colonies, and those without capsules R-forms due to the rough colonies, e.g., *Salmonella*. Some *Acetobacter* spp. form cellulose which binds the cells in a pellicle (Schlegel 1985). Xanthan and curdlan are gelling agents produced with the aid of the capsule component of *Xanthomas campestris* and/or *Alcaligenes faecalis* ssp. *mycogenes* (DFG 1987). Capsulated bacteria caused difficulties especially in the beverage industry where they complicate the filtering process. The polysaccharide is often successfully dissolved before filtration by using a suitable enzyme preparation.

carbohydrates (CH)
$(CH_2O)_n$. The most important nutritional component apart from protein and fat. It is a source of energy and carbon. Natural occurring CH-compounds are degraded by MO. The smaller part is assimilated and the larger part converted in the presence of O_2 into CO_2, H_2O and energy. → Respiration. Examples are sugar, starch, cellulose and pectin.

carbon dioxide
CO_2. Colour- and odourless. Asphyxiating gas. 1.8 g/l; 1 g is soluble in 300 ml of water at 0°C; in 700 ml of water at 25°C; air contains 0.03 vol%. The "MAK" value is 5000 ml/m^3, corresponding to 9000 mg/m^3 air. It is a product of organic material combustion, volcanoes, → respiration and several→fermentation processes (→ swelling/blowing). It collects in cellars, etc., due to the fact that it is heavier than air. Caution! Suffocation danger. Unstable carbonic acid is formed in a watery solution: $CO_2 + H_2O \rightarrow H_2CO_3$. – CO_2 in the atmosphere is assimilated and fixed by green plants through photosynthesis. It is also utilised by several C-heterotrophic MO in the production of, e.g., fatty acids and propionic acid. It is reduced to CH_4 in

methane production. Several MO grow anaerobically if the required atmospheric CO_2 exceeds 3%. Growth mostly ceases when the levels rise to between 50 – 80%. Not only the shelf life of products can be extended by increasing the CO_2-concentration, but it is also useful in → CA-storage. The oxygen still present in fresh, cooled, vacuum packed meat is utilised, thereby increasing the CO_2 concentration to about 20 – 30%, which inhibits the growth of aerobic Gram – organisms responsible for spoilage. – EC no. E 290. Extraction of components contained in food with supercritical CO_2, e.g., caffeine and spice ingredients, leads to the destruction of several MO, but obligate anaerobes and endospores survive.

carboxyl group
-COOH. Acidic group of organic acids, substituted to alkyl-, phenyl- or other groups: e.g., CH_3-COOH (acetic acid). Release of the carboxyl group by MO → decarboxylation.

carcinogens
An agent which can cause cancer, probably by interacting with the DNA of the target cell. Examples: → aflatoxins; → nitrosamines, etc.

cardboard
Used in the production of packaging material with a stable form and which can be piled up. Unpainted cardboard boxes have coarse surfaces and they allow the adhesion of MO better than when painted. The surfaces of synthetically coated boxes can be sterilised → packaging goods sterilisation. Boxes must be stored in dry conditions away from the production rooms so as to prevent contamination by fungi.

caries
Tooth decay. Bacteria in the tooth film or plaque cause breakdown of the tooth enamel through demineralisation and constant disturbance of the underlying dentine by secondary infection. Lactic acid bacteria are mainly responsible for the initial decay if enough mono- and disaccharides are present in the food rests.

carotinoids
Class of yellow and red pigments, representing highly unsaturated aliphatic and alicyclic hydrocarbons and their oxidation products. Produced by several MO and may cause colouration of foods and synthetic utensils. Examples: → *Brevibacterium linens, Rhodotorula* spp., etc. → Pigments.

carrier
Food legislation generally prohibits the handling of foods by individuals that harbour one of the pathogens *Vibrio cholerae, Salmonella* spp. and *Shigella* spp. These individuals show no clinical signs of the disease, but may shed or excrete the pathogens for up to 6 months after the disappearance of clinical symptoms in the case of salmonellosis. Being important reservoirs, such individuals may transmit these pathogens to other healthy persons, either by direct contact or by food. Constituting a health risk, such individuals should not be involved in any commercial food processing or handling operation, or have any contact with commercial food products. Furthermore, their employment is not allowed in the kitchens of restaurants, canteens, hospitals, children's or infants' homes, or community welfare (Hahn and Muermann 1987; D'Aoust 1989).

carry-over-effect
"Transmission". The transfer or transmission of deleterious substances, taken up by the animal via its feed, directly or after chemical conversion, in products from animal origin. Examples: → ochratoxin A in pig kidneys and blood plasma; → aflatoxin M in milk and milk products.

carry-over-principle

Food or→ enzymic preparations that have been stabilised with→ preservatives in approved concentrations, may not be mixed/added into other food products for which these preservatives/additives have not been approved. This would result in a dilution of the substance below its effective concentration in the end product, e.g., preserved liquid rennet for the manufacture of cheese. In such cases the preservatives have not to be declared on the package.

casein

A phosphoprotein; major milk protein (ca. 80%); as micelles in the milk, each of which consists of aggregated molecules of the 3 main casein types: α, β and kappa-casein. The latter acts as the stabilising part in the micelle, and can be cleaved by enzymes such as pepsin (E.C. 3.4.23.1) or rennin (chymosin) (E.C. 3.4.23.4) into a soluble C-terminal and an insoluble N-terminal fragment. The coagulation of casein by the action of rennet is an important step in the manufacture of most cheese types. Coagulation may also result from acidification to a pH of ca. 4.6, which approximates the iso-electric points of the caseins. Lowering of the pH to 5.3 – 5.3 by acid production of starter cultures affects the solubilisation of calcium and phosphate out of the micelles, resulting in the destabilisation and coagulation of the milk, as is practiced during the manufacture of fermented milk products. – Sweet coagulation ("clotting") may cause spoilage of drinking milk, and may be due to the survival of proteolytic bacteria or their enzymes after→ pasteurisation (Krämer 1987; Teuber 1987).

catalase

An enzyme which is extracted from liver or produced with the aid of MO, e.g., *A. niger, Micrococcus lysodeikticus* (syn.: *M. luteus*). The common name is catalase; the systematical name is hydrogen peroxide: hydrogen peroxide oxido-reductase (E.C. 1.11.1.6). It catalyses the cleavage of hydrogen peroxide into water and oxygen. It is used to remove hydrogen peroxide from food, e.g., mayonnaise,→ gluconic acid (GDCh 1983).

catalase test

A positive or negative catalase reaction is a simple and important step in the→ differentiation (identification) of bacteria. Several Gram + bacteria are catalase – and the larger majority of Gram – bacteria is catalase + . One drop of 5% H_2O_2 is added to a bacterial colony; if gas bubbles appear they are indicative of the presence of catalase + organisms. $2H_2O_2 \rightarrow 2H_2O + O_2$. A liquid culture can be centrifuged and the sediment used, or 10% hydrogen peroxide solution can be added directly to the suspension. O_2-production can be measured quantitatively with an oxygen electrode.

caustic soda solution

Hot NaOH solution used to clean and disinfect tanks and pipelines. No corrosion danger for stainless steel (Wallhäusser 1988).

cell

Smallest unit in biology with the ability of independent replication into identical units and to remain viable for longer or shorter periods.

cell division

The basis of→ multiplication in bacteria. With few exceptions always binary fission. It occurs in two steps: (a) separation of the nuclear material and (b) cytokinesis – the division of a cellular body into a progeny of two cells. The cellular count (microbial population) per g or ml is important in the hygienic evaluation of foodstuffs. In the production of protein the→ biomass (dry mass/ml) is an important criterion. In the first case the→rate of division or→generation time is of interest, in the second the growth rate or doubling time. The difference lies in the method of detection, only cells able to divide will allow a (viable) →bacterial count (cfu), whereas the biomass is made up of both viable and nonviable cells which can be collected by harvesting (e.g., by centrifugation) (Schlegel 1985). – For fungi different criteria exist since they do not multiply by binary fission, but by apical extension, bifurcation and branching. Budding is predominant in the yeasts. Growth rate or doubling time can only be determined with reference to the biomass (→fungal count).

cell wall

The cell walls of bacteria and→cyanobacteria consist of a basic framework of peptidoglycan (murein) surrounding the cell like a net. It is permeable to water, salts and substances with smaller molecular weight and on the inside it is adjacent to the semi-permeable cytoplasmic membrane. The thickness of the murein network probably determines the reaction in→Gram staining. The murein sacculus – a giant molecule – is formed by the N-acetylglucosamine and N-acetyl muramic acid molecules which are alternatively linked with a β-1,4-glucosidic bond into heteropolymeric chains, which are again linked via lactyl groups with pep-tide bonds to the amino acids. This rigorous framework is substituted with proteins, polysaccharides and lipids.→ Lysozyme hydrolyses the β-1,4-glucosidic bonds of murein, by which bacteria are destroyed. Present, e.g., in tears or egg albumen (Schlegel 1985). – Chitin is an important substance in the supporting framework of the cell walls of fungi and yeasts. It consists only of N-acetyl glucosamine molecules bound to each other with a β-1,4-glucosidic bond. In addition, chitin may form compounds with cellulose, chitosan, β-glucan and mannan, typical

for particular fungal groups. Fungal control with chitinase or other enzymes dissolving the cell wall has not been proven feasible yet (Müller and Loeffler 1982).

cellar fungi

Cladosporium herbarium or *C. cellare* may be found on the walls of old storing cellars with a high humidity as well as on flasks and corks. They can also grow on steel in such a manner that it resembles dark cobwebs or thick cottonwool coats. The fungi utilise the volatile organic substances, e.g., esters, alcohols, organic acids, etc., in such extreme environments.

cellar slime

Slimy growth of→*Aureobasidium pullulans* (pullulan production) and often→ *Cladosporidium* spp. and *Rhodotorula* spp. and unicellular blue and green algae on the walls in old wine cellars with high humidity. The volatile air-borne acids dissolve in the slimy layer and are then metabolised by the MO.

cellulase

Enzymes which hydrolyse cellulose to cellobiose; catalyses the hydrolytic cleavage of the β-1,4-glycosidic bonds in cellulose. Found in plants, bacteria and moulds. Commercially produced with the aid of MO, especially fungi. Trivial names: β-glucanase, endo-1,4-β-glucanase. Systematic name: 1,4-(1,3; 1,4)-β-D-glucan-4-glucanohydrolase (E.C. 3.2.1.4). – Applications: degradation of cellulose, e.g., for eliminating turbidity, improvement of the digestibility, and the production of plant extracts. – MO used for production: *A. niger, A. oryzae, Rhizopus delemar, R. oryzae, Sporotrichum dimorphosporum, Trichodermareesei, T. viride, Thielavia terrestris.*

cellulose

Plant polysaccharide consisting of β-1,4-linked glucose units. Unbranched chains with M_r between 300,000 and 500,000. Enzymatically hydrolysed to the disaccharide cellobiose ($\rightarrow$ cellulase). The chains are arranged in microfibrils and stabilised by interchain hydrogen bonding. Covalently bonded into hemicelluloses and lignin in wood. Distribution: most abundant organic substance on earth (ca. 50% of the CO_2 of the atmosphere). Stabilises the plant cell wall and also that of the Oomycetes. Produced as extracellular polysaccharide by *Sarcina ventriculi* and $\rightarrow$ *Acetobacter xylinum*. Degradation: $\rightarrow$ cellulase. In nature, e.g., by rumen bacteria of herbivores, and in soil by moulds. – For the industrial production of glucose from cellulose appropriate cellulase preparations are not available yet; $\rightarrow$ Natick process.

cephalosporins
Group of chemically related antibiotics, produced by *Cephalosporium* spp. Mechanism of action: inhibition of cell wall synthesis, especially of Gram-pos. bacteria, similar as for the penicillins.

Cephalosporium $\rightarrow$ *Acremonium*.

cereals
Basic foodstuff with a high content of CH (40 – 78%), protein (8 – 25%) and lipids (1.5 – 6%), as well as minerals, vitamins and fibre. Water content should be < 13% after harvest; drying otherwise necessary. Av. bacterial population after harvest $10^5 - 10^7$/g, depending on locality, climate, cultivar, soil and fertilisation. Moulds $10^3 - 10^6$/g; yeasts $10^2 - 10^3$/g (Kunz 1988). – The "internal" population between epidermis and pericarp is composed mainly of *Alternaria, Aspergillus, Cephalothecium, Cladosporium, Fusarium, Helminthosporium, Mucor, Penicillium, Rhizopus,* etc. The surface microbes on the epidermis are more heterogeneous, and include bacteria (*Erwinia, Flavobacterium, Pseudomonas, Proteus, Xanthomonas, Enterobacter, Serratia,*

Micrococcus, Staphylococcus, Streptococcus, Enterococcus, Pediococcus, Lactobacillus, Bacillus, Clostridium, etc.), yeasts (*Candida, Cryptococcus, Hansenula, Pichia, Trichosporon, Torulopsis, Rhodotorula, Sporobolomyces,* etc.) and moulds (*Alternaria, Aspergillus, Aureobasiium, Cladosporium, Fusarium, Helminthosporium, Hyalodendron, Penicillium,* etc.) (Spicher and Baumgart 1986).

certified milk
$\rightarrow$ Raw milk from livestock under constant microbiological control; the refrigerated product may be distributed without previous $\rightarrow$ pasteurisation (Teuber 1987).

cfu
Colony forming units. A new and more correct term for $\rightarrow$ microbial counts. Single cells and cellular aggregates (e.g., *Sarcina, Staph.,* etc.) are cultivated, e.g., by the $\rightarrow$ pour plate and other similar methods, to obtain single colonies. Evaluation is within the fluctuation range given by the plating method. The deviation is even greater with fungi and mycelium forming yeasts $\rightarrow$ fungal (mould) count (Baumgart 1990).

Chaetomium
Mould genus of the Ascomycetes. – Strongly cellulolytic, and typically found in the inner part of grains and therefore also in flour. Distribution mainly in soil and, through contact, also on plants. Moderate spoilage potential for foods, but larger on materials and utensils. Some spp. produce the antibiotic chaetomine.

chain formation
Appears frequently by *Str., Lb.* and other bacteria as well as by several yeasts, e.g., *Brettanomyces*. The cells do not separate after division and/or budding. Chains and aggregates are responsible for the wide fluctuations in $\rightarrow$ microbial counts when cultured.

chalky spoilage

"Kreidekrankheit" (Germ.). Microbiological damage to bread and baked products caused by white dusty to crumbly coats of *Moniliella sauveolens, Endomycopsis fibuligera* ("Kreideschimmel"), *Candida* spp. and other yeasts (Baumgart 1990; Kunz 1988).

characteristics

Genetically determined features of MO partially used for → identification and → differentiation of a group, genus or sp., contained mostly in dichotomous "determination keys". Easily determinable characteristics of bacteria are morphology, staining properties (→ Gram-staining), as well as enzymatic, metabolic and ecological properties. Similar schemes are available for yeasts. For fungi mainly morphological and ecological characteristics are used (Baumgart 1986, 1990).

cheese

Fermentation product of pasteurized milk. The shelf life is longer than that of the raw material. Unripe cheese is fermented with butter or sour cream cultures without rennet (→ cottage cheese; quarg) in order to acidify the casein. The milk used in the manufacturing of ripened cheese is coagulated by the addition of rennet or an approved → rennet substitute. The lactic acid produced by cheese starters leads to the release of calcium and phosphate from casein. In this way different structured gels are formed, resulting in various cheese types. The dry matter is also important. Various types of cheese are known, e.g., hard cheese which is partially manufactured from unpasteurised milk; → semi-hard cheese and → soft cheese with mould growth on the surface and/or throughout the cheese, or with surface bacterial growth (Teuber 1987). → Raw milk cheese.

cheese defects

Bitter taste due to rennet with high pepsin levels or proteolytic MO producing peptides with a bitter taste. Mycotoxin production by undesired moulds. Colour changes due to pigmented bacteria, and undesirable small holes may be formed by early gas production in hard cheese. → Early blowing and → late blowing (Kunz 1988; Teuber 1987).

cheese processing suitability

The suitability of milk used in the cheese making process must be controlled before the addition of the starter culture or → rennet. The criteria are: the period required for curdling at a temperature of 30°C following the addition of rennet, so as to sort out batches of slow-curdling milk; it must be free from antibiotics and/or other preservatives contained in the fodder; preripening of the milk at 8 – 10°C with a mesophilic acid producer in order to adapt the culture to the new medium. Such control mechanisms have advantages in spring and autumn when the feeding ration is changed (Teuber 1987).

cheese starter cultures

Starter cultures used in the manufacturing of various cheese types. Factors contributing to the type of cheese are: moisture content; composition of the starter culture; temperature treatment from the onset until the completion of the ripening process; ventilation; period allowed for the ripening process.

chlamydospores

Generally refers to a thick-walled, asexually-produced resting spore (see Figure 11, page 174) of some fungi; mono- or multicellular. Survives unfavourable conditions – after the thallus has died off – and serves the survival of the sp. rather than to promote its distribution, as is the case with → conidia. Distribution: *Fusarium, Mucor, Candida*, etc. Control (measures) more difficult than for other vegetative organs (Müller and Loeffler 1982).

chloramphenicol
Chloromycetin. Antibiotic produced by *Streptomyces venezuelae*; of the four stereo-isomers only the D(-)-*threo*-C shows antibiotic activity. Inhibits protein synthesis on the 70S ribosomes of prokaryotes and mitochondrial ribosomes of eukaryotes. Synthetic production since > 30 years. Clinical application limited because of relative toxicity; more common usage in veterinary medicine.

Chlorella
Genus of unicellular coccus shaped, nonmotile green algae, found in fresh water, on moist soils, bark, etc. In association with other MO, especially moulds→ lichens. *C. vulgaris* is used for the commercial production of→single cell protein.

chlorine
Cl. Pungent smelling, yellow-green gas with strong microbicidal action. Commercially distributed in grey steel cylinders. "MAK"-value 1.5 mn/m^3 air. Important application for the decontamination of→ drinking water and swimming pools with 0.3 – 0.6 ppm free, available chlorine. Strong oxidising agent; forms the strongly antimicrobial hypochlorous acid with water. Highly (unspecific) reactive with organic compounds; may be partly inactivated by heavily polluted water.→ Halogen compounds; → disinfectant (Wallhäusser 1987).

Chlorophyta
Green algae: division of the algae. Ca. 10,000 spp. found in fresh and sea water. Mono- or multicellular. Some spp. used for the production of→single cell protein→ *Scenedesmus*.

chocolate
Finely dispersed mixture of cocoa, milk powder, sugar and different other ingredients such as nuts, almonds, raisins, coffee, etc. During processing at 70 – 80 °C most MO are killed by the heat and the→radicals formed. The a$_w$ of 0.37 – 0.50 serves as reliable safety factor to prevent growth of MO. – Salmonella infections transmitted by chocolate have been reported. For microbiological examinations see Baumgart (1990) and Speck (1984).

cholera
Acute, infectious human disease, mainly transmitted by faecal-polluted water. Oral intake of→ *Vibrio cholerae* (biotypes *cholerae* or *eltor*) is followed by typical symptoms within 6 h to 5 d (incubation period): profuse dehydrating diarrhoea ("rice water stools"). Most severe form (cholera gravis) may pass > 1 l diarrhoea stool/h that may result in circulatory collapse and eventual death. Therapy directed towards replacement of fluid and mineral losses; tetracycline treatment (500 mg orally, every 6 h for 3 days). – Rice water stools may contain up to 10^9 *V. cholerae* cells/ml. In addition to water, the organism may also be transmitted via fish, dried fish, shrimp and other seafood, green salads (lettuce), ice (from contaminated water) added to cold drinks, "washed fruit", etc.→Excretors rare. Endemic in many tropical countries; risk of infection increases with mass tourism in the tropics. It is recommended that only heat-treated food be consumed in endemic areas (Madden et al. 1989; Sinell 1985).

Chromobacterium
Genus (*incertae sedis*) of Gram-neg., motile, facultatively anaerobic rods; strong proteolytic, nitrite +, catalase +, oxidase + (Bergey 1984). *C. violaceum* and *C. fluviatile* produce violet-pigmented colonies. Distribution: soil, surface water (especially in autumn), etc.; opportunistic pathogens in man and animals. Opt. growth temp.: ca. 30 °C; min. 4 °C; opt. pH 7 – 8.

Chrysosporium

Mould genus of the Hyphomycetes; an-amorphous of *Anixiopsis* (Ascomycetes). Universally distributed; found regularly on dried fruit and cocoa beans. Growth at rel. low a_w values: e.g., min a_w for *C. xerophilum* and *C. fastidium* 0.70. During prolonged storage of cereals with water content of 15 – 16% damage by *C. inops*.

CIP

(1) Collection de l'Institut Pasteur (28 Rue du Docteur Roux, F-75724 Paris Cedex 15, France). (2) Cleaning-in-place process for the→cleansing and→disinfection of closed systems; has been proven successful for pipelines involved in the processing of mayonnaises, dressings, milk and other liquids and semi-fluid products. The CIP system may be programmed for carrying out the sequence of steps automatically. The operation is initiated by rinsing at elevated temperature using a fat and/or dirt solvent, followed by a→disinfectant; final rinsing with sterile (hot) water and drying.

Circinomucor→*Mucor.*

citreoviridin

Neurotoxic mycotoxin (yellow pigment) produced by *Penicillium citreoviride* a.o. *P.* spp. on polished rice stored under humid conditions. Consumption may cause acute cardiac beriberi, a fatal disease symptomised by ascending paralysis, convulsions and respiratory arrest. Known in Japan for centuries (Reiss 1981).

citric acid

A mono-tricarboxylic acid, present in many fruits (e.g., *Citrus* spp.), milk (ca. 2.4 g/l) and some vegetables. Key metabolic intermediate in many MO, and produced and excreted by several moulds. Use as oxidation and acidifying agent in the food industry, especially for beverages and confectionery. EEC regulations require declaration and purity standards as for other food additives. EC nos.: E 330 (citric acid), E 331 (sodium citrate), E 332 (potassium citrate), E 333 (calcium citrate). – Important role in cell metabolism as starting point of the→tricarboxylic acid cycle (Krebs cycle), in cell energy generation, and in the coordination of several other metabolic pathways, e.g., activation of acetyl-coenzyme A carboxylase (key enzyme in fatty acid biosynthesis), and in amino acid biosynthesis. – Metabolised by different→LAB accompanied by the production of→diacetyl (butter aroma). – Used in differential and selective microbiological media, e.g., Citrate Agar and Citrate-Azide-Tween-Carbonate Agar (CATC-Agar) for differentiation of→Enterobacteriaceae (Baumgart 1990). Other uses include applications in the food industry, as softeners for synthetic materials, and as substitute for phosphoric acid in detergents and washing powders. – Annual world production ca. 350,000 ton of citric acid from molasses (yield up to 60% of the sugar used). Mainly strains of *Aspergillus niger* used. Until the 60's stationary surface culture methods were in use for citric acid production; yields were 200 – 220 g/l of nutrient substrate.→Submerged fermentation processes are presently in use.

citric acid cycle→tricarboxylic acid cycle.

citrinin

Antimycin. Nephrotoxic, carcinogenic mycotoxin (yellow pigment) produced by *Aspergillus* and *Penicillium* spp., e.g., *P. roqueforti*, that prefer reduced temperatures. Antibiotic action against many bacteria and moulds, also vertebrates (→mycotoxins). May be found in: mouldy bread, cereals, peanuts, etc. Often accompanied by ochratoxin A. – Fluorescent under (longer wavelength) UV (360 nm), even in high dilution.

citrus fruit

Main microbiological damage/spoilage during transport and storage of citrus fruit generally caused by → *Penicillium italicum* and → *P. digitatum*. Cross infection of the former also to other fruit types. *Diplodia natalensis* and *Phomopsis citri* of importance in humid areas; *Alternaria citri* more typical of semiarid regions. Soft rot caused by *Geotrichum candidum*; the smell attracts flies by which the organism is further distributed. Control measures: → diphenyl in packaging material, or → orthophenyl-phenol.

Cladosporium

Genus of the fungi of the → Deutero-mycetes, forming septate mycelium and dark-coloured conidia. Because of black to dark-coloured yeast-like cells (melanin production) → black yeasts (black moulds) (Ellis 1971; de Vries 1967). Septate or aseptate ellipsoidal conidia in chains; so-called good-weather spores that may be transported during hot spells in summer by air over long distances. Includes sapro-trophs and pathogens of plants and animals; also responsible for myco-aller-goses, especially in connection with dust from hay and grains (Schachta and Jorde 1989). Mycoses of humans and pets caused by some spp. (Gedek 1980). – May settle on all types of foods, packaging materials, fruit cartons in storerooms, etc.; may destroy materials; → wall mould. – Some spp. "specialised" on certain products: *C. butyri* on margarine, *C. suaveolens* on butter, *C. cladospoides* and *C. shaerospermum* on old tires, *C. herbarum* and *C. resinae* on synthetics. Non-fastidious: grows in sparse medium containing minerals with nitrate and glucose or sucrose (Czapek-Dox-agar).

Cladosporium herbarum

Found on all types of plant products. Growth conditions: opt. 24 – 25 °C; min. – 7 to – 5 °C; max. 30 – 32 °C; a_w: min.

0.85 – 0.86, opt. 0.95 – 0.96; pH 3.1 – 7.7. Germination of conidia and growth from 0.25% O_2 in the storage room atmosphere. – Damage to meat, vegetables, fruit juices, spices, cereals, nuts, fruit, pastas, etc. Mycotoxins not known.

clarification

(1) Beer: "Läutern". Process step in beer brewing to clarify the → wort and separate the malt and hop solids. The clear wort is cooked to kill all vegetative MO.

(2) Wine: Treatment of (e.g.) wine after fermentation or prior to filling; precipitation of tannins and dispensed particles including some MO. Clarification aids include bentonite or gelatine. Special requirements for gelatine specified in (e.g.) German Regulation of 1983, with reference to MO: aerobic MO max. 10,000 g; coliforms negative in 0.1 g; *E. coli* and *Cl. perfringens* negative in 1 g (Krämer 1987).

class

Latin: classis. → Systematics.

clavatin → patulin.

Claviceps purpurea

Belongs to Ascomycetes. With *C. paspali* typical parasites on grasses, rye, barley, and more recently associated with wheat and triticale; to a lesser extent also on oats and maize. Infection cycle begins in spring-time when ascospores are distributed by wind to susceptible grasses where they germinate and infect the ovaries. A thick mycelium is formed, and colourless conidia, suspended in a honeydew secretion, are produced (the *Sphacelia* stage). Insects, attracted by the honeydew, disperse the conidia to the ovaries of healthy plants. The mycelium gradually develops into a tough structure, the → sclerotium (*Secale cornum*, ergot:

the overwintering stage). In the following season the sclerotia give rise to stromata and ascospores. – In the ripening ears of rye the sclerotia are typically black to purple-black and hard, and become elongated and larger than the grain; on wheat they are brownish and only slightly larger than the grains. Early treatment of the field with → fungicides may prevent excessive damage. With modern milling technology the sclerotia are separated from the healthy grains on basis of their density and size. This preventative measure is not taken for grains sold directly "from the field". – The alkaloid content of sclerotia presents a health risk and is of concern for health authorities in most countries. Presence of alkaloids in flour or bread may cause a type of food-poisoning called → ergotism (Roth et al. 1988). Acc. to EEC Commission regulation no. 1569/77 of 11.07.1977, a maximum content of 0.05% is allowed. Detection of the alkaloids after extraction with HPLC [Klug, C.: *MvP-Hefte* 2 (1986), Max von Pettenkofer Institute of the German Health Authority,

Berlin]. Animal experiments indicate that the ergot alkaloids are reduced and detoxified by the bread making process to ca. 40 – 50% of the original level [Wolff, J. et al.: *Z. Ernährungswiss* 27 (1988) 1 – 22].

clavicin, claviformin → patulin.

cleaning
Most important basic operation in food processing plants, transport vehicles, industrial kitchens and for the people working there (→ personal hygiene) in order to produce hygienically acceptable products with good keeping qualities. An important prerequisite for → disinfection. The success of cleaning depends on the composition of the cleaning agent, the kind of dirt, the condition of the surface and the application of the cleaning process (Figure 4). In addition to manual or automatic scrubbing procedures, a choice of three methods for application to accessible surfaces are generally available: high pressure cleaning (30 – 70 bar), steam cleaning (100°C; 5 – 10 bar) and foam

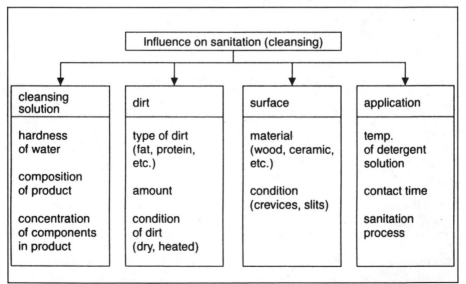

Figure 4 Factors influencing the sanitation process (source: Krämer, J.: *Lebensmittel-Mikrobiologie*. Stuttgart: Ulmer Verlag, 1987; p. 237, modified).

cleaning followed by rinsing. → CIP procedures are used for closed systems, e.g., dairies (Krämer 1987; Sinell 1985).

cleansing agents
Surface active organic compounds (surfactants), used for combined→cleaning (sanitation) (→ detergents) and→ disinfection operations. May be cationic, nonionic, anionic or amphoteric substances. Action in part destructive, and partly→ inhibitive. Cause→ degeneration of many bacteria when applied at suboptimal levels.

clone→ pure culture.

Clostridium
Genus with at least 80 spp. (Section 13, *Bergey's Manual* 1986) of endospore forming anaerobic, Gram-positive, rod-shaped bacteria. Most spp. motile by peritrichous flagella. Few strains aerotolerant. Dissimilatory sulphate reduction absent (→ *Desulfotomaculum*). Typical products of sucrose fermentation: butyric acid and acetic acid, CO_2, H_2 and for some spp. also CH_4. Importance spoilage organisms of vacuum packaged and canned foods. Some spp. are pathogenic or toxinogenic:→ *Cl. botulinum* (botulism), *Cl. perfringens* (gas gangrene, and type A major clostridial food poisoning organism) and *Cl. histolyticum, Cl. tetani* (tetanus). Nitrogen fixation by some spp. Distribution: main habitat is soil, and also GI tract of man and animals. – Thermophilic clostridia may cause spoilage of tropically preserved foods, e.g., *Cl. thermosaccharolyticum*: opt. growth temp. 55 – 62°C; produces lactic and butyric acids, CO_2 and H_2, and causes hard→ swell spoilage of canned foods. Transmitted via sugar.

Clostridium botulinum
Causative agent of→ botulism food poisoning (→ incubation period). Natural dis-

tribution in soil, in sediments of surface waters and coastal waters. May be found on all food products that have been in contact with soil. – Straight or slightly curved rods, Gram +, motile, strictly anaerobic, ovoid endospores arranged centrally or subterminally, distending the sporangium. Growth: min. 10°C (type E and saccharolytic type B strains 3°C), opt. 30 – 37°C; pH >4.5. a_w >0 – 93; O_2 <2.7%. Destruction: pasteurisation for vegetative cells; moist heat (>90°C) for endospores (F_o 2.5 – 3; D_{110} 2 min; D_{121} 0.21 min). – Toxins produced in protein rich substrates; 8 types (A to G) distinguished, differ in their distribution, heat resistance, degree of toxicity, pH resistance and proteolytic ability. Type A more typical of USA (and mainly associated with canned vegetable products) and type B of Europe (mainly meat products); type E associated primarily with seafood (psychrotrophic), and types C and D of consequence for farm animals. Home-canned foods are presently primarily incriminated; increased risk caused by the "bio" trend. Incrimination of industrially processed products due to specific failures, e.g., insufficiently cured bacon/hams and sausages (too low conc. of→ nitrite curing salt); canned green beans, spinach, asparagus, maize (sweet corn) (insufficient heat treatment); incorrectly smoked and underprocessed trout and fish products [Review: Hauschild, A. H. W.: *Food Technol.* 36 (1982) 95 – 104]. – Toxins heat labile: 95 – 100°C/2 min core temperature for inactivation in product. – Recognition of toxins by their lethal action in mice and neutralisation with specific antisera (bioassay) which still is the most specific and reliable method for identification of botulinal toxins (Hauschild 1989).

Clostridium perfringens
Formerly *Cl. welchii*. Five types (A – E), based on production of four extracellular

toxins (α, β, and iota). The α-toxin produced by all strains, and practically all outbreaks caused by type A strains; however, determination of serotype important for epidemiological investigations. Gram +, immotile, rel. large rods, either single cells or palisades; subterminal endospores. Normal inhabitant of the human gut. Growth: min. 12 – 15°C, opt. 43 – 46°C, max. 52°C; pH 5 – 8.5; max. O_2 conc. 1.5%; salt < 6%. Destruction of vegetative cells by pasteurisation, and moist heat > 90°C (D_{90} up to 145 min; D_{100} up to 60 min). – Infection dose around 10^6 vegetative cells ingested with meals. Increase (ca. 3 divisions) in GI tract, and endospore production; symptoms occur within 8 – 24 h after ingestion. Toxin (protein with MW 36,000) produced intracellularly between cell wall of sporangium and endospore, and is released during digestion of walls. Usual symptoms: diarrhoea and severe abdominal pain; infrequent: nausea, fever and vomiting, lacking appetite. Death uncommon. Treatment usually not necessary, and complications rare. – Contaminating endospores, e.g., from → spices, will germinate in food after heat shock, resulting from processing. Rapid proliferation in food wastes at room temp., and may reach infection dose within a few hours. Health risk especially related to poor management of restaurants, canteens and → airline foods (on board catering) (Labbe 1989).

Clostridium sporogenes
Gram +, motile rods, single and pairs; ovoid endospores, subterminal and distending the sporangium. NO_3^- reduction, but no accumulation of NO_2^-. Acid and gas production from different CH. Deamination of amino acids with the production of short-chain fatty acids (Stickland reaction), causing sensoric (aroma and taste) changes. – Distribution: soil, dung, hay, milk. Growth: 18 – 45°C; pH 5.5 – 8.9; a_w

> 0.95; O_2 < 12%. Destruction: pasteurisation for veg. cells; moist heat > 90°C for endospores (D_{95} 4.5 h; D_{110} 8 min; D_{120} 15 s). – Spoilage of underprocessed canned foods (gas production, proteolysis). Spores commonly found in milk during summer and in hay used for feeding. Causative agent of butyric acid production without gas formation, in semi-hard and hard cheeses; white putrefactive spots in Emmental type cheese ("stinker").

Clostridium thermosaccharolyticum
Motile rods, singles or pairs; ovoid endospores, terminal, distending ("drumstick"). Non-haemolytic, strongly gas producing in deep culture; produces acetic, butyric and lactic acids, in addition to ethanol, H_2 and traces of tartaric acid from glucose. Growth temp.: opt. 55 – 62°C, min. 35 – 40°C, max. 68 – 70°C. Destruction of endospores: D_{120} 3 – 4 min. Distribution: soil, blown cans, sugar, effluents from sugar refineries. Spoilage of canned products containing vegetables and CH, with pH > 4.5; → swell (blowing) by CO_2 + H_2; spoilage in addition by souring and acid smell. Causing losses during sugar processing when extraction is performed < 70°C (Krämer 1987).

Clostridium tyrobutyricum
Gram +, motile, large rods, anaerobic; subterminal endospores, ovoid, distending. Profuse production of CO_2 and H_2 during fermentation of lactate and different sugars. Distribution: soil, silage, cow dung, milk. Growth: 10 – 42°C; pH 4.6 – 7.3. Destruction: vegetative cells by pasteurisation; endospores by moist heat > 90°C (D_{90} 14 min; D_{120} 0.01 min). – Feeding of silage results in higher raw milk spore numbers during winter, rendering such milk unacceptable for Emmental cheese processing. – Causing late blow-

ing of semi-soft and hard cheeses; gas production 4 – 5 weeks after manufacture leading to swelling/blowing of cheese loaves, especially by H_2 (which is practically insoluble in water). As a precaution 20 g of → nitrate may be added to 100 l of milk in the cheese bath; addition of 500 units of → lysozyme/ml has also been approved. These precautions, however, are not allowed for Emmental manufacture (Reuter 1987).

CO_2 → carbon dioxide.

coagulase test
Proof for toxinogenic *Staph. aureus* strains with rabbit plasma at 37°C. Method: see Baumgart 1986, p. 101; Speck 1984, p. 421 and DIN 10163 1984.

cocci
Spherical to ovoid bacterial cells. They may occur single, in pairs (diplococci), chains (*Streptococcus*), clusters (*Staph.*) or regular tetrads (*Sarcina*). Gram + or Gram – .

cockroaches
Blatta americana, B. orientalis, etc. Commonly found in food processing plants, canteens, restaurants, etc. Carriers of MO and difficult to control.

cocoa
Cocoa beans are embedded in the sugar containing mucilage in the pods of *Theobroma cacao*. The beans are piled up in wooden boxes in layers of more or less 1 m thick and spontaneous fermentation occurs. The beans are mixed and turned daily. Bacteria and fungi develop in the adhering mucilage at 30 – 40°C (Table 9). Through oxygen supply and/or utilisation the microbial succession pattern changes regularly. Ethanol is produced anaerobically by yeasts and *Zymomonas mobilis*, which is autoxidised to acetic acid by acetic acid bacteria under aerobic con-

ditions and decisively contributes to the aroma. The use of *Saccharomyces chevalieri* (5 × 10^6/g) as starter culture has proved successful, but is seldom used. During fermentation toxins are apparently not produced whilst pathogenic MO do not multiply (DFG 1987). As soon as the water containing carbohydrate-rich mucilage is utilised the beans are dried in the open air. The cocoa butter is pressed from the cotyledons and the pressed cakes are ground. Contamination through bird dung is unavoidable, but the MO are killed by suitable heat processing (80°C) in the production of chocolates and beverages. Microbiological control (Baumgart 1990).

COD → chemical oxygen demand.

coffee
Seeds from *Coffee arabica* and *C. ro-*

Table 9 Bacteria and yeasts associated with the spontaneous fermentation of cocoa beans (DFG 1987).

Bacteria	Comment
Acetobacter roseus	prod. of organic acids
Acetobacter aceti	
Acetobacter rancens	
Lactobacillus fermentum	
Lb. delbrueckii ssp. *lactis*	
Lb. delbrueckii ssp. *bulgaricus*	
Streptococcus salivarius ssp. *thermophilus*	
Bacillus stearothermophilus	
Zymomonas mobilis	prod. of ethanol

Yeasts	Comment
Saccharomyces chevalieri	pectolysis; ethanol prod.
Torulopsis candida	
Candida norvegensis	
Kluyeromyces fragilis	
Pichia membranaefaciens	

busta. The fruit, so-called "coffee berries", are covered by a leathery peel and both seeds are enclosed in the slimy and hygroscopic mesocarp (fruit flesh). The fruits are either dried and the beans mechanically released from the covering or they are carefully squeezed out and the sugar containing mesocarp is weakened through fermentation. The spontaneous microbial population, especially *Erwinia dissolvens,* but also other pectinolytes, is able to decompose the covering within 12 – 24 h to such an extent that it can be washed away with water. The green beans are dried to obtain a water content <14% and a corresponding a_w of 0.75 (Rehm 1985).

cold chain
Linkage of the inevitable extrinsic changes in the situation of chilled, frozen or deep-frozen products between production, transport and delivery to the consumer. Interruption of the cold chain, with the possible effect of influencing the guaranteed microbiological quality and shelf-life should be avoided under all circumstances. Common weak spots in the cold chain are all connected to handling outside the cold storage rooms, or transport in insufficiently refrigerated trucks, strikes at the borders or in the transport business and insufficient or defective refrigeration facilities in shelves and containers in retail outlets.

cold fermentation
Fermentation in wine production taking place at a temperature between 6 and 10°C. Today it is hardly used, but it was favoured for a long time. Cold fermentation yeasts are preferred in the brewing industry.

cold rooms
Refrigerated rooms difficult to disinfect, since temperature of the floors, walls and other objects in the room significantly in-

fluences the required application period of the → disinfectant.

cold shock
Several bacteria exposed to an abrupt fall in temperature from 37°C to 0°C lose the ability to divide due to damage to the cytoplasmic membrane. It is more pronounced in Gram – than Gram + organisms (Ayres et al. 1980). Several → *Fusarium* species need to be chilled to <8°C (temperature during night time in autumn) in order to induce toxin production.

cold smoking → smoking.

cold sterilisation
Decontamination process without heat. (1) Filtration of liquids or gases; → bacterial filter. (2) Radiation with β-, γ- or UV rays. (3) Treatment with → microbicidal disinfectants which are only effective on smooth surfaces.

cold storage
Method to extend the shelf life and for the transport of living plant material, e.g., fruit and vegetables at a temperature between ca. 5 and 15°C depending on the product. The respiration is slowed down ($\rightarrow Q_{10}$ value) and the possibility of infection or reproduction of → mesophiles is minimised.

coli
Coli bacteria. Trivial name for → *Escherichia coli*.

coli enteritis → enteropathogenic *E. coli*.

coli titre
Method for the quantitative assessment of → coliforms in nutrient broth, using → Durham tubes for detecting gas production (Baumgart 1990). The value of this method is doubted since the validity of

"coliforms" as→indicator organisms of faecal contamination is questioned. However, some legal regulations still refer to tolerance levels for "coliforms", and the procedure therefore still has some value in some→official methods for the microbiological examination of milk, dairy products, lactose and drinking water.

coliforms
Loosely refers to any Gram-negative rod-shaped enteric bacterium.→Enterobacteriaceae. Designation more specifically for any Gram−, facultatively anaerobic asporogenous rods fermenting lactose with the production of acid and gas at 37 °C within 48 h (35 °C incubation temp. used in the USA; some workers specify 32 °C for dairy products);→coli titre as indication of faecal pollution. Specification of medium and temperature are critical for interpretation. Although→E. coli is generally referred to as indicator organism of faecal pollution, other genera of the Enterobacteriaceae may produce similar test reactions (e.g., Aeromonas, Citrobacter, Enterobacter and Klebsiella). In addition, a coliform is considered by some water microbiologists to be cytochrome-oxidase-negative, and to be able to grow anaerobically and aerobically in the presence of bile salts or other surface-active agents. In Europe the total →Enterobacteriaceae group concept is preferred, using brilliant green-oxgall broth (with glucose substituted for lactose) for enrichment, and violet red bile agar with 1% glucose for colony recovery. For some products the detection of coliforms is still prescribed as method for assessing the relative→health risk and hygienic processing conditions (→official methods) of a food. May still have some value for internal plant control as indication of the presence of ubiquitary "dirt" bacteria during processing and handling; also as indication of recontamination after heat processing. − Commercial dehydrated

culture media: Lactose Broth (Merck 7661, Oxoid CM 137, Difco 0004) and Lactose Peptone Broth acc. to Eijkman (Merck 7655, Oxoid CM 451, Difco 0017).

collection→culture collection.

Colletotrichum
Mould genus belonging to the Deuteromycetes; partly an anamorph of *Glomerella,* spp. of which are plant parasites. Produce brown or black spots on fruit and vegetables or bitter rot. C. circinans produces dark green or black spots on onions, C. lindmuthianum smudge on green beans and peas, C. musae crown rot of bananas, and C. gloeosporioides→ anthracnose of avocado, mango and papaya, especially in humid regions. − Production of→mycotoxins has not yet been detected (Müller 1983a).

colony
Visible form of bacterial or fungal growth on a solid or gel-like substrate→cfu. Very small colonies (<0.5 mm) are called pinpoints. The colony (macroscopic) morphology is often a useful characteristic for→identification.→Fungal colonies are substantially larger. Ideally an individual colony should develop from a single spore.→Conidia may contribute to a striking colour, or melanin production in the hyphae may render dark brown to black colour.

colony count→cfu.

colouration→pigments.

commensal
Literally "table partners". MO on the outer and/or inner surfaces of the host do not only use nutrients from the host, but also supply metabolic poducts in return, e.g., vitamins. It may be difficult to distinguish it from parasitism.→Skin flora;→oral flora; →enteric flora.

commercial starter

Amount of starter culture required for a production process. Propagated either from a liquid, lyophilised or frozen concentrated culture. Depending on the product to be manufactured, from 0.1% to 5% of inoculum is used, with a viable cell number ranging from 10^6 to 10^8/ml. The initial cell number should at least amount to 10^6/ml so as to reduce the risk of contamination.

commercial sterility

In the absolute sense a controversial term; practically referring to a condition of a heat processed commodity following commercial appertisation.

compressed yeast→ baker's yeast.

concentration changes

(1) Beneficial concentration differences in the cell are established and maintained through the active transport of nutrients and/or metabolites. Energy is used in this process. (2) Changes in the concentration of inhibitors of all kinds in the substrate (→ecosystem), either through dilution or concentration, lead to a change in the growth conditions of the MO present. Some species have improved chances to multiply as the competitors are inactivated or killed. → Succession.

concentration gradient

The uptake of nutrients in the cell gives rise to a concentration gradient directly around the cell towards the cell wall and/or cell membrane. Continuous diffusion tends to "stabilise" the situation. Important process for nutrition in stationary systems; of less significance in systems in which continuous flow is maintained (e.g., pipelines, hosepipes). Stratification of microbes, ranging from aerobes through micro-aerophiles to anaerobes, is due to the oxygen concentration gradient underneath the free surface.

condensation

If the humidity in enclosed spaces (cold storage chamber, refrigerators, ripening rooms, plastic bags, jam bottles, etc.) is higher than 90%, water condenses on surfaces of which the temperature is lower than that of the air or environment. In the range of $0-1°C$ a temperature drop of $2°C$ is sufficient. The higher the temperature the higher the required difference. Up to 30°C natural difference may be found between night and day temperatures, resulting in condensation on a canvass roof. Such continuous or intermittent wet surfaces are suitable living spaces (niches) for MO and especially fungi (→surface fungi). The respiration heat of fruit and vegetables can lead to sufficiently large temperature differences, if the cooling and circulation facilities are insufficient in the store.

condensed water

Present in rooms with foodstuffs. It is not pure distilled water but is enriched with nutrients from the atmosphere. Aroma substances originating from plant and animal products, or other metabolic products dissolve in the water film or drops and act as nutrients and sources of energy for the colonising MO. Dripping condensed water promotes the expansion of MO in these rooms.

confectionery

Products prepared from different sugars and numerous different ingredients, e.g., milk, cream, eggs, honey, fat, cocoa, dried fruit, gelatine, agar, almonds, nuts, etc. In spite of the typically low a_w value, xerophilic yeasts and fungi may grow under specific conditions. Examples of a_w values: fondant $0.70-0.84$; fruit jellies $0.59-0.74$; marzipan $0.65-0.70$; turkish honey $0.60-0.70$; liquorice $0.53-0.66$; soft caramels >0.48. Microbiological examination (Baumgart 1990; Speck 1984, pp. 700-718).

conidia

→ Dormant forms of fungi, produced in large numbers. They are coloured, have thick resistant walls, are formed asexually by transversal fission or budding and are distributed by wind, water and animals, especially insects. Conidia possess one or more than one nuclei. → Heterokaryosis. The form, size, colour and mode of conidiogenesis are important characteristics in the classification of fungi. Most of them survive passage through the GI tract of animals without any reduction in viability.

conidiophore

Conidia forming organs of fungi. Their characteristic forms are important criteria for the identification of genus and species (Baumgart 1986; Reiss 1986).

conserves

Food with a long shelf life packaged in material to prevent contamination and microbial spoilage as long as the protective material remains intact. Food conserved in tin, glass or plastic containers is usually heat treated and free from MO able to multiply in the food. Typically, such products are considered "commercially sterile", however, they are not microbiologically sterile in an absolute sense → sterility. This implies that following a change in the pH, O_2-access or dilution (without recontamination) viable → sporeformers may develop (Heiss and Eichner 1984). It is essential that the period of time between enclosure and heat processing (conservation) be kept as short as possible so as to prevent MO present from producing toxins and to ensure that the food in consequence be treated sufficiently; → D value; → F value. Sterilisation by means of ionising radiation is called → radappertisation.

consumer's milk → pasteurised milk; → drinking milk.

contact contamination

Refers to transmission of contaminating (infecting) MO either by direct contact (e.g., between food products) or indirectly (e.g., by hands, utensils, etc.). In the preparation of poultry (either deep-frozen or fresh) in the household, knives, hands, plates and surfaces may become contaminated with → salmonellae, which then may be transferred to other non-heat processed foods (e.g., vegetables, salads). Probably the most common cause of → salmonellosis. → Cross contamination.

contaminated air

Food processing areas or rooms that are either under positive pressure or convectively aerated, may contain large numbers of viable MO, in addition to dust and off-odours. A strong release of microbes into the air may especially be expected in malteries, mills and bakeries. Contamination risk of nearby processing plants should be minimised by purification of the → air supply, e.g., for cheese ripening rooms and for meat processing areas.

contaminated atmospheres → contaminated air.

contamination

"Pollution"; impurity. Mixture with MO or other noxious elements. Often erroneously used as synonym for → infection. Raw material is usually contaminated from the environment (soil, water, air). Internal contamination is scarce if collected hygienically or if the products are not damaged, e.g., milk from a cow with → mastitis; growth of fungi from the scars on fruit (oranges, peaches); or from the root in maize. The occurrence is higher with root crops due to injuries by harvesting. For the importance of the different sources of contamination see Table 10.

Table 10 Contamination sources of food raw material groups (acc. to Sinell, 1985, modified).
  1 = minor contamination, preventable by good hygiene
  2 = moderate contamination, prevention at least partly possible
  3 = major contamination; prevention possible under careful management
      and processing procedures

| | Contamination Sources | | | | |
| | Interior | Exterior | | | |
Food	Excretions	Body Surface Animal, Man	Water	Soil	Air (Dust)
Fresh meat, poultry	1	2	1	1	
Offall	1	2	2	1	1	
Milk	1	1	2	1	1	1
Shell eggs	3	2	1	2	1
Fish and aquatic animals	1	1	3		
Seeds, grains, peels	1			2	3
Fruit	1	1	1	2	3
Leafy vegetables	1	1	1	2	3
Roots, bulbs, onions	2	2			3	2

contamination precautions
Are the most important steps to minimise or avoid contamination during processing. Fundamental rules are published by the CAC (Codex Alimentarius Commission) as well as in the→ "LMBG" for Germany and subsequent Legal Regulations for Hygiene (compare FAO/WHO Codex Alimentarius Commission, 1983, *Vol. A*, Rome).

continuous fermentation
Methods to culture MO continuously. The most important examples are, e.g., the production of→ baker's yeast or→ single cell protein, including the use of algae. Wastewater treatment works also make use of this principal. The system may be automixed. If biomass is to be recovered the overflow could either be centrifuged or circulated over a rotating filter from where the concentrate is collected.

continuous heating→ continuous
pasteurisation.

continuous pasteurisation
Formerly practised pasteurisation method (63°C/30 min) used to ensure the destruction of all pathogens in milk. → Pasteurisation. For heat-sensitive products, e.g., liquid egg (→ egg products) a similar method is applied.

control
The safety of foodstuffs from a health viewpoint begins in the field or shed and is dependent on continuous control of every step in the production chain including transport, storage, handling and processing until it reaches the consumer. In addition to all precautions taken in the production procedure (in-plant control), the official regulations must be observed. The intrinsic (in-plant) process control is more important than the official control in

preventing the precocious spoilage of the product and to avoid food poisoning. Data of the internal control analyses are earlier available, mostly in time to prevent dispatch of defective products. The process control, the crucial step in the procedure, must be adjusted optimally to the processing steps of the product and depends on the premises, technical outfit, temperatures and production flow rate up to delivery. Thermographs are efficient in controlling critical periods of heating or heat treatment and can be presented in official control records. Microbiological control of → air entering in the system, the raw materials and packaging materials complete the steps (→ GMP directions). Precise book-keeping of single results is a valuable aid in the tracing of vulnerable spots in the production process, during critical periods (mostly after public holidays) and of unreliable employees (Sinell 1985). The official control is mostly aimed at the end product before delivery. Sampling methods (→ sampling plans) are adapted to seasons and different regions. The methods for microbiological examinations of several product groups are compiled and described comprehensively in the → official examination methods (§35, "LMBG"). Further control methods are given in Baumgart (1990) and Speck (1984).

cooked cheese
Sour milk quarg with spices, cream or butter fat which is ripened until the mixture becomes glazy. It is heated at 72°C for a period of 45 – 60 min in a water bath. After it is molten and stirred, 2% table salt and caraway are added. Due to the spices it has a limited shelf life.

cooked sausages
Mainly cooked ingredients are used and it can only be sliced when it is cooled, e.g., "Leberwurst" (liver sausages), "Blutwurst", "Presack", "Zungenwurst",

"Schwartenmagen", frankfurters, braunschweiger, bologna, corned beef, etc. When it has been cooked sufficiently it is much less perishable. As the products are not free from bacteria they should be stored and marketed at a temperature of < 5°C (Krämer 1987; Sinell 1985).

cooked (cured) sausages (meats)
"Brühwurst" (German). Mainly refers to mildly heated cured meats that are either heated in the final container (sausages in casings), or those that are skinned or peeled and portioned after heating (e.g., frankfurter and bologna types). Production from fine or coarsely ground ("bowl cutter") raw meat, with addition of salt or nitrite curing salt, water (as ice) and up to 10% of → plasma. In some countries addition of up to 10% meat substitutes, e.g., soya isolate, is allowed. Mild heating causes coagulation of the proteins. Remains firm, even after additional heating. Final pH ranges from 5.6 to 6.4, depending mainly on initial pH of the meat, and the possible addition of polyphosphates which serve to improve the consistency/stability. A wide range of product types known that are typical for specific regions and countries. Limited shelf life because of → a_w ranging from 0.93 to 0.98; hot smoking (50 – 120°C) contributes to increased keeping quality, although → pocket formation could be detrimental to quality. Main → spoilage association, especially of vacuum packaged, refrigerated sausage products, are LAB. Recontamination or survival after mild heating. May cause sensory defects, e.g., souring syneresis (in hermetically packaged products) and slime formation. Greening caused a.o. by *Lactobacillus viridescens* and *Lactobacillus minor*, resulting from H_2O_2 production and its reaction with myoglobin. Shelf life at refrigerated storage (4 – 7°C) up to 3 weeks, depending on pre- and post-packaging processing conditions, as well as

maintenance of the cold chain. Limited shelf life of sliced and packaged bologna types.

core rot
"Kernhausfäule". Rot affecting apples and pears during storage. It starts around the kernel and brown lesions develop in the fruit flesh. Eventually the fruit is affected completely. White or pink mycelia develop like cottonwool in the cavities. The species usually involved are *Fusarium* spp., *Trichotechium* spp. and *Alternaria alternata,* rarely other species. Infection is usually through the kernel or scarred blossom (Müller, G. 1983a).

core temperature
The region which is lastly affected during temperature changes in an enclosed system, e.g., boxes, flasks and heat exchangers, and is of decisive importance from a microbiological point of view. It applies typically to heating but also to chilling. It is therefore necessary that the core region be maintained at the desired temperature for the required time period in order to guarantee that all MO are killed and/or inactivated. This may lead to the over-treatment of the peripheral zone. The core temperature is reached quicker in moving systems, e.g., rotating autoclave ("retort sterilisers") than in stirring systems. → F value; → heat sterilisation.

cork taste
Cork taste or cork "taint" error in wine of which the flasks are closed with natural cork. About 2% of all the flasks are affected resulting in annual losses of about 60 million DM in D. Changes in the taste and the smell are most probably caused by sesquiterpenes produced by the fungi growing on the cork, e.g., *P. roqueforti, P. frequentans, Cladosporium herbarum,* etc., and which dissolves in the wine [Heimann, D. et al.: *Lebensm.-Rundsch.* 79 (1983) 103; Dittrich 1987].

corn steep liquor
Steeping water rich in organic substances resulting from the production of starch from maize. Concentrated to a syrup-like consistency and is then added to culture media for cultivation of fungi. Raw material for the technical production of biomass.

Corynebacterium
Bacterial genus of the order Actinomycetales: Gram +, irregular, non-sporulating rods (Bergey 1986), psychrotrophic. Several spp. are pathogens of man and animals, e.g., *C. bovis* (→ mastitis), *C. diphtheriae* (diphtheria). Found in raw milk, milking machines and dairy utensils, brines (Krämer 1987; Kunz 1988).

cottage cheese → fresh cheese.

counting chamber procedures
Direct method of microscopical counting especially for yeasts. It is not possible to distinguish between dead or viable cells. → Bacterial count. A THOMA – counting chamber is usually used. A thick slide is smoothly ground and etched with a grid of known dimensions. The depth of the chamber is 0.01 mm under the cover slip. Bacteria can also be counted in similar chambers with a depth of 0.02 mm, but phase contrast microscopy should be used. It is problematic and the results are only approximations for use in routine laboratory procedures (Baumgart 1990).

cresol
Methylphenol. (1)→ Disinfectant with good antimicrobial properties, especially against fungi. Disinfectant for hands (lysol soap) and abrasive disinfection (Wallhäusser 1988). (2) Degradation product of proteins, formed in addition to other flavour intensive substances, e.g., phenol and indole, during microbiological proteolysis.

critical point analysis
Control of critical points in the production

procedure in order to detect coincidental defects. It is part of → GMP. More or less identical to quality control systems developed in the USA: "Hazard Analysis Critical Control Point" (HACCP). – Critical points prior to end product control include contamination with dangerous MO during production and processing, and enrichment of MO during this period. Continuous control of pH, temperature, hygiene and the colonisation of the raw material or end products are important. Biological features (MO present, nutrients available, etc.), → intrinsic factors, ecological factors (temperature, pH and a_w-change during the process) (→ extrinsic factors) should be considered. Basic prerequisites are the devising of a sampling plan and the continuous training of personnel involved.

cross contamination
A contamination chain in the production and processing of food where uncontaminated parts are contaminated by others or transmitted by a vehicle, e.g., the dividing of labour in an abatto where certain steps are not mechanised and are to be done by hand; mechanical defeathering of poultry in a continuous system; → spin chiller for the prechilling of poultry; wash basins in the production of → mixed salads. Refrigerators for domestic use and in restaurants allow the contact of highly contaminated fruit, vegetables and eggs with hygienically extremely sensitive products, e.g., leftovers, delicatessen salads, sliced cold meats, etc. This process is enhanced in the limited space, by air movement created by repeated opening and closing of the refrigerator. A special case is "Schmierinfektion" (SGLH 1982); → contact contamination.

cross resistance
The resistance of MO against → antibiotics and other substances with similar chemical structures or mechanism of action, e.g., tetracyclines, streptomycin and mac-

rolide-antibiotics. Antibiotics are worldwide not allowed in food preservation due to the phenomenon of cross resistance. In this way the risk of inducing resistant strains is minimised. → Nosocomial infection; → penicillin.

cryophile → psychrophile.

Cryptococcus
Genus of non-fermentative imperfect yeasts of the Basidiomycetes (→ Deuteromycetes); teleomorph: *Filobasidiella*. *C. neoformans* (cryptococcosis, also associated with bovine mastitis) and *C. bacillisporus* are pathogens. Distribution: soil, faeces of birds, rinsing water of fruit, and in frozen fruit). – Cryptococcosis (torulosis, European blastomycosis) is a severe disease of man and animals. Infection by inhalation of dust contaminated with the mould, and more indirectly via food and feed. Pulmonary infection mild to severe, depending on immunological resistance. Other spp. related to *Rhodotorula* have been isolated from fresh seafood.

cultivar
A commercial variety or cultivation form of particular MO species, e.g., mushrooms or specific starter cultures. It can be compared to the common type names of ornamental plants.

cultivated mushrooms
Apart from the well known → mushrooms (champignons), other edible species forming vegetative bodies are also being commercialised as far as possible. The special experience in several East Asian countries is presently adopted and made use of in Europe. Known species and partly commercially available species grow on similar substrates as the conventional mushrooms: e.g., *Volvariella volvacea* (padi straw mushroom), *Lepista nuda* (purple wood blewit) and *Stropharia rugosoannulata*. *Coprinus comatus* (ink cap

fungi) is commercially available as "instant culture" in the horticulture business under the name "Spargelpilz" ("asparagus mould") without supplying any closer substrate description. *Pholiota mutabilis* ("Stockschwämmchen"), *Pleurotus ostreatus* (oyster fungus) (→ allergens) and *Lentinus edodes* (shii-take) are cultivated on wood (Botticher 1974).

culture
General term referring to MO cultured "artificially" in the laboratory or for commercial purposes. They are cultured and stored in large containers usually under sterile (axenic) conditions. A distinction can be made between→ pure cultures,→ mixed cultures,→ starter cultures, etc.

culture characteristics
The surface, form, colour, edge and size of a MO→ colony cultured under standardised conditions (culture medium, temperature, time, etc.). In liquid media turbidity, sediment, flocculation and film formation can be distinguished. The constant characteristics are useful in the identification, differentiation and description of genera and species (Baumgart 1990).

culture collection
Collection of species and isolates of pure cultures kept in the laboratory and used for scientific or commercial purposes. Preservation is by means of refrigeration, or the culture is kept in special containers with liquid nitrogen, or it is freeze-dried and then stored in vacuum sealed glass tubes. – Species and isolates with special characteristics can be purchased from the large culture collections comparable to the botanical and zoological governmental collections, e.g., DSM ("Deutsche Sammlung von Mikroorganismen und Zellkulturen") in Braunschweig-Stöckheim, ATCC ("American Type Culture Collection") in Rockville, MD, USA, CMI ("Commonwealth Mycological Institute") in Kew, GB, or CBS ("Centraalbureau voor Schimmelcultures") in Baarn, NL.

culture medium
A liquid or solid nutritious medium with a known and/or constant composition with specific characteristics used for the reproduction of MO. Several culture media for specific purposes are described and some are commercially available, e.g.,→ fermentation,→ selection and→ differentiation.

culture mould
→ "Noble mould". Trivial name for→ *P. caseicolum,*→ *P. roqueforti* and→ *P. nalgiovensis* that are available as commercial starter cultures and are used in the production of soft cheeses and mould-ripened salami.

culture test tube
Test tube, eprouvette. Reagent glass-like culture container manufactured from thick glass without a rim; different diameters and sizes are used for the cultivation and storage of MO. It can be closed with cellulose, cottonwool, metallic caps with different constructions and spring clamps.

cultured creamery butter→ butter.

curd
Casein fraction of milk that is coagulated with the aid of rennet and starter cultures (souring) during the production of cheese.

cured meat products
Generally refers to meat products "cured" with a mixture of salt and nitrite/nitrate; sugars and spices/flavourings may also be used depending on the product type. Raw cured products (bacon), cooked (heat-processed) meat products (cooked hams, bellies, shoulders, "Kasseler", tongue, etc.),→ fermented sausages (dry and semi-dry

sausages) (Germ.: "Rohwurst") → cooked sausages. Ca. 90% of the processed meat products are cured to increase their shelf life (preservation) and improve sensory properties (flavour, colour), and especially for safeguarding against food-poisoning bacteria, mainly → *Clostridium botulinum* (Bacus 1984; Krämer 1987).

curing aids
CH such as sugars or corn syrup containing glucose and maltose, in addition to dextrin, support the rapid growth of the desired → curing microorganisms. → Nitrate is reduced to → nitrite which inhibits → proteolytes and clostridia. – Sodium ascorbate serves to maintain a low Eh value (→ redox potential) which serves to stabilise the colour and to prevent the formation of → nitrosamines from nitrite and amines (Krämer 1987; Sinell 1985).

curing microorganisms
Microbial population selected by the presence of → nitrite and → salt in the → brine, dominated by *Staphylococcus* spp., micrococci, *Lactobacillus* spp., *Enterococcus* spp., *Pediococcus* spp., *Vibrio* spp., etc., including halotolerant yeasts. By their metabolic action the pH is reduced from 7 to ca. 5, thereby preventing the proliferation of strong → proteolytes (Bacus 1984; Krämer 1987; Sinell 1985).

curing salt → nitrite curing salt.

curling effect → griseofulvin.

Curvularia
Mould genus of the → black moulds (Hyphomycetes), with weak allergenic potential. *C. lunata* is used for the hydroxylation of 11-deoxycortisol to hydrocortisone (cortisol) and for the recovery of hormones (Müller and Loeffler 1982).

cyanobacteria
Large group of photosynthetic pro-

karyotic organisms, differing from other bacteria by the possession of chlorophyll a and by oxygenic photosynthesis; no chloroplasts or mitochondria, and no flagella although gliding motility sometimes observed. Gram-negative cell wall sensitive to → lysozyme. Unicellular or filamentous. Occur in different types of habitats, ranging from fresh to brackish and even hypersaline waters. Eutrophication of reservoirs, e.g., after heavy storms, favour the rapid development of cyanobacteria (water bloom). Some spp. are toxic, e.g., *Anabaena flosaquae, Aphanizomenon flosaquae, Lyngba majuscula, Microcystis aeruginosa, Oscillatoria* spp., *Schizotrix calcicula,* etc. They produce "phycotoxins", that have been identified as cyclic peptides, and synthesised from D- and L-amino acids of varying sequences (Botes, Theis and Carmichel: in Steyn and Vleggaar 1986). – Fish that have fed on toxic cyanobacteria may constitute a health risk because of the heat resistance of these toxins (e.g., Ciguatera poisoning in the USA, "Haff sickness").

Cyanophyceae → cyanobacteria ("bluegreen bacteria").

cyclopiazonic acid
→ Mycotoxin produced by *P. caseicolum* and *P. griseofulvum*, and other spp. of the genera *Penicillium* and *Aspergillus*. Formed during mould growth on food (Table 31, see page 195, *Penicillium*). May be found in maize, peanuts, Camembert cheese (max. 0.37 mg/kg). Toxic for chickens, but no definite health risk for humans, even after consumption of large amounts of soft cheese with mould growth (Reiss 1986).

cytochalasins
Group of secondary fungal metabolites with cytostatic activity; classified as → mycotoxins. Produced by different moulds,

e.g., *Aspergillus, Helminthosporium, Metarrhizium, Phoma, Rosellinia* and *Zygosporium.* Induce morphological changes and multikaryosis in viable animal cells. Inhibit the addition of monomers to actin microfilaments, and cell division. Found in mouldy grains and their products (Betina 1984).

cytoplasm
Protoplasm; plasma. Complex cellular content in which (a.o.) the→nucleus (nuclei for moulds and other eukaryotic cells) is embedded. Surrounded by a semi-permeable membrane (cytoplasmic membrane, plasma membrane, protoplast membrane), which controls the active transport of nutrients and metabolites between the inner part of the cell and the external environment. The plasma membrane is bordered externally by the cell wall.

cytotoxicity
Toxicity towards human or animal cell cultures. Typical of some→ mycotoxins, e.g., → trichothecenes or→ cytochalasins.

Czapek's medium
Czapek-Dox medium. Nutrient medium for cultivation of several moulds and soil bacteria. Contains mineral salts and sucrose or glucose as C and energy source. Nitrogen provided as→ nitrate; other minerals: K_2HPO_4, $MgSO_4$, KCl and $FeSO_4$; pH ca. 7.3. Commercially available: BBL 11140 or Merck 5460. For some mould genera it is used parallel to→ malt extract agar or→ wort agar for species diagnosis, e.g., penicillia (Raper and Thom 1968).

D value

D_{10} value; decimal reduction time. The time required at a given temp. (e.g.,70 °C = D_{70}) for the reduction of the number of viable cells or spores/endospores of a specific organism by 90% (or 10-fold). The time is quoted in minutes or seconds; the temperature is indicated by a subscript. The D value serves as measure of the → heat resistance of a MO (Figure 22, see page 264, TDT curve). Also serves as basis for the calculation of a minimum safety margin for heat sterilisation in practice; → F_o value (Heiss and Eichner 1984). – Also used in → radiation treatment for → decontamination or → sterilisation (appertisation). Indicates the doses in kGy (= τD_{10}) required for the reduction of the number of viable cells or spores of a given organism by 90%: e.g., *Salmonella* spp. 0.20 – 0.45 kGy; *Cl. botulinum* (endospores) 1.50 – 2.50 kGy (Classen et al. 1987).

deamination

Enzymatic cleavage/removal of an amino group from → amino acids and → nucleic acids during the spoilage of foods, with the release of → ammonia and fatty acids: e.g., oxidative deamination of L-aspartate to ammonia and fumarate (catalysed by aspartase). Especially Gram – bacteria are involved in such → putrefactive processes that may have a detrimental effect on the sensory properties of a product (Krämer 1987; Kunz 1988).

death phase

The stage during which the viable number of cells in the microbial population declines; on certain occasions the → biomass may still increase, reflecting continued overall → growth by this measure. The death phase typically represents the time period of microbial growth following the stationary phase. It is characterised by a strong increase in metabolic products, e.g., organic acids, and may be regarded as "self-intoxication" (environmental pollution!). Cells isolated from this phase may show selective increase in resistance against inhibitory substances present.

Debaryomyces

Yeast genus belonging to the Endomycetales. Cells round to ovoid, some yellow to red pigmented, often with intracellular oil drops; practically no fermentative ability; produces a surface skin. Found in milk products, meat, fermented tobacco; may settle on epidermis of man and animal (Baumgart 1986; Kunz 1988).

decarboxylation

(Enzymatic) cleavage of the → carboxyl group from a carboxylic acid, → amino acid or → nucleic acid (e.g., during food spoilage) to produce → biogenic amines (from amino acids) and CO_2: e.g., histidine → histamine; ornithine → putrescine; lysine → cadaverine. Many Gram – bacteria are capable of this type of spoilage; especially protein rich products may thus be rendered inedible and toxic (Krämer 1987; Kunz 1988). – Decarboxylation also typically part of (e.g.) → tricarboxylic acid cycle.

decimal reduction time (DRT) → D value.

decontamination

Sanitation, disinfection. Drastic reduction of the microbial population in wastewater, effluents, etc.: e.g., abattoirs, rinsing water, water from spin chillers, etc., or on surfaces. Also used when referring to destruction of a significant part of the contaminating MO in certain (dehydrated/low moisture) food commodities, e.g., →

radurisation of spices. Not identical with→sterilisation or→pasteurisation by heat.

deep freezing

General method to extend the shelf life or keeping quality of several products characterised by the fact that the temperature should not exceed − 18 °C until the product is delivered to the consumer. Only some MO are killed when frozen, and pathogens may survive. The microbial numbers slightly decrease during storage, depending on the time period. From a microbiogical point of view, slow thawing is preferable since it allows the destruction of more MO when the critical range of − 4 to +6 °C is passed slowly. Pretreatment methods, e.g., blanching of vegetables and→pasteurisation of cream, may contribute significantly to a reduction of the bacterial count or to the destruction of pathogens. – Particularly critical procedures include washing and pre-cooling in chilled water because of possible→cross contamination;→salmonellosis; ("contact infection");→spin chiller.

deep-frozen products

These products should be free from pathogens if they are to be consumed without heating ("ready-to-eat" foods). Examples are:
(1) Fish (Matjes fillet, fish pies, smoked trout fillets)
(2) Crustaceans, shellfish and molluscs (shrimps, crabs, snails, mussels, calamari and hind-legs of frogs) either cooked or ready-to-eat
(3) Meat products ("Brotzeitteller", cooked liver sausage, vienna sausages, krakauer, etc.)
(4) Pastries with moist fillings
(5) Cakes, especially with moist fillings
(6) Small cakes and breakfast pastries (fruit pies, filled eclairs)
(7) Mousses

(8) Dairy products (butter with herbs, quarg creams, cream desserts, cream flakes and portions)
(9) Other products (muesli with berries, muesli, shepherd salad and other salads, sauces and dressings)
The requirements for all these products are that they can be consumed with or without heating or reheating after thawing. – It may be critical when such products are kept at temperatures higher than 6 °C for longer periods, since→opportunists may reproduce rapidly, in addition to quality deterioration resulting from increased enzymatic activity and simultaneous microbiological spoilage.

degeneration

Morphological and/or metabolic changes; e.g., caused by harmful substances to LAB (→disinfectants;→antibiotics;→nitrite, etc.), or suboptimal temperatures. The performance of→starter cultures may thus be strongly reduced. Typically experienced with culture collections (Kunz 1988).

degradation

Enzymatic cleaving (catabolism) of high molecular or water insoluble substances into small units that are usually water soluble. Being mostly monomolecular in size, these fragments can be transported into the microbial cell where they are metabolised for energy generation and the synthesis of new cell material.→Extracellular enzymes have a key function in the degradation of non-assimilable macromolecules (primary substrate) outside the cell, thus ensuring the nutrition of the microorganism. Degradation is the basis, both for the mineralisation of organic matter in the ecosystem, and for the microbial spoilage of foods. Examples:→starch is broken down to glucose;→protein to amino acids, and→fat to fatty acids and glycerin;→biological degradation ("biodegradation").

degree of acidity

Important parameter, in addition to pH value, for the monitoring of different→fermentations, and for end product quality control. Potentiometric procedures for end product titration are the methods of choice; the results, however, are given in different units, for historical reasons: (a) degree of acidity of milk and milk products acc. to Soxhlet-Henkel (°SH) corresponds to 1 ml of 0.25 n NaOH per ml of sample; (b) degree of acidity of→sourdough and bread (Sr°) corresponds to 1 ml of 0.1 n NaOH per 10 g of sample. – In other areas of food production the amount of titratable acidity is given in mol or mmol sodium hydroxide solution per ml or g of sample.

dehydrated nutrient medium

Standardised nutrient media in powdered or granulated form with preparation instructions commercially available for almost all purposes in the food microbiology. It is diluted with tap or deionised water and then autoclaved. Produced by, e.g., BBL, Difco, Merck, Oxoid.

dehydrated vegetables

Dried leaf and root vegetables which have been cleaned, washed, shredded, blanched and occasionally even cooked before mechanical dehydration. Drying at 55–60° does not guarantee that the products are free from MO. It is processed further for use in powdered soup. Care should be taken that instant soups are free from MO since these are not cooked, and often only lukewarm water is used for preparation on camping sites (Krämer 1987).

dehydrogenases

Metabolic enzymes that remove hydrogen from a reduced substrate.

Dematiaceae→black moulds.

Dematium

Obsolete name for black yeasts, especially referring to the genus→*Aureobasidium*.

Demeter pipette

Free flow pipette, subdivided in 1–2 compartments each of 1 ml and 0.1 ml. Called after the famous microbiologist Demeter. Application especially in routine milk examinations, e.g., for inoculation of agar plates for the determination of the→viable count; time- and labour-saving; for→dilution series only steps of 1:100 are required.

demineralised water→ion exchanger.

denitrification

Nitrate reduction. Respiratory metabolism by some bacteria by which NO_3 or NO_2, as terminal electron acceptor, is reduced to gaseous products (nitrous oxide or N_2) or ammonia. The gases are highly insoluble in water and escape as "air bubbles" from the system. The first step involves reduction of nitrate by nitrate reductase; in the second step, nitrite reductase is involved. The following sequence may occur:

$$NO_3^- \rightarrow NO_2^- \rightarrow NO \rightarrow N_2O \rightarrow N_2$$

Biological method for the removal of→nitrate from→drinking water, after addition of methanol as energy source and a mixture (mixed population) of nitrate reducing bacteria. – Genera that contain denitrifying spp.: *Alcaligenes, Bacillus, Chromobacterium, Flavobacterium,→Paracoccus, Pseudomonas*. The application of some strains for the reduction of nitrate in vegetable juices as infant food, is investigated; applications for drinking water are being practised already.

deoxynivalenol→*Fusarium*.

dermatophytes

Medical designation of fungi that infect skin and keratinised tissue of man and animals (dermatomycoses), e.g., athlete's foot.

destruction

During food processing operations MO may be typically killed by thermal energy. In a→disinfection operation some MO may be lethally damaged by chemical compounds:→disinfectants. When the destruction process of a homogeneous population is followed over time under constant temperature and component concentration, an exponential plot is obtained for the decrease in viable→population, suggesting a "single-hit" mechanism. For a mixed (heterogeneous) population the plot follows an irregular course. The complete destruction (or removal) of all forms of life is termed→sterilisation. Pasteurisation refers to the killing or removal of all pathogens. The number and type of survivors will decisively influence the keeping quality (shelf life) of a product (Krämer 1987; Kunz 1988).

destructive disinfection methods

Disinfection processes by which MO are destroyed, i.e., that are microbicidal. Opposite:→inhibitory disinfection methods. Physical: heat (sterilisation), burning, radiation sterilisation (→radappertisation). Chemical, in order of decreasing effectivity: alkylating compounds (→ethylene oxide), strong oxidising agents (peracetic acid, H_2O_2, halogens), ion aggressiveness (strong acids or alkalis), aldehydes, phenols, tensides, antibiotics [Mrozek, H.: *Arch. Lebensm. Hyg.* 31 (1980) 91 – 99].

Desulfotomaculum

Relatively small genus of Gram –, anaerobic endospore-forming bacteria, distinguished from *Clostridium* spp. by their sulphate reducing ability. Chemo-organo-trophs; C- and energy sources: lactate and pyruvate. Found in soil and the rumen. Growth: opt. 55°C, min. 27 – 30°C, max. 70°C; pH 5.9 – 6.4. Endospores may survive in low-acid canned products (D_{120} 2 – 3 min). H_2S production may cause precipitation of black heavy metal sulphides; responsible for "sulphur stinker" spoilage defects (Kunz 1988; Mitscherlich and Marth 1984).

detergents

Cleaning agents. Support the mechanical cleaning of surfaces by chemical action, in order to remove dirt particles, e.g., fats and proteins, which may negatively influence or inactivate the subsequent→disinfection process. Distinguishing is made between:
– alkaline detergents
– alkaline detergents containing carbonate
– alkaline detergents containing chlorine
– alkaline detergents containing aluminium
– acidic detergents
They should not be corrosive. Several preparations (sanitisers) may also contain a disinfectant (Kunz 1988).

Deuteromycetes

Fungi Imperfecti; Deuteromycotina. "Artificial" category for fungi with no (known) sexual stage, but may also include the asexual (anamorphic or conidial) stage of→Ascomycotina and→Basidiomycotina with a known sexual (teleomorphic or perfect) stage, which, however, are only formed under specific ecological situations (see Table 11). Generally refers to asexual fructification without change of the karyophase. Common property is the formation of→conidia and, in some, also→sclerotia and→chlamydospores. Ca. 30,000 spp. included, representing most fungi associated with food or food spoilage, e.g., *Aspergillus, Penicillium, Fusarium, Colletotrichum*, etc., and also

Table 11 Main fruiting bodies known of the most important Deuteromycetes (acc. to Domsch et al. 1980; Müller and Loeffler 1982; Reiss 1986).

Anamorph Imperfect (Asexual)	Teleomorph Perfect (Sexual) Form
Acremonium	Nectria
Aspergillus	Emericella
	Eurotium
	Chaetosartoria
	Dichlaena
	Fennellia
	Petromyces
	Sartorya
	Sclerocleista
Basipeptospora	Monascus
Bipolaris	Cochliobolus
Botrytis	Sclerotinia
	Botryotinia
Cladosporium	Mycosphaerella
Curvularia	Cochliobolus
Cylinrocarpon	Nectria
Fusarium	Nectria
	Gibberella
	Calonectria
	Plectosphaerella
Geotrichum	Dipodascus
Helminthosporium	Cochliobolus
Monilia	Neurospora
	Sclerotinia
Myrothecium	Nectria
Paecilomyces	Byssochlamys
Penicillium	Dichlaena
	Eupenicillium
	Hamigera
	Talaromyces
	Penicilliopsis
	Trichocoma
Rhizoctonia	Pellicularia
Scopulariopsis	Microascus
Stilbella	Nectria
Tubercularia	Nectria
Verticillium	Cordyceps
	Nectria
	Torrubiella

including anascosporogenous yeasts, e.g., *Candida*. Many produce→mycotoxins (Müller and Loeffler 1982).

dew point
Temperature in an enclosed space or on a wall at which air or gasses condense at a specific rH. An important criterion to bear in mind, since products in waterproof packaging or glass flasks, e.g., jam, may undergo considerable acceleration of microbial spoilage when stored in refrigeration and ripening rooms or cellars.

dextranase
Trivial name for dextran hydrolysing enzymes. Systematic name: 1,6-α-D-glucan-6-glucanohydrolase; E.C. 3.2.1.11. Catalyses the endohydrolytic cleavage of 1,6-α-D-glucosidic bonds in dextrans. Application: degradation of dextrans in raw sugar juices. MO for production: *Aspergillus* spp., *P. funiculosum*, *P. lilacinum*, *B. subtilis*, *Klebsiella aerogenes* (GDCH 1983).

dextrans
Group of extracellular, high-molecular weight polysaccharides produced by certain MO, e.g., *Leuconostoc mesenteroides* and some other→LAB. Excessive slime production by *Lc. mesenteroides* may cause severe losses in the sugar industry; may also complicate filtration procedures in the beverage industry.

dextrose→ glucose (D-glucose).

diabetics beer
→ Beer produced with *Sacch. diastaticus* which ferments dextrins and raffinose in addition to glucose and sucrose. Low amount of residual sugars (Kunz 1988).

diacetyl
CH_3-CO-CO-CH_3; butane-2,3-dione. "Butter aroma", e.g., in→cultured creamery butter ("ripened cream butter"). Produced a.o. by *Lactococcus lactis* ssp. *diacetylactis*, *Leuconostoc lactis* and *Lc. mesenteroides* ssp. *cremoris* from citric acid (present in milk: ca. 2.4 g/l) via oxalic

acid and pyruvate during controlled souring/ripening of cream. These spp. are contained in several cheese→ starter cultures. Threshold value for diacetyl in butter: 0.5 mg/kg for taste, and 0.1 to 10 mg/kg for aroma. – Diacetyl-producing LAB deleterious to beer quality; exception "Berliner Weisse" (top-fermented wheat beer) (→ beer). Especially *Pediococcus* spp. are considered detrimental for beer because of production of acetoin (CH$_3$-CO-CHOH-CH$_3$) that is autoxidised by molecular O$_2$ to diacetyl that causes flavour and taste defects. Taste threshold value ca. 0.2 mg/l (Priest and Campbell 1987). Acetoin also produced by *Enterobacter aerogenes* but not by *E. coli*. – *Pediococcus* spp. may also cause off-flavours and tastes in wine, with diacetyl levels from 1 to 50 mg/l (Dittrich 1987).

diarrhoea
Diarrhea; dysentery. Refers to passage of fluid stool with a high frequency. Most typical symptom associated with→ food poisoning/food-borne infection.

diatoms
Group of algae consisting of > 10,000 spp. Cell wall typical siliceous structure (SiO$_2$); general requirement for silicon. Unicellular. Found in fresh, brackish and marine waters, and in moist terrestrial environments. Natural deposits of fossil diatom frustules (diatomaceous earth, diatomite, kieselguhr) are mined and find many industrial applications as kieselguhr, a.o. as filtering material (e.g., in Berkefeld filters), and as mild abrasive, e.g., in tooth pastes.

diauxie
Occurs when a MO, in presence of two C sources, preferentially metabolises one till depletion, before metabolisation of the second source is started. A "switch-over" between the two phases is typically characterised by a lag phase during which *de novo* synthesis of the newly required enzyme(s) is initiated: e.g., in presence of sorbitol, glucose is completely fermented before aldose-reductase is synthesised to enable the cell the glycolytic degradation of sorbitol (Bruchmann 1976; Schlegel 1985). Analogous examples for N sources, e.g., ammonia and nitrate.

diethylpyrocarbonate
C$_2$H$_5$-O-CO-O-CO-C$_2$H$_5$. Preservative of beverages, e.g., wine and beer. Effective against yeasts and moulds, but no longer permitted in USA and several European countries because of cleavage into toxic ethylcarbamate (urethane) in addition to ethanol and CO$_2$. Presently, dimethyldicarbonate (Velcorin) finds application as alternative; spontaneous degradation into→ methanol and CO$_2$; half life at 10 °C ca. 40 min, at 20°C ca.15 min. Methylcarbamate that may be formed is harmless. Dimethyldicarbonate used in conc. of up to 200 mg/l at room temp. (Classen et al. 1987).

differential medium
By contrast to→ selective media, different spp. and even groups of bacteria may grow on a solid differential medium, on which they may be distinguished by different growth forms and reactions in the medium. Based a.o. on ingredients (e.g., blood, egg yoke, Fe^{3+} ions) and pH indicators. Dehydrated culture media for different applications, mainly for bacteria, are commercially available (e.g., BBL, Difco, Merck, Oxoid).

differentiation
Heterogeneous microbial populations are typically found in foods, and several direct and indirect methods are available for determination of the→ total cell count. Differentiation of these groups or spp. of MO is often desirable, and an important aid in the interpretation of quantitative results; generally, however, additional proce-

dures and substantial experience are required. Typically, the first step involves the isolation of→pure cultures. (1) Bacteria: the classification system of *Bergey's Manual of Determinative Bacteriology* (1986 Edition) forms the basis for the→identification of families, genera and spp. Additional information is also available in *The Prokaryotes, 2nd Ed.* (Balows et al. 1992). The "natural" (phenetic) system as originally suggested by Von Linné has found application in the classification of plants and animals but not for MO, for which still too little information is available on the relationship between phylogenetic and phenotypic characteristics. The main differentiation criteria for bacteria involve →morphology,→metabolic properties,→chemical composition of the→DNA and the→cell wall, and, in some instances, pathogenicity towards experimental animals. Numerous procedures have been published and are being generally accepted and applied (Balows et al. 1992).→Differential media and time and labour saving "multitest" systems have been developed for simplification of identification procedures such as the determination of the "sugar fermentation spectrum" (e.g., commercial systems available: API system, Enterotube, Oxifermtube, etc.). Novel, computerised systems involve (a.o.) the gaschromatographic determination of the fatty acid profile, or the poly-acrylamide electrophoretic determination of the cellular protein profile (Page) of bacteria. (2) Yeasts: typical characteristics are determined for a→pure culture, isolated, e.g., from beer, wine, non-alcoholic beverages, confectioneries, etc.; simplified determination keys are available for the differentiation of phenetic groups, representing the main spp. typically associated with the respective product groups. Typical criteria include: sexual, morphological, cultural and physiological properties. Several specific

and→differential solid media are available for yeasts, also as commercial dehydrated products. (3) Fungi: morphological and cultural properties constitute the main criteria for differentiation. Physiological characteristics as additional criteria are reliable only in exceptional cases, and involve colour (pigment), aroma, and pathogenicity towards plants, animals and man. Production of→antibiotics and→mycotoxins are not reliable criteria; the practical implications may, however, render additional examinations unnecessary. – Generally, pure cultures are grown on specific growth media and at different temperatures (under artificial light or in darkness), and subsequently investigated microscopically as first indication of possible group or genus relatedness. In addition to macroscopical properties (colony shape, colour, surface and elevation), the fine structure and surface properties of the reproductive organs, and size and shape of spores or conidia, may be determined microscopically, and serve (in some cases in combination with infectivity, etc.) as criteria for the recognition of a sp. For fungi a relatively limited number of→differential media are available, although several basal media are offered commercially. For differentiation keys and additional literature (see Beuchat 1987; Reiss 1986; Rhodes 1979; Spicher: in Baumgart 1986).

diffusion

Molecules are soluble, at least in part, in liquids (water) or solids (lipids, synthetics). These molecules may be solubilised upon contact with these phases, resulting in a slow migration in crystalline lipid, and more rapid in oil or water, away from the point of contact and towards decreasing concentration gradient. This movement between other molecules is energy independent. The rate of diffusion is dependent, a.o., of the viscosity of the medium, the temp., etc. Gas inclusions constitute impenetrable hurdles. – This phenomenon has important consequences for food systems. Examples: absorption of O_2 from the air or CO_2 from → CA storage; diffusion of solubilisers, softeners and other constituents from packaging materials, or the escape of aromatic compounds through the packaging material. Diffusion of microbial toxins from the colony (point of production) into the food. At 25 °C → patulin diffuses 10 cm from the mycelium growth in apple juice-agar-gel, within 12 days. Even large molecules such as enzymes (proteases) migrate, e.g., during ripening of cheese into the protein matrix.

diffusion test
Examination of water soluble diffusible substances on agar plates or in test tubes. Typical application in the → antibiotic sensitivity test which relies on the diffusion of antibiotics into agar; also used for determining the presence and/or effectivity of other inhibitory substances such as preservatives and disinfectants, by means of the → disc diffusion test, the → agar well test or → cylinder test (Baumgart 1986). – Examination of enzymes or enzymatic activities of MO (→ exoenzymes) on buffered gels; turbidity caused by fat droplets (e.g., tributyrin) cleared by lipases; proteolytic activity on medium containing casein (skim milk). Clear zones indicate activity, whilst the diameter around a colony or depth in a test tube, may serve as quantitative measures.

digester → biogas.

dikaryophase
Phase during the life cycle of fungi, especially → Basidiomycetes, characterised by a binucleate cell or mycelium.

dilutant
Solution used to prepare the first dilution and → dilution series. Its osmotic value corresponds more or less to that of the cells. "Physiological" salt solution (Ringer's solution) used in medicine, is not optimal for microbiological purposes. The following solution has been proven more satisfactory: 0.1% casein peptone (tryptically digested), 0.85% NaCl and pH 0.70 after sterilisation. For the → drop culture method this solution is prepared with addition of 0.08% agar.

dilution series
For → microbial counts determined by the pour plate, surface plate and drop culture methods. The first dilution with the smaller and well mixed sample is diluted further in steps of 1:10 in order to obtain 30 – 300 colonies on each plate. See Figures 13 and 14, pages 208 and 209 (→ pour plate method). Details in Baumgart (1990), or Smith (1981). – For determination of the minimal inhibitory concentration (MIC) (e.g. for the assessment of new disinfectants; the determination of the concentration of an antibiotic in milk; the estimation of residual antibiotic activity in → enzyme preparations) a dilution series of 1:1 is to be preferred. A new sterile pipette should be used in principle for each step so as to prevent transfer of bacteria or inhibitory substances present on the inner wall to the next dilution step.

dilution streaking
Important method to obtain → pure cultures. A small part of a colony from a bacterial count plate is streaked with an inoculation needle on a plate, without disrupting the nutrient surface. For this purpose the plate must be dried after solidification, prior to streaking. For details see Baumgart (1990), Smith (1981).

dimethylcarbonate → diethylpyrocarbonate.

dimorphism
Ability of some fungi (e.g., Mucorales and *Penicillium*) to grow in either of two distinct vegetative states, depending on environ-

mental conditions. Phenomenon initially observed by Pasteur in 1876 in relation to fermentation and respiration during changing conditions of aerobiosis. Reduction of the O_2 and increase of the CO_2 partial pressure (e.g., in hermetically sealed packages) induces the change of the mycelial state to the (unicellular) yeast phase. Determination of the viable mould count results in the detection of exceptionally high numbers of mould colonies arising from the single "yeast-like" cells under the aerobic conditions existing on the plate. As a result of the original environmental conditions (e.g., high rH and reduced O_2), typical mycelia may not have been detectable in the sample (e.g., cereals from grain silos). The presence of high numbers of single cells (as opposed to typical mycelia) results in a higher contamination frequency, and thus has severe practical consequences. – Other changes in the→biotope, e.g., rising or lowering of the temp., reduced levels of inhibitory substances, etc., may result in dimorphism [Beemann, W.: *Zbl. Bakt. II. Abt.* 136 (1982) 369 – 416].

diphenyl
Biphenyl; phenylbenzol; E 230. Fungistatic agent for the protection of the skin of citrus fruit against infection by moulds (*P. italicum, P. digitatum, Diplodia natalensis, Phomopsis citri, Alternaria citri, Phytophthora* spp.). Paper or other packaging materials are impregnated with an alcoholic solution of diphenyl because of its activity in the gas phase (Wallhäusser 1987). → Preservatives. LD_{50} (rat, rabbit, cat), p.o. 2.5 – 3.5 g/kg B.W. ADI value: 0 – 0.125 mg/kg B.W. (Classen et al. 1987).

dipicolinic acid
Pyridine-2,6-dicarboxylic acid. Typical constituent (as Ca salt) of bacterial endospores; up to 15% of the DM. With the exception of *P. citreoviride* not reported for any other organisms, neither for the→ vegetative forms of the endospore-forming bacteria. Some relationship seems to exist between calcium-dipicolinate content and heat resistance (or loss thereof) of endospores.

diplococci
Diplococcus (sing.). Coccus (round) shaped bacterial cells remaining in pairs after binary fission; → cocci. Typical of some *Streptococcus, Enterococcus, Lactococcus* and *Leuconostoc* spp.

Diplodia
Genus of the Deuteromycetes (Sphaeropsidales), → black moulds. Base rot of citrus fruit, avocado, papaya, mango, pumpkin and melons. Wet rot of peaches. – Prevention: dipping into warm water at 45 – 55°C; sorting out of infected fruit (Beuchat 1987). – Cereals contaminated with *D.* may be toxic to young chicklets.

Dipodascus → *Geotrichum.*

direct microscopic examination
Rapid microscopic method used for assessing the composition of the microbial population of a food, or possible contamination of a starter culture. Gram-positive and Gram-negative bacteria and their cellular morphology, bacterial endospores, yeasts and moulds can be distinguished. The results may be indicative of processing failures (e.g., survival of endospores in heat processed foods; large numbers of cocci in processed foods may indicate the presence of enterotoxigenic *Staphylococcus aureus*). – A film of blended food sample is prepared on a microscopic slide; after air-drying it is fixed with moderate heat (alternatively 1 – 2 min with methanol), and stained. Films of food with high fat content should be defatted with the use of xylene and methanol successively. For more complete description, see FDA (1984).

disc diffusion test

Agar disc diffusion test. Laboratory method for determination of the presence of antibiotics in milk, enzyme preparations, etc. Filter paper discs with constant thickness and structure, typically 9 mm in diameter (e.g., from Schleicher and Schüll, Dassel; Millipore), are dipped into the liquid sample or a dilution series and, after allowing excessive liquid to drip off, placed onto the surface of an agar plate which is inoculated with the organism under test. After incubation microbial growth turns the medium turbid, except for areas (zones) around discs containing an antibiotic to which the organism is sensitive. The antibiotic diffuses into the medium, and the size of each zone may be measured as indication of the relative antibiotic concentration; the absolute concentration is determined with the use of a calibration curve, e.g., in "penicillin equivalents". → Diffusion test.

disinfectant

Chemical agent for → disinfection; should not be harmful to human or animal tissues and preferably not to materials or utensils used in a food processing operation; should not impose any spoilage risk on product. Effective → cleaning operation vital, prior to application of disinfectant, e.g., for removal of food residues (proteins, fat, CH or metal ions). – Commercial products listed in Table 12; applications in food processing plants/ kitchens, for the removal/destruction/inhibition of undesired MO from rooms, working surfaces, machines, pipelines, tanks, etc. – Differentiation between → destructive and → inhibitive disinfectants (see Figure 5), that may be selected acc. to purpose of application. Should be microbicidal rather than microbistatic (e.g., by dilution, short contact time, wrong pH, presence of organic matter, etc.); may result in → selection of undesired MO. "Overdosing" may be harm-

ful to the staff ("MAK" values), the utensils (corrosion) and the environment, in addition to increased costs. – Wide spectrum of activity desirable; most disinfectants are antibacterial and some also antifungal and antiviral, however, with little or no effect against endospores. Main groups of disinfectants: → quaternary ammonia compounds (quaternaries or "quats"), → amphoteric compounds, → halogens (chlorine, iodine), → oxidative disinfectants, → aldehydes, → phenols, → alcohols (Wallhäusser 1988; Krämer 1987).

disinfection

Sanitation or decontamination process for the removal, destruction or inactivation of undesired MO on inanimate objects, causing, e.g., spoilage or infection. Applied in food processing plants, kitchens or shops for the prevention or reduction of → cross contamination. Effective only after initial → cleaning (sanitation) procedure. Commercial → disinfectants are available for this basically important operation in food plants.

disintegration

Disintegration of solid and semi-solid foodstuffs, after the sample has been taken and weighed, to enable the preparation of a dilution series for determination of the → bacterial count by plating procedures. A 10 g sample with 90 ml → diluent is disintegrated into small pieces in a "Stomacher" (model Colworth), Waring blender or "Ultra Turrax". The first dilution is made with 1 or 10 ml of the supernatant after the coarse particles have been discarded. It should not be kept for more than 1 h at 0 – 5 °C before transfer to the plates (Baumgart 1990). For determination of → mould counts special care should be taken to standardise the complete procedure with a view to (e.g.) length of the cutting period, the number of revolutions and the sharpness of the blades, since errors are possible.

Table 12 Disinfectants for the food sector assessed acc. to the guidelines of the "DVG"; version of
24. 6. 87 (Reuter, G.: *Deutsches Tierärzteblatt* 8/87).
A = foods of animal origin (except milk), including commercial canteens
B = milk and milk products

Name	Manufacturer/ Distributor	Active Components	Area of Application
CALGONIT SAC	Joh. A. Benckiser GmbH	active chlorine	A
CALGONIT SAN	Ludwigshafen	organic acids	A
CALGONIT SAQ		QAC's	A
CALGONIT SAQ-10		QAC's	A
CONTROL	DuBois Chemie GmbH Dietzenbach	QAC's	A
DARASAN 209	Teroson GmbH	QAC's	A
DARASAN 214	Heidelberg	QAC's	A
DARASAN 7056		organic acids	A
Disinfectant sanitiser, liquid	Diversey GmbH Kirchheimbolanden	QAC's, complex former	A
Divosan aktiv plus		per-acetic acid, hydrogen peroxide	A
Divosept DR 75		QAC's, aldehyde	A
FINADET® DES 115	Deutsche Fina GmbH Frankfurt	QAC's	A
FINK-Antisept G	Fink Chemie GmbH Hamm	QAC's	A
G 447 Budenat® LM	BUZIL-Werk/Memmingen	QAC's	A
Hexaquart® F	B. Braun Melsungen AG Melsungen	org. and inorg. acids tensides	A
Hexaquart® L		QAC's	A/B
HOWALIN D 075	Howa Reinigungschem-ikalien GmbH, Köln	QAC's	A
KAMASEPT OF	Dr. Nüsken Chemie GmbH Kamen	QAC's, glutaraldehyde, glyoxal	A
LYSO® rapid	Schülke & Mayr GmbH Hamburg	ethanol, n-propanol glyoxal	A
Melsept® SF	B. Braun Melsungen AG Melsungen	QAC's, glutaraldehyde, glyoxal	A
neoquat S	Dr. Weigert Chemische Fabrik, Hamburg	QAC's	A
NÜSCODOS	Dr. Nüsken Chemie GmbH	org. acids	A
NÜSCOTAN	Kamen	QAC's	A
P3-steril	Henkel KGaA	QAC's	A/B
P3-topax 65	Düsseldorf	active chlorine	A/B
P3-topax 66		active chlorine	A/B
P3-topax 91		QAC's	A

(continued)

Table 12 (continued).

Name	Manufacturer/ Distributor	Active Components	Area of Application
Quartasept B/F	Schülke & Mayr GmbH	QAC's	A
Quatohex®	B. Braun Melsungen AG	QAC's, biguanids	A
SABONA N	Sabona GmbH Feldkirchen (W.)	organic acids	A
Sokrena food	BODE CHEMIE GmbH & Co., Hamburg	QAC's	A
SU 303	Lever Industrie	organic acids	A
SU 388	Mannheim	paracetic acid	A
SU 3181		QAC's	A
SU 3182		QAC's	A
TEGO® 51	Th. Goldschmidt AG	amphotensides	A/B
TEGO® 90 E	Essen	amphotensides, QAC's	A/B
TEGOL® 2000		amphotensides, QAC's	A/B
TEGOTOP®		amphotensides, QAC's	A/B
TOLO® 440	TOLO Chemie GmbH & Co. KG, Osnabrück	QAC's	A
VENNO-DA 20	VENNO GmbH	QAC's	A
VENNO-OS	Norderstedt	org. acids	A

dissimilation
Catabolism. Degradation of organic substances (CH, fat, protein) into inorganic compounds (CO_2, H_2O, NH_3, H_2S, etc.), with the generation of energy (formation of ATP by substrate and respiratory chain phosphorylation). → Assimilation.

distillery yeasts
Specific strains of Saccharomyces cerevisiae ("dust yeasts") used for the production of ethanol and fruit brandies. These strains remain dispersed in the fruit mash and rapidly ferment the available sugars (ca. 72 h).

division rate
The doubling of a bacterial population per time unit (hours or minutes). → Reproduction.

DNA
Deoxyribonucleic acid. Acc. to the model of Watson and Crick (Alberts et al. 1986) DNA represents a double helix of deoxyribonucleotides, each of which contains one of the bases adenine, thymine, guanine or cytosine in alternating order (macromolecule). Contains the genetic information in all cells and some viruses concentrated in the→nucleus of eukaryotes and nucleic equivalent (genome) of→prokaryotes. – The high conc. of purins in microbial cells (> 6% of the DM) complicates the processing of → single cell protein for human nutrition.

dormant forms
Dormancy; hypobiosis. A state during which the organism or spore (→endospore) exhibits minimal metabolic ac-

tivity. Cryptobiosis: ultimate state of dormancy when no physical or chemical changes can be detected. Different kinds of resistant forms known for MO that serve for survival and (for some kinds) distribution during harsh conditions. → Sclerotia → spores and → conidia for moulds; For → cyanobacteria endo- and exospores, as well as cysts are known. Dormant forms may pass the GI tract of animals in that state, by contrast to → vegetative forms.

double vision

Neuroparalytic disturbances, paralysis of the motor nerves of the eye ball, resulting from the action of → botulinus toxin. Eyesight impairments also include slow pupillary reaction to light, photophobia, ptosis, etc., and serve as final warning signal for intoxicated person to consult a physician. → Botulism.

drawing (off)

Separation from the wine of sedimented

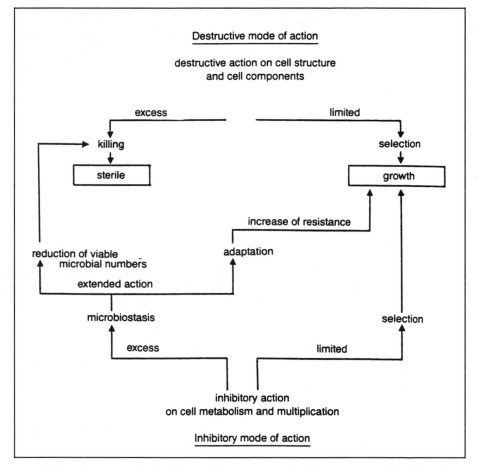

Figure 5   Reaction pattern of disinfectants relative to mechanism of action [source: Mrozek, H.: *Arch. Lebensm.-Hygiene* 31 (1980) 91 – 99].

yeasts and other settled particulate matter after completion of fermentation.

dried fruit
Products with a water content of 18 – 25% corresponding to an a_w value of 0.70 and may thus be considered microbiologically "safe". If no water is absorbed from the atmosphere, it can be stored for a longer period of time. It is often pasteurised at 80 – 85°C with steam, following dehydration at ca. 75°C, so as to ensure that it is free from pathogens. Without these measures the products are heavily contaminated in general (Krämer 1987).

dried milk → milk powder.

dried yeast → baker's yeast.

drier
Modern, technologically refined equipment used for the extraction of water from food. The product should not reach a temperature > 50°C to ensure that optimum quality is retained. Few MO are killed. Grated or shredded pieces are commonly dried on bar frames, conveyor belt driers or in spinners. Liquids and mashy products are dried in sprayers. A more effective reduction in the MO count is achieved during rotating drum drying since higher end temperatures are reached. Freeze drying effectively prevents severe damage to MO and the product, and it is therefore suitable for use in the production of dried conserves for → starter cultures and → enzyme preparations (Heiss and Eichner 1984; Kröll 1978). – The development of driers with adapted technology for developing countries is a most important prerequisite for → post harvest protection of crops.

Drigalski spatula
Drigalski triangle. Triangular glass rod used for the even distribution of inoculation (measured volume of serial dilution) on agar plate surface. → Surface plating.

drinking milk → consumer's milk → pasteurised milk.

drinking water
Important raw material in the food industry. Only water with the quality of drinking water may be used; exception: boiler water. The microbiological quality is defined in the Regulations on Drinking Water of 22.5.86. §1 deals with the fundamental requirements that the water must be free from any pathogens, to which it does not comply if E. coli is detected in 100 ml of water. Coliforms should not be present in 100 ml (limit value). This prerequisite is satisfied when 95% of a minimum of 40 samples have been found negative (Hahn and Muermann 1987). Microbiological examination procedures are referred to in the "Official Collection of Examination Methods" ("Amtliche Sammlung von Untersuchungsmethoden") acc. to §35 "LMBG" for the detection of:
- E. coli and other coliforms
- faecal streptococci
- Pseudomonas aeruginosa
- sulphite reducing spore-forming anaerobes
- colony count

Bacterial genera found in water are given in Table 13. – In addition to physical methods (filtration, UV radiation and heat treatment), chemical additives are also approved (with reference to residual limits) in the Drinking Water Purification Regulations, and include a.o. Cl, sodium hypochlorite, magnesium hypochlorite, calcium hypochlorite, chlorine dioxide, ammonia and ammonium salts for reduction of the bacterial count. Ozonisation and treatment with → silver are also allowed (Kunz 1988).

drop culture
Drop plate method. Method to determine

Table 13 Bacteria occurring in drinking water and their significance [acc. to Dott et al.: *Zbl. Bakt. Hyg. B.* 182 (1986) 449 – 477].

Genus	Possible Effect on Drinking Water Quality
Acinetobacter	possible antagonist of indicator bacteria
Actinomyces	taste and flavour
Arthrobacter	production of colour and/or slime
Bacillus	nitrate reduction, corrosion potential, antagonist of indicator bacteria
Beggiatoa	"red" water, sulphur oxidation
Crenothrix	"red" water, iron bacterium
Desulfovibrio	"black" water, H_2S production, corrosion
Enterobacter	microbial growth
Escherichia	indicator bacterium
Flavobacterium	opportunist, possible antagonist of indicator bacteria
Gallionella	"red" water, iron bacterium
Klebsiella	microbial growth, opportunist
Legionella	potential pathogen
Leptothrix	"red" water, iron bacterium
Methanomonas	methane oxidation
Micrococcus	nitrate reduction, corrosion, antagonist of indicator bacteria
Mycobacterium	potential pathogen
Nitrobacter	nitrate reduction, corrosion, slime production
Nitrosomonas	nitrate reduction, corrosion, slime production
Nocardia	potential pathogen
Proteus	possible indicator antagonist
Pseudomonas	opportunist, indicator antagonist
Serratia	opportunist
Sphaerotilus	"red" water
Streptococcus	indicator of fecal contamination
Streptomyces	smell, taste

the bacterial count with a small expenditure on materials. Selective media are preferably used for in plant (intrinsic) → process control. The plate is divided into 4 or 6 segments at the bottom (Figure 6), clearly marked, the bottom covered with 12 ml nutrient medium and dried at 50 °C for 1 h; 0.05 ml of the suspension from the → dilution series is carefully dropped and streaked out over a surface of 2 cm on each segment with the point of the pipette. It is incubated with the cover downwards after the liquid is absorbed by the nutrient agar (Baumgart 1990). Count-ing is done with the aid of a magnifying glass. This method is suitable for fungi only in exceptional cases since the distances between the colonies are too small. It must be evaluated under the stereo-microscope while still in the → "star" stage.

dry rot

Occasionally used for fungal spoilage of stored fruit with *Botrytis cinerea*. In the long run it causes mummification of the fruit. The term is also used in the case of spoilage caused by *Fusarium* spp. in potatoes.

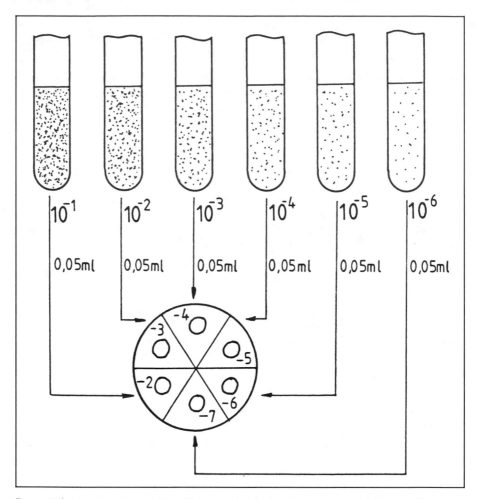

Figure 6   Schematic representation of the procedure for the drop culture method (source: Baumgart, J.: *Mikrobiologische Untersuchungen von Lebensmitteln.* Hamburg: Behr's Verlag, 1990).

drying

The oldest method to extend the shelf life of foodstuffs. Originally drying was done outside under suitable weather conditions, e.g., raisins, fish, rock fish, hake. Later it was done inside the house, making use of warmer air and smoke, e.g., smoked sausages or meat. Presently performed in technologically refined → driers. The microbiological destruction rate is low. If the water is extracted rapidly, the product soon reaches a range where it is safe and spoilage will not take place. This level depends on the composition of the product (→ water activity, → water requirements). Drying only makes sense if the product is packed and/or stored in such a way that no atmospheric moisture will be absorbed and the a_w value remains < 0.70.

duck eggs

Higher incidence of → *Salmonella* contamination than → eggs of domestic chickens. Increased risk of → cross contamination in the household, or transmission via → egg products into other foods. May not be used for the manufacturing of mayonnaise, scrambled eggs, ice cream, confectioneries, etc. Should be designated specifically as "duck eggs" for commercial distribution.

Durham tubes

Small glass tubes (ca. 5 mm × 30 mm), that are autoclaved upside-down inside a normal-size culture tube, containing, e.g., violet-red-bile-lactose broth. Gas produced by the fermentation of lactose in this substrate is partly retained in the Durham tube, and serves as positive reaction, e.g., in the presumptive test for → coliforms.

dysentery

Refers to inflammatory disease of the intestine (mainly colon) typically caused by *Shigella* spp. (*Sh. flexneri, Sh. boydii, Sh. dysenteriae* and *Sh. sonnei*); shigellosis. Typical → food-borne disease. Apart from the bacterial (*Shigella*) type of dysentery (also called bacillary dysentery), the disease may also be caused by an amoeba, *Entamoeba histolytica* (amoebiasis or amoebic dysentery), which attacks the mucosa of the large intestine, after ingestion of the organism in the cyst form.

ear moulds

Moulds that affect the flowering parts of cereals (e.g., barley, oats, rye, wheat) and develop on the ears or grains. The extent of development/growth is influenced by factors such as climate, crop rotation, the type of grain, the use of pesticides and the dipping of the seed. – Of interest are those types that are able, or suspected, to produce toxins, that may exert harmful effects on farm animals. Economic losses may be substantial. – Ergot (→ *Claviceps purpurea*) is definitely toxic, but is removed practically completely during purification before milling (rye, wheat, barley). – Uterotropic metabolites, causing abortion in cattle and sheep, have been suggested to be produced by *Ustilago maydis* (= *U. zeae*). – Common bunt or dwarf bunt may be caused by *Tilletia* spp., especially *T. controversa*, and may be the cause of reduced weight increase of pigs, cattle and poultry; *T. caries* (= *T. tritici*) may cause off-odours ("fishy smell") resulting from trimethylamine or even histamine. – Black stem rust is caused by *Puccinia* spp., especially *P. graminis*, which releases clouds of spores during harvesting, and causes allergic reactions when inhaled by sensibilised persons. – → *Alternaria* spp. are typical producers of → mycotoxins; in feeding trials 78% of the strains isolated from cereals have been found toxinogenic. – Most → *Fusarium* spp. produce toxins, and a high infection rate of plants (via roots or leaves), resulting in development in ears, is typically related to low frequencies of crop rotation.

early blowing

Cheese spoilage caused by *E. coli*, coliforms and *Klebsiella* spp.; introduced by recontamination of the cheese milk. Metabolic activity produces formic acid, H_2, and CO_2. Indication of hygienic deficiencies during processing (Teuber 1987).

East Asian specialities

Fermentation products from rice, soya, cassava, etc., are traditionally prepared in East Asia in small home industries by spontaneous contamination or inoculation. Starter cultures are used in commercial production, also for export to Europe. *Mucor* and *Rhizopus* are important in tropical regions, and *A.* and/or *Neurospora*, partly in combination with lactic acid bacteria, in Japan and China. Rice and cassava consist mainly of carbohydrates and are enriched with proteins and vitamins through fermentation which also contributes to improvement of the taste. Soya is hydrolysed, and becomes more digestible and also tastes better; → sufu. – The best known examples are: "Angkak" in China prepared from rice with → *Monascus purpureus* as colourant and seasoning. "Miso" in Japan is prepared from soya and/or rice or barley with *A. oryzae*, *Schizosaccharomyces rouxii*, *Torulopsis* spp. and lactic acid bacteria and is used as seasoning or a base for soups. "Shoyu" is a Japanese soy sauce prepared with *A. oryzae*, *Torulopsis* and lactic acid bacteria. "Tape" is prepared from cassava or rice with *Chlamydomucor oryzae*, *Mucor* spp., *Rhizopus oryzae* and yeasts in Indonesia where it is eaten fresh as the main source of food in some regions. "Tempe (h)" is prepared from whole soy beans with *Rhizopus microsporus, Rh. arrhizus, Rh. oryzae, Rh. stolonifer*, etc. It is roasted or fried in oil and used as substitute for meat in Malaysia and Indonesia. "Saké" (rice wine) is prepared in Japan from rice with → *A. oryzae* (amylases) and yeasts (DFG 1987; Reiss 1986; Steinkraus 1983).

ecological characteristics
Stable characteristics of MO which change characteristically if the parameters of the ecosystem change, e.g., temperature, O_2-concentration, light, light-darkness periods, pH, a_w value, available nutrients, etc. Variations of these parameters give useful information for → identification or → differentiation of MO (Baumgart 1990).

ecology
Study of the interaction between organisms and the environment (ecosystem) resulting in a stable or labile equilibrium. By changing the parameters of the ecosystem during production, e.g., increasing the temperature, → psychrotrophs are inhibited and → thermophiles are promoted. Knowledge of the exact ecological parameters and the specific MO possibly involved in a specific process, will contribute to optimisation or correction of a procedure (Table 14). The more information available on the ecological demands of MO, the better the possibility of processing a product for improved shelf life and hygienic quality.

ecosystem
The sum of environmental characteristics (abiotic and biotic factors) and the organisms present. Several parameters of such systems can be changed or modified by processing technological steps so as to influence the growth and reproduction potential of the MO in a positive or negative way. The technologist has the task of influencing the ecosystem in such a way that MO have little or no opportunity for growth or reproduction.

ecotypes
Ecosystem related varieties of microbial spp. with selectively developed characteristics (indigenous MO). They are well con-

Table 14 Transfer of ecological data for food spoilage and food poisoning microorganisms in a food processing sequence.

1)	Identification of the organism(s) and/or toxin	(laboratory)
2)	Conditions for growth and toxin production of MO	(literature and consultation of specialists)
3)	Occurrence of MO in raw materials, in the processing plant and processing and handling steps	(laboratory)
4)	Parameters of processing and handling temp./time, pH, a_w, O_2, etc.)	(technical plant management)
5)	Comparison of data from 1) and 4) for the identification of critical points	(head of laboratory)
6)	Possible consequences for 4) with reference to sensory quality	(discussion with all parties involved)
7)	Examination and evaluation of the modified process, including critical control points	(technical management)

ditioned to the specific habitat and extremely difficult to control. An ecotype does not represent a separate systematic unit.

ectotoxins→ exotoxins.

Edam cheese
→ Hard cheese with evenly distributed gas holes produced by gas forming aroma bacteria (*Lactococcus lactis* ssp. *diacetylactis; Leuconostoc mesenteroides* ssp. *cremoris*). Ripening at 13 – 15 °C for 5 weeks; packaging in laminated foil for prevention of surface growth by, e.g., *Brevibacterium linens* (Teuber 1987).

"Edelfäule" (German)→ noble rot.

"Edelschimmel Kulmbach"→ *Penicillium nalgiovensis.*

Edwardsiella
Genus of the→ Enterobacteriaceae. Gram – , rod-shaped, usually motile and resistant to colistin. Opportunistic pathogen causing diarrhoea (symptoms reminding of salmonellose). Transmitted via faecal contaminated foods and water, e.g., *E. tarda*. Distribution: intestines of animals; also found on fresh and frozen fish (Sinell 1985).

egg products
For processing,→ eggs are broken out (hand or machine), examined for visual signs of spoilage, and either homogenised as whole egg, or the yolk separated from the white. As primary source of microbial contamination, the shell needs to be washed before breaking. Other sources of contamination include contaminated/spoiled eggs, and equipment (breaking utensils, pipes, holding tanks, etc.). Homogenised (liquid) egg and especially egg yolk may easily spoil, and are pasteurised at 64 – 65 °C for 2 – 3 min, primarily to destroy salmonellae.

Chilled storage at 5 °C will inhibit growth of possible survivors. For shipment and long term storage, liquid egg can be kept at < – 18 °C; freezing, however, will not free the product of salmonellae. Shelf life may also be increased by addition of up to 12% salt or sugar (for use in confectioneries) to lower the a_w. The egg pulp may also be dried before storage and distribution; prior to the drying process, however, the free carbohydrate (primarily as ca. 0.6% glucose in the DM of egg white) should be removed by fermentation (spontaneous or by added starter culture, e.g., of *Enterococcus, Lactococcus* or *Saccharomyces* spp.) so as to prevent Maillard reactions during processing. Sampling procedures specified by the USDA basically distinguish between three types of samples for detection of the presence or absence of *Salmonella:* surveillance, confirmation and certification; furthermore, the USDA requires the sampling and testing for *Salmonella* of each lot of dried egg product. The following specifications are to be met by all egg products, liquid, frozen or dried:
– aerobic plate count, < 25,000/g
– "coliforms", < 10/g
– yeasts and moulds, < 10/g
– *Salmonella*, negative by prescribed sampling and testing procedures (e.g., absent in 25 g)
→ Lysozyme can be produced from fresh egg white (ICMSF 1980; Bergquist et al. 1984).

eggs
Eggs (from chickens/poultry) are surrounded by shell consisting of Ca- and Mg-carbonate; O_2 exchange through thousands of micropores, 20 – 30 μm in diameter, important during incubation until hatching. The cuticle, a protein coat on the outside of the shell, becomes partly damaged during cooling after the egg has been laid, and cannot completely prevent passage of MO into the egg. Just inside the shell, the outer coarse membrane and

the inner fine membrane act as further barriers against penetration of MO into the egg white (albumen). The latter basically consists of water and protein, in addition to some important antibacterial factors: e.g., lysozyme (muramidase), conalbumin (metal chelating agent), avidin (binds biotin), ovomucoid (trypsin inhibitor), etc., as well as an unfavourable pH of 9.1 to 9.6. Antibacterial activities mainly against Gram+ bacteria. During storage the a_w value drops gradually due to water loss, and the pH increases from 7.3 to 9.6, as a result of the escape of CO_2. – Mechanical damage (cracks) of the egg shell makes the egg more suscep-tible to penetration of MO from the environment, especially those associated with the faeces of poultry, e.g., Enterobacteriaceae, pseudomonads, *Alcaligenes, Aeromonas, Proteus, Micrococcus,* bacilli and moulds, and may cause spoilage at an early stage. Common types of spoilage: green rot (*Pseudomonas* spp.), colourless rot (*Acinetobacter-Moraxella,* → *Achromobacter*), red rot (→ "coli-forms", *Serratia*), black rot (*Pseudomonas, Proteus, Escherichia* and other Enterobacteriaceae with strong proteo-lytic properties) (Ayres et al. 1980; Krämer 1987; Sinell 1985). The egg yolk provides a rich nutrient substrate for MO that may have penetrated the previously men-tioned barriers. In addition, the yolk may be infected by, e.g., mycobacteria and *Salmonella* via the blood system.

Eh value → redox potential.

elevators
Potential (and sometimes important) source of contamination in food process-ing plants. Water condensate and other water residues on the floor create favour-able conditions for mould growth; the co-nidia may be distributed to other levels through air movement and the draught caused by the elevator shaft.

ELISA test
"Enzyme Linked Immunosorbent Assay". Serological method for the specific detection of (e.g.) microbial toxins in serum, faeces, urine, food, etc. Gener-ally: immunoassay for specific antibodies or antigens. Commercial test kits are avail-able, e.g., for detection of *Staphylococcus* enterotoxin at a level of ca. 1 ng/ml (Baum-gart 1986). Test procedures also stan-dardised for botulinus toxins, *Cl. perfringens* toxins, aflatoxin M_1, etc.

Emericella → *Aspergillus.*

Emmental cheese
"Swiss cheese". → Hard cheese type with 62% DM, manufactured from raw milk, and sold only after at least 3 months ripen-ing. Maturing involves at least 4 weeks ripening at 20 – 28 °C to allow → propionic acid fermentation, responsible for typical large holes and taste. – Starter culture (0.1 – 0.2%) contains a.o. *Streptococcus thermophilus* and *Lb. helveticus* and small numbers of propionic acid bacteria; ren-neting at 30 °C. The mixture of curd and whey is heated at 53 °C; syneresis is accel-erated and many Gram– bacteria and yeasts are killed. Pressing (moulding), brining, drying, ripening and maturing at 12 – 14 °C. – Prevention of → late blowing, only milk obtained from → herds not receiv-ing silage is used; → *Cl. tyrobutyricum* (Teuber 1987).

endemic
Refers to a disease that prevails or often recurs in a country or an area, or among a particular group(s) of people. When a large number of individuals in a com-munity are attacked in a short time, an → epidemic exists.

ENDO agar
→ Selective medium for the detection of → coliforms, and → differential medium for → *E. coli.* Addition of sodium sulphite

and basic fuchsin prevents/inhibits the growth of Gram+ bacteria; lactose promotes the development of the desired bacteria. Crystallised fuchsin appears as metallic sheen on colonies of *E. coli*. Commercial dehydrated culture media, e.g., BBL no. 11199, Difco no. 0006, Merck no. 4044, Oxoid no. CM 479.

Endomyces

Genus of the perfect fungi belonging to the Ascomycetes. Hyphae may be divided into "square" units: arthrospores. – *E. lactis* is the teleomorph of → *Geotrichum candidum*.

Endomycopsis

Genus of ascogenous yeasts; septate hyphae regularly produced. Presently grouped under the genus *Saccharomycopsis*.

endospore staining

Due to their thick wall the endospores of *B.* and *Cl.* spp. are difficult to stain by simple procedures. A modification of the Ziehl-Neelsen stain or the Schaeffer-Fulton stain are however used for successful spore staining. A heat-fixed smear on a microscope slide is flooded with saturated malachite green solution and either left at room temperature for 10 min or briefly heated to near boiling and then washed in running water. This may be followed by counterstaining of the vegetative cells with 0.5% safranin solution. The endospores stain green and vegetative cells red (Baumgart 1990). → Endospores.

endospores

"Bacterial spores"'. Extremely resistant survival forms produced by bacteria (genera *Bacillus, Clostridium, Coxiella, Desulfotomaculum, Sporolactobacillus, Sporomusa* and *Thermoactinomyces*) under conditions of nutrient limitation. Formation intracellularly in a → vegetative cell (sporangium), and contains practically the total DM of the latter, but only 10% of its volume. Destruction by → microwaves ineffective because of small/compact size and low water content. – Sporulation represents a complex differentiation process, taking at least 7 h. It is initiated by formation of an axial DNA filament followed by cytoplasmic membrane invagination to form an asymmetric septum by which the cell is divided into two protoplasts. The smaller spore protoplast (forespore, prespore) is gradually engulfed by the plasma membrane of the larger protoplast, followed by peptidoglycan deposition to form the cortex. A lamellate protein structure, the spore coat, then develops externally to the outer membrane of the forespore; contains ca. 15% Ca-dipicolinate (→ dipico-linic acid) in DM which plays a part in the heat resistance of the endospore. Staining of the spore coat and the exosporium is difficult → spore staining. – Typically, sporulation is induced by nutrient deficiencies or accumulation of certain metabolites, e.g., thorough cleaning (sanitation) without subsequent → disinfection; drying out does not normally lead to sporulation. – Under favourable conditions (availability of water and nutrients) germination gives rise to a vegetative cell, leaving behind the spore coat. Activation, e.g., by heat ("heat shock") (→ pasteurisation; → tyndallisation), may accelerate the germination process (Schlegel 1985; Alberts et al. 1986). – Comparable to endospores of bacteria are → ascospores, sporangiospores, etc., of moulds. By contrast, → exospores of bacteria are found only among the Actinomycetes as conidial organelles.

endotoxin

Cell wall component of Gram – bacteria. Type specific → lipopolysaccharide responsible, e.g., in salmonellosis, for typical symptoms. Heat resistant, and is released and active after the death of the cell.

enrichment

Procedure or process by which a given group or type of MO in a mixed population is favoured. Minor groups of MO such as toxinogens or pathogens may represent a minor proportion of a mixed population. A specifically formulated enrichment (mostly liquid) medium contains substances that will suppress the growth of major groups and encourage or allow the growth of the required organism(s), e.g., toxinogens or pathogens of importance in the safety evaluation of a commodity. A → differential medium may then be used for their selective detection. Enrichment procedures are used for the detection of *Campylobacter, Cl. perfringens,* Enterobacteriaceae, *Salmonella* and *Yersinia* (Baumgart 1990). → Official methods. − Enrichment of yeasts and moulds may be achieved in → Sabouraud or → Czapek-Dox broths, either in static or shake culture. The pH of the medium may be adjusted to 4.5 − 3.7 with lactic or tartaric acid, or chloramphenicol and/or a tetracycline used to inhibit the growth of bacteria.

Entamoeba histolytica

Protozoan sp. causing amoebic → dysentery (amoebiasis). Monokaryotic, unicellular; invasive forms (penetrating and developing in the gut mucosa) 20 − 50 μm diam. Highly motile in fresh faeces. → Permanent (dormant) forms (cysts) may undergo two nuclear divisions to form a mature cyst with four nuclei in the gut of a symptomless host; not detectable in the stool of patients with clinical amoebiasis. Vegetative forms (trophozoites, precysts) are killed rapidly after shedding, and are not infective in contrast to cysts. The latter may survive up to 1 month, and contaminate drinking water, vegetables, salads, etc. Sensitive to > 40 °C or < − 4 °C, and to desiccation. These amoebas form → proteases that enable their penetration into the gut mucosa (referred to by the designation "histolytica"). May be transported to the liver where severe damage is caused, and, if not treated, may cause death. Persisting diarrhoea, interrupted by "normal" stool, may occur within weeks or even years after the infection. → Endemic disease in tropical and subtropical countries; may be contracted by tourists.

enteritis

Inflammation of any part of the GI tract; → gastroenteritis; colitis.

Enterobacter

Genus of the → Enterobacteriaceae, some representatives of which formerly have been designated → *Aerobacter.* Typical inhabitants of the GI tract, and also found in sewage and soil. Some spp. typically associated with plants (e.g., *Ent. agglomerans*). A representative of the → "coliforms".

Enterobacteriaceae

Family of Gram − , facultatively anaerobic rod-shaped bacteria with respiratory or fermentative (chemoorganotrophic) metabolism. Acid and gas (CO_2) production from CH; nitrate reductase + , catalase + , cytochrome oxidase − . Typical representatives include → *Escherichia, Edwardsiella,* → *Klebsiella,* → *Citrobacter,* → *Salmonella,* → *Enterobacter,* → *Hafnia,* → *Serratia,* → *Proteus,* → *Yersinia,* → *Erwinia* (Bergey 1984). The name refers to the fact that many representatives are typical inhabitants of the GI tract; this family is therefore considered as → indicator organisms of faecal pollution. Several are food spoilage organisms, some are phytopathogens. − Detection a.o. on violet red neutral red bile dextrose agar (VRBD agar); commercially available: BBL no. 95300, or Merck no. 1406. For detection and evaluation methods for foods, see Mehlmann (1984).

enterococci

Formerly the serological group D streptococci, represented a.o. by *Ent. faecium, Ent. faecalis, Ent. hirae, Ent. gallinarum,* that may be present in the human and animal intestine. Characterised by relatively high resistance towards environmental factors; growth usually at 10°C and 45°C, at pH 9.6, and in 6.5% NaCl. May be pathogenic, causing septicemia in humans and domesticised animals. Food poisoning (rare) through the production of→ biogenic amines. May possibly be involved in the→ secondary fermentation of some→ hard cheeses, produced from raw milk.

enteropathogenic *Escherichia coli*

EPEC. May generally refer to any strain of *E. coli* causing enteric disease, but more specifically includes only strains responsible for diarrhoeal illness without producing either heat-labile (LT) or heat-stable (ST) enterotoxin (enterotoxic *E. coli* or ETEC). In addition it is not invasive of the intestinal mucosa, and thus also distinguished from "entero-invasive *E. coli*" (EIEC). Infants are mainly infected by EPEC, causing severe out-breaks of "infantile gastroenteritis" or "coli-dysentery", e.g., in nurseries. About 15 serotypes known (e.g., O26, O55, O114, O128) with higher frequency in tropical countries; generally low health risk for adults, although may cause illnesses such as traveller's diarrhoea, "Montezuma's revenge", etc. Transmission: humans→ food→ humans, with salads and vegetables as common vehicles. Important factors: personal and kitchen/canteen hygiene. – Attachment by→adhesins to the mucosal wall where the brush border microvilli may be destroyed; invasive forms (EIEC) may penetrate the mucosa to cause a dysentery type of diarrhoea. ETEC strains adhere to the mucosal surface and cause severe diarrhoea, with high fluid losses, reminding of→cholera. Incubation time 6 to 36 h; the illness may last from 1 to a few days (Doyle and Padhye 1989; Sinell 1985).

enterotoxemia

Food poisoning by→*Clostridium perfringens* types C and D; not typical example of→food poisoning or→food-borne infections.

enterotoxins

Refers to a microbial toxin which, either upon ingestion or production in the GI tract, causes disturbances of the human intestinal tract.→Food poisoning, caused either by→endotoxins or→exotoxins.

enzyme preparations

Enzyme mixtures that are extracted or partly purified from biological materials, →biomass. May be diluted with typical food components such as starch, lactose, cellulose powder, etc., in view of the desired concentration for practical application, before commercial distribution. Technical developments enable commercial producers to meet enzyme specifications. Reference to enzymes in, e.g., food legislation normally concerns commercial enzyme preparations, the use of which may be restricted by economic factors, safety regulations, stability, etc. – →Amylases,→amyloglucosidase,→cellulase,→dextranase, → galactosidase, → glucanases, → glucose-isomerase, → glucosidases, → hemicellulase, → invertase, → catalase, → lipase, → lysozyme, → pectinases, → proteases, → pullulanase. – Most important areas of application: proteases for detergents (35%), amylases for starch hydrolysis (30%), rennet substitute (chymosin) for cheese manufacture (15%), pectinases for fruit and vegetable processing (3.5%), and invertases for the production of fructose (3%).

enzymes
Formerly called "ferments"; important components of all living cells. Proteins or proteins in combination with prosthetic groups or coenzymes (e.g., CH, lipids and/or metals such as Fe, Mo, Cu, etc.) that act as highly specific and effective biocatalysts in the cell metabolism. By decreasing the activation energy an enzyme acts to increase the rate of a reaction. Enzymes may be extracted and purified from plant or animal materials and from MO, for use in analysis, research and technology. The sensitive structure and specific activity are retained. Every enzyme has a unique classification number or E.C. number (E.C. = Enzyme Commission), consisting of four figures that reflect the nature of the substrate, the reaction catalysed and possible coenzymes involved, with the last figure referring to a serial number, e.g., β-galactosidase (lactase): E.C. 3.2.1.23.

epidemic
Outbreak of an infectious disease, e.g., → salmonellosis or → dysentery, that has been transmitted by food or water to a group or population within a region or area. Increased contamination frequency also via diseased persons. Incidence in a given population exceeds that typical of other populations.

eprouvette
Obsolete designation for → culture tubes.

ergosterol
Steroid regularly found in fungi but not in plants. May serve as criterion for mould contamination. Absorption maximum in UV at 283 nm is characteristic and may have practical advantages to culture methods for the routine examination of food and feed products for moulds. Even nonviable hyphae, spores or conidia (resulting from the use of preservatives or heat treatment) may be detected by this method, thus reflecting previous metabolic activities in the product, including possible → mycotoxin production [Müller and Lehn: *Arch. Anim. Nutr. Berlin* 38 (1988) 1 – 14; Zill et al.: *Z. Lebensm. Unters. Forsch.* 187 (1988) 246 – 249].

ergot → *Claviceps purpurea.*

ergotism
Mycotoxicosis caused in humans by the ingestion of → sclerotia (ergot) of *Claviceps* spp. that had developed on rye and wheat. Sclerotia ingested with grazing, hay, and fodder grasses may be harmful to farm animals. The mould overwinters by means of the hard and resistant sclerotia which contain a number of toxins, called ergot alkaloids. These alkaloids may be divided into two classes, the clavine alkaloids and the lysergic acid derivatives, some of the latter which are clinically useful, e.g., in the treatment of migraine (with ergotamine), to control postpartum bleeding (with ergometrine), and Basedow's disease. – Flour containing 0.1% ergot is not harmful for humans; 1% however is considered as toxic (Gedek 1980). – The disease typically causes a prickling sensation in the skin of hands and feet; other symptoms may include numbness, diarrhoea, vomiting, and (in severe cases) neurological symptoms, or gangrene and blackening of toes and fingers (St. Anthony's fire). Well known disease in Europe since the Middle Ages. The most recent outbreaks date back to 1952 in Southern England and Southern France. Isolated cases are reported with increasing frequency, and are related to recent trends towards "natural" foods and the direct purchasing of grains from the farm (Roth et al. 1988).

Erwinia
Genus of the → Enterobacteriaceae. Gram – , motile (except for *Er. stewartii*) large rods, producing acid but little or no

gas from CH. Strongly pectinolytic. Causes plant diseases and storage rots. Spoilage of carrots and celery by *Er. carotovora* through solubilisation of the pectin lamellae, resulting in soft rot; cross contamination of healthy products. *Er. carotovora* var. *atroseptica* (*Er. phytophthora*) and *Er. carotovora* var. *carotovora* may also cause blackleg of potato plants, and storage rot of potato tubers during storage at elevated temperatures. *Er. dissolvans* is mainly involved in the moist fermentation of → coffee (Müller 1983a).

Escherichia
Genus of the → Enterobacteriaceae, represented only by one sp., *E. coli*. Gram −, aerobic to facultatively anaerobic short rods. H_2 and CO_2 produced from glucose in 1:1 ratio. Found in the colon of vertebrates and the faeces of virtually all animals. Growth: $2.5 - 45\,°C$; pH $4.7 - 9.5$. Survival: few days up to several years, depending on product (substrate). Numerous data available in literature; → Mitscherlich and Marth (1984). Killing: pasteurisation. − *E. coli* is regarded as → indicator organism in food and water; → official methods. Apart from harmless strains typical of the gut microbial population, pathogens may be found, e.g., → enteropathogenic *E. coli*. → Detection: → official methods; Baumgart (1990), Speck (1984). An identification method based on the presence of β-D-glucuronidase and tryptophanase (indol production) has been developed recently. The "MUG" medium contains 4-methyl-umbeliferyl-β-D-glucuronide that turns intensively bright blue fluorescent at 360 nm in presence of the former enzyme; the indol produced by the second enzyme is detected with → Kovac's reagent. − For rapid routine identification different "Test Kits" are commercially available, e.g., Bactident® from Merck. The traditional "IMVIC" test is still one of the most important biochemical procedures for the iden-

tification of *E. coli*; ca. 95% of the strains give the IMVIC pattern + + − − (indole production, methyl red indicating acid production from glucose, Voges-Proskauer for detection of acetoin formed from glucose, and utilisation of citrate as sole source of C) (biotype 1), and 5% as − + − − (biotype 2) [Hahn: *Arch. f. Lebensm.-Hyg.* 38 (1987) 33 − 68; Doyle and Padhye 1989].

ethanol
C_2H_5OH. Ethylalcohol; "*spiritus vini*"; trivial name: alcohol. Colourless liquid with b.p. $78.3\,°C$. "MAK" value: $1900\ mg/m^3$ of air. LD_{50}-LD_{100} for adults p.o. 3.3 ml of 96% ethanol/kg of body weight (Classen et al. 1987). Anaerobically produced by yeasts and some strains of bacteria from, e.g., sugars. Apart from → *Saccharomyces cerevisiae* it may be industrially produced − for non-food and non-pharmaceutical purposes − from the juice of *Agave americana* with the bacterium → *Zymomonas mobilis*. A thermophilic process is presently being developed, in which a bacterium "AKO-1", isolated from the bottom of Lake Kivu (East Africa), will be applied for the fermentation of different sugars, including xylose from wood, to ethanol at $73\,°C$ (Gottschalk et al. 1986).

ethylcarbamate → diethyl-pyrocarbonate.

ethylene oxide
(Oxiram). A gaseous disinfectant that may be used for chemical sterilisation of inert, non-food materials. Approval for food uses withdrawn. High toxicity on inhalation and towards tissues.

eukaryotes
Organisms of which the nucleus is surrounded by a double-layered membrane, called a "unit membrane". The genetic information in the nucleus is dispersed

over a number (> 1) of → chromosomes. Karyogamy by mitosis that guarantees the identical replication of the genetic material (genome). Eukaryotic MO are represented by the → fungi (moulds) (including the yeasts), → algae and → protozoa. Opposite: → prokaryotes (Schlegel 1985).

Eupenicillium → *Penicillium*.

Eurotium → *Aspergillus*.

eutrophic
Characterisation of surface water with a high nutrient content giving rise to elevated organic pollution. Opposite: → oligotrophic.

excreta
Excrete. May refer to the "discharge" from a living system, e.g., MO, with a specific purpose or role in cell metabolism, e.g., → extracellular enzymes, or end products of basic metabolism such as CO_2, H_2, ethanol, lactic acid, and other → organic acids, that cannot be assimilated further. The secondary metabolic products such as → exotoxins, → mycotoxins, → pigments, → aroma compounds, extracellular polysaccharides (slime, etc.) have special importance, and may bring about desired or undesired changes in a food; → sensory properties.

exoenzymes
Biochemical designation for enzymes that cleave long chain molecules, from the free ends, into smaller components. In microbiology reference is made to → extracellular enzymes that are being excreted from cells or are attached to their surfaces.

exospore
→ Survival body produced outside fungal cells, e.g., basidiospores of the Basidiomycetes, and → conidia. Reduced water content and stable/tough outside wall or layer. Distribution typically by wind.

exotoxins
Toxins (poisons) from bacteria or fungi, that may often be synthesised only after the exponential growth phase, and may be excreted from the cells into the surrounding medium (food). Examples: → *Clostridium botulinum*, → *Staphylococcus aureus*, → mycotoxins.

expansin → patulin.

extracellular digestion
→ Extracellular enzymes. Important method by which animals and some plants degrade non-absorbable nutrient materials outside the cell: e.g., diastase in the mouth, pepsin and trypsin in the intestines, proteases at carnivorous plants. Practically all MO.

extracellular enzymes
Enzymes of MO excreted from the cell into the medium for the purpose of degrading macromolecular substances (starch, pectin, cellulose, protein) or water insoluble materials (fat, lipids) extracellularly. The smaller units of the original (non-absorbable) molecules may then be assimilated. – Such enzymes may remain active, even after removal or inactivation of the MO, and cause changes (spoilage) in the medium (food), e.g., rancidity in frozen products.

extrinsic factors
Factors external to a food that may be applied (e.g., by the processor) for extending the shelf life or keeping quality of a food, e.g., → temperature, → preservatives, → CA storage, etc. Several important food processing procedures are based on these factors. Opposite: → intrinsic factors.

F

F value

From °F (Fahrenheit). F_o is the time (in minutes) required at 121.1 °C (250 °F) for the inactivation or destruction of all MO in a population. – In the sterilisation of canned products the initial number of bacterial endospores (most heat resistant form) is not known. A hypothetical number of $10^{12} \rightarrow$ Clostridium botulinum spores per g or ml is therefore assumed, so as to prevent botulism under all circumstances. A $\rightarrow$ D value of 0.21 min at 121.1 °C has been determined for these endospores. The reduction of botulinus endospores at this temperature by 12 log units is called the F_o concept (12D concept). F_o amounts to $0.21 \times (\log 10^{12} - \log 1) = 2.52$ min. For more resistant endospores, e.g., of $\rightarrow$ Cl. sporogenes with $D_{121} = 2$ min, or $\rightarrow$ B. stearothermophilus with $D_{121} = 4$ min, F_o values range between 12 and 20 min; these organisms, however, are not involved in food poisoning, and the 12D concept therefore will only have relation to C. botulinum spores. In practice a 5D concept is aimed at with regards to the mentioned spoilage organisms (Heiss and Eichner 1984). For examples of F_o values determined from practice, see Table 15.

Table 15 F_o values for some canned vegetables (acc. to Kunz 1988).

Vegetable	F_o Value in Min.
Beans, lentils	3.0 – 6.5
Peas	4.0 – 6.5
Cabbage	1.5 – 3.0
Celery	3.0 – 12.0
Carrots	3.0 – 10.0
Asparagus	2.5 – 3.5
Cauliflower	3.0 – 8.0
Potatoes	4.0 – 10.0

facultative

Optional or alternative, e.g., facultatively anaerobic refers to a MO that preferably grows aerobically, but which can also grow under anaerobic conditions.

faecal indicators $\rightarrow$ indicator organisms.

faecal population

Inhabitant microbial population of the GI tract of animals; excreted with the faeces. $\rightarrow$ Enterobacteriaceae, $\rightarrow$ lactobacilli, Bacillus and Clostridium spp., enterococci, Eubacterium, Bacteroides and Bifidobacterium spp. A large proportion of the facultatively anaerobic groups have significance in food spoilage and food-borne infections. – Faecal contamination may imply a health risk to the consumer, and may be considered an indication of deficiencies in processing plant hygiene; easily detectable groups or spp. have been selected as $\rightarrow$ indicator organisms of possible faecal contamination; $\rightarrow$ gut microbes, $\rightarrow$ coliforms.

family

Lat.: Familia; $\rightarrow$ systematic unit; taxonomic rank.

farmer's lung

Allergic pulmonary disease; extrinsic alveolitis. Caused by the inhalation of spores and mycelial fragments of certain actinomycetes and fungi, associated with mouldy hay. High infection risk during harvesting, and also during handling of feeds in association with dust development.

fat

Basic component of foods. Energy reserve in animals, plants and MO, e.g., yeasts. Mainly triglycerides (triacylglycerides); esters of glycerol with saturated or unsaturated fatty acids.

fatty acids

Released by microbial or product-specific→lipases from triglycerides in foods; causing the sensory defect→rancidity. Intensive smell or taste common (e.g., acetic acid, butyric acid, etc.). Partly excreted as metabolites by MO under anaerobic and microaerophilic conditions (lactic acid, propionic acid, etc.).

fecal contamination→contamination.

fermentation

Generally refers to a wide range of microbial processes in industry, involving both respiratory and fermentative metabolism. Strictly spoken fermentation refers to (anaerobic) energy-yielding intracellular processes (metabolism) in which the released hydrogen is not transferred to an exogenous electron acceptor, but to an organic hydrogen acceptor, e.g., NAD + 2H → $NADH_2$, which is then transferred to the typical metabolic intermediates or special acceptors with concomitant regeneration of NAD. Oxidation of any intermediate is balanced by the equivalent reduction of other intermediates in the fermentation pathway. Reduced compounds excreted into the (food) substrate (→sensory properties). Several important biotechnical processes are typically fermentative (→fermentation products). Energy yield (substrate level phosphorylation) lower than for oxidative (→aerobic respiration) processes. – Anaerobic respiration, by which nitrate or sulphate are reduced in absence of oxygen, is not an example of fermentation;→denitrification (Schlegel 1985).

fermentation products

Metabolic end products that may be produced by MO in foods under oxygen limitation; may be detrimental (spoilage) or favourable to→sensory properties. Examples: lactic acid (*Lactococcus, Pediococcus,* homoferm. lactobacilli); lactic acid + ethanol (acetic acid) + CO_2 (*Leuconostoc,* heteroferm. lactobacilli); propionic acid (*Propionibacterium*);→biogas (Gottshalk 1986; Rehm 1985).

fermentation trap

Glass rods in various forms, partly filled with liquid (water), used for sealing fermentation vessels. Escape of CO_2 is allowed, simultaneously preventing the entering of air (O_2), dust, MO and insects (fruit flies).

fermenter

Fermentor;→bioreactor. (Stainless steel) vessel for aerobic or anaerobic fermentation, either for batch or continuous culturing of MO, for production of→biomass and/or metabolic products, e.g.,→organic acids,→enzymes, etc. Different sizes, varying from 1 l to 500 m^3; shape mainly vertical, cylindrical and closed. Should be sterilisable, in addition to regulation of pH, temp., O_2, nutrients, foam, etc., so as to provide optimal conditions for the growth of MO (Kunz 1988).

field mushroom→mushroom.

filtering materials

When→bacterial filters are used in food processing plants, care must be taken to avoid having materials detrimental to health (asbestos, etc.) enter the food or the factory air.

filth test

Microscopic examination for the possible detection of contaminating materials and residues, that may be an indication of defective raw materials or processing failures, with possible health consequences, e.g., hairs, excreta of rodents and birds, fragments of insects and feathers, etc. (Sinell 1985).

fish

Provides as food excellent growth substrate for bacteria (75 – 85% water, 15 – 19% protein, < 1 – 8% fat, < 1% CH). → Ocean fish represent the major part for human consumption (in Germany 113,000 t vs 12,180 t fresh water fish in 1986). Spoilage during refrigerated storage mainly by *Pseudomonas, Alteromonas* and the *Moraxella-Acinetobacter* group; at elevated temperatures also *Bacillus, Micrococcus, Proteus* and coryneform spp. (→ cold chain). Initial degradation is indicated by a darkening of the gills followed by increasing off-smell (trimethylamine). Microbial → decarboxylation of amino acids (e.g., histidine) may produce toxic levels of biogenic amines (e.g., → histamine), especially in tuna. Fresh fish will spoil most easily, followed by smoked fish, fish marinades, frozen fish and fish preserves (Krämer 1987; Kunz 1988).

fixed bed fermenter

→ Bioreactor with small carrier particles, containing (immobilised) MO or enzymes on a support or solid substrate. The liquid substrate slowly flows upwards through the reactor bed or layer, enabling the production of (e.g.) ethanol from sugar.

flagellum

Plural: flagella. Filament responsible for movement in majority of motile bacteria; thread-like appendages that occur either singly (monotrichous), in groups or tufts of 2 or more (polytrichous/multitrichous). Arrangement also typical for a group or sp., e.g., distribution over the cell surface (peritrichous), monopolar (on one end), bipolar (on both ends), or subpolar (near one pole). Motility often considered as important phenotypic characteristic, without referring to arrangement, etc., of flagella. Detection in routine laboratory difficult; dark field microscopy or special staining procedures required.

flat sour

Type of microbial spoilage of canned fruit and vegetables; contents have been soured without gas production, whilst the ends appear flat and not swollen (→ swell). Lactic acid production, e.g., by *Bacillus stearothermophilus* under tropical conditions, and *B. coagulans* (*B. thermo-acidurans*) at room temperature. Contamination primarily via soil residues on the raw materials, blanching utensils, and via sugar or starch (Müller 1983a; Krämer 1987).

Flavobacterium

Genus of Gram – , aerobic to facultative anaerobic, rod-shaped bacteria, psychrotrophic, non-motile. Cytochrome-oxidase positive. Yellow to orange, non-fluorescent, water insoluble pigments produced (Bergey 1984). Most spp. are strong proteolytes. Associated with spoilage of meat, poultry, fish, milk and dairy products. Distribution: fresh and ocean waters, soil.

flour

Basic foodstuff produced from grain by mechanical milling, after it has been cleaned (→ *Claviceps*) and conditioned. The bacterial count lies between $10^3 – 10^5$/g. The water content should not be more than 13 – 15% corresponding to an a_w value of 0.7. The flour keeps fresh at this water content as long as it does not absorb water from the atmosphere. If the water content increases fungal growth may be possible, followed by souring, and other microbial defects including bitterness and rancidity (Kunz 1988).

fluidised bed fermenter

Special bioreactor for→ solid state fermentation. Small solid state particles (< 1 mm) are whirled in a turbulent stream of air saturated with water, above a sieve. The MO and substrate are pseudo-homogeneously distributed; the MO being in equilibrium with the moisture content of the aeration gas. Volatile metabolites, e.g., ethanol, are recovered from the escaping air. In this way optimal conditions are created for transfer of components; continuous process control is possible and allows reduction of waste problems (Kunz 1988).

foam controlling agent → paraffin.

food conserves

Also referred to as semi-preserved foods (Germ.: "Halbkonserven"). Foods packaged in cans, tins, bottles, synthetic containers, that are preserved by slight reduction of the a_w and/or the pH (e.g., with acetic acid/vinegar), by hermetical sealing, and moderate heat treatment. Shelf life several weeks depending on storage temperature (ca. 10°C). Example: certain fish specialities, e.g., Bismarck herring, etc.

food legislation

The first harmonised legislation "to regulate the handling and trade with foodstuffs and food products" of the German Empire was approved on 14 May 1879, and was aimed at "protection of the consumer against health hazards and economic losses". Health hazards mainly refer to pathogenic MO or their → toxins (→ food poisoning). The foodstuffs and food products regulation ("Lebensmittel- und Bedarfsgegenständegesetz" or → LMBG) of 15 August 1974, mandatory from 1 January 1975, forms the present basis for the microbiological examination of different groups of food commodities; these methods are described, collected and continuously amended and published under §35 LMBG by the Federal Health Department → official methods (Hahn and Muermann 1987). See also Furia (1977) for information on the *Federal Register* and the functions of the FDA and USDA in the USA.

food poisoning

General term for human diseases caused by the ingestion of contaminated food. By virtue of the different elicitors (causative agents) distinction can be made between:
(1) Infection: food or water act as vehicle or transmitter of pathogenic bacteria, protozoa or viruses, e.g., → *Salmonella*, → *Shigella*, → *Entamoeba* and → hepatitis.
(2) Intoxication: toxins in the food are produced by MO, e.g., → *Staph. aureus*, → botulism, → mycotoxins.
(3) Massive → contamination of food occurs with saprophytes or otherwise harmless bacteria, e.g., → *Cl. perfringens* and → opportunists.
(4) Chemical toxins are produced by MO or intrinsic enzymes of the food, e.g., → histamine, biogenic amines and → nitrite.

The contamination of foodstuffs can be traced back to the raw materials, production or processing in combination with the faultive treatment. See Table 16. To trace the cause of the food poisoning, stool and vomit specimens of the patient as well as the suspected food must be examined microbiologically. The physician should be consulted. The main causes are insufficient cooling (70%), insufficient heating, reheating and food being kept hot (45%), insufficient hygiene (30%). Multifactorial causes are often involved and therefore the total in Table 16, adds up to more than 100%. 65% of all cases of food poisoning can be traced back to large kitchens, restaurants and catering services, 31% to household kitchens and only 4% due to commercially produced foodstuffs → airline foods (Knothe et al. 1978; Krämer 1987; Sinell 1985).

food sprouts → sprouts.

food vinegar → vinegar.

Table 16   Causes of food poisoning according to results following the study of 1152 cases of food poisoning recorded during a period of 15 years in the USA (Sinell 1985).

Causes	Proportion in %
Insufficient cooling	46*
One or more days between preparation and consumption	21
Infected person	20
Incorrect heat treatment or canning	16
Insufficient heating	16
Insufficient reheating	16
Use of contaminated ingredients or raw foodstuffs	11
Cross contamination	7
Insufficient cleaning of utensils	7
Foodstuffs with a questionable origin	5
Use of food rests	4

*The total adds up to more than 100% since more than one cause was documented in several cases.

foot and mouth disease

Aphto-virus. Communicable zoo-anthroponosis through raw milk or in rare cases drinking water.

formaldehyde

HCHO; methanol; methylaldehyde, formol or formalin. Pungent smelling gas; in 30 – 40% water solution; soluble in water and ethanol. Excellent → disinfectant for the treatment of surfaces. Active against bacteria, fungi and some viruses, but low effectivity against endospores. Direct application near foods not allowed because of mutagenic effects. No definite proof yet of carcinogenic properties; because of strong irritation of the mucous membranes cocarcinogenic effects cannot be excluded. – "MAK" values for air: 1 ppm (1.2 mg/m^3). Found in: fruit and vegetables ($4 - 20$ mg/kg); in 1 cup of coffee $3 - 7.5$ mg → cigarette smoke. It is formed by the degradation of → hexamethylenetetramine (Classen et al. 1987).

formates

→ Salts of → formic acid.

formic acid

HCOOH. Salts known as formates. Metabolic product (in addition to other → organic acids) of fermentation by Enterobacteriaceae, several Bacillus spp. and other bacteria. Some species split HCOOH to H_2 and CO_2. Weakly smoking liquid with a stinging smell and etching to the skin. Commercially mainly available in 25% concentration. – Approved as preservative under EC no. E 236; sodium formate E 237; potassium formate E 238; declaration obligatory. Growth/multiplication of most MO inhibited but not killed at concentrations approved for foods (Wallhäusser 1987). – LD_{50} for mouse (p.o.): $1 - 2$ g/kg BW; lower toxicity of salts. For man 10 g (p.o.) is dangerous and $50 - 60$ g may be lethal (acid etching, severe acidosis, hemolysis). → ADI value $0 - 3$ mg/kg BW (Classen et al. 1987).

fractionated sterilisation

Tyndallisation. Heat on three consecutive stages (days) at ca. 100 °C. The germination of endospores is induced by the → heat shock; the following heat treatment will then kill the vegetative cells.

freeze drying

Lyophilisation. Elegant but expensive process for moisture (water) removal by sublimation from a deep-frozen product or specimen. Long-term preservation for MO,→starter cultures,→enzymes and certain foodstuffs. Packaging in air- or water vapour-tight materials/containers. Skim milk, lactose or glutamate can be used as protective substrate for bacterial suspensions or mould conidia.

freezing

Temperatures <0°C used for food preservation; shelf life extension. Ice formation results in the reduction of available water for the MO. Physiological water depletion, in addition to the low temperature, is the main growth limiting factor. Freezing point reduction affected by salts, organic substances and colloids (Table 17); only below this temperature microbiological stability can be maintained. The temperature range − 2.5 to − 7.5°C however,

Table 17 Freezing point reduction, with reference to distilled water, for different foodstuffs (acc. to Heiss and Eichner 1984).

Product	Temperature in °C
Meat	− 0.6 bis − 1.2
Fish	− 0.6 bis − 2
Egg albumen	− 0.45
Egg yolk	− 0.65
Tomatoes, raspberries	− 0.9
Cauliflower	− 1.1
Onions, peas, strawberries	− 1.2
Peaches	− 1.4
Apples, pears	− 2
Plums	− 2.4
Sucrose solution (1 mol)	− 2.65
Salt (NaCl) solution (1 mol)	− 3.45
Cherries	− 4.5
Nuts	− 6.7

cannot be considered as "safe" with regard to moulds over a period of >30 weeks. – Reducing of the temp. below zero will gradually lower the proportion of "free" water. At − 18°C (deep freeze temp.) a constant value is reached; the proportion of frozen water amounts then to 65% for white bread, 88% for beef, and 95% for strawberries. Generally it can be assumed that ca. 0.4 g of water per g (DM) of protein or starch will not be transferred into the frozen state. – The lethal effect of ice crystal formation to MO has not been proved. During the deep freezing of meat and vegetables 30 – 70% of the microbial population survives, and 5 – 10% for fruit. Gram − bacteria are damaged more than Gram + . Different→Q_{10} values during freezing and thawing are probably a more important factor in the destruction of microbial cells than ice crystal formation. Animal and plant cells may however be mechanically damaged during ice formation, causing the typical losses of juice during thawing; responsible (a.o.) for cross contamination in household (Heiss and Eichner 1984).

freezing rate

Rapid freezing either prevents ice crystal formation (by encouraging vitrification) or results in the formation of smaller ice crystals in and between the cells. A slower freezing rate will increase the cell damage, e.g., by extracellular ice formation and intracellular dehydration, and results in greater moisture (juice) losses from the cell during thawing. Slow thawing will result in a higher lethality rate, and may be caused by longer continuation of imbalance in intermediary metabolism.

fresh cheese

Quarg, quark, related to cottage cheese. Manufactured from pasteurised milk, originally only with added starter cultures, presently also with some→rennet added, so as to enhance the coagulation of the casein, and to stabilise the gel. Relatively low temperature process at 20 – 30 °C for 15 – 22 h when a butter culture or *Lactococcus lactis* ssp. *cremoris* culture is used; "hot" souring with *Streptococcus salivarius* ssp. *thermophilus* and *Lactobacillus bulgaricus* (yoghurt culture). Cheese baths of up to 50,000 l; cutting at pH 4.8, and stirring till pH of 4.6 is reached; removal of the whey; thermisation at 60 °C, separation and kneading. Cream, spices, canned and fresh fruits, etc., may be added. Aseptic filling and packaging. – For cottage cheese a starter culture of *Lactococcus lactis* ssp. *cremoris, Lactoc. lactis* ssp. *diacetylactis, Lactoc. lactis* ssp. *lactis* and *Leuconostoc cremoris* is used, with a small amount of rennet added. The curd is sliced at pH 4.7 into particles of 1 – 2 cm, stirred at 42 – 60 °C for 1 – 3 h; the whey is separated, drained and washed with bacteria-free water and ice water; additives may be used as for quarg. High spoilage potential; cold chain should be maintained < 5 °C. Spoilage mainly by yeasts and moulds as a result of recontamination. Ascospores of → *Byssochlamys* sp. and/or → *Monascus purpureus* from mouldy silage; may survive pasteurisation and thermisation (Teuber 1987).

fresh milk → pasteurised milk.

fresh water fish
Fish harvested from lakes, rivers, dams or fresh water aquaculture. High spoilage potential as ocean fish. The marketing routes are, however, shorter in general, and the incidence of spoilage and losses is lower. Abuse of smoked fillet may enable the growth of *Cl. botulinum* (type E). This explains why vacuum (anaerobic) packaged smoked fish is not approved for distribution/marketing in some states of the USA.

Frings acetator
"Frings trickling generator". Wooden vats filled with beechwood shavings or chips, the surfaces of which are covered with → acetic acid bacteria (*Acetobacter* spp.). The ethanol containing liquor (e.g., apple cider) trickles through the shavings down to a reservoir from which it is pumped continuously to the top. Air is introduced from the bottom and rises through the shavings, thereby supplying O_2 to the system and removing excessive heat. Additional cooling may be necessary so as to maintain the temp. at 29 – 35 °C. The process is continued until the circulated liquid contains 10 – 12% of acetic acid and with a residual ethanol content of < 0.3% (Krämer 1987; Rehm 1985).

frozen meat
Deep frozen (< – 18 °C) storage of meat carcasses as halves or quarters. Keeping quality primarily depends on fat proportion. Quality principally dependent on initial microbial population at freezing. Aerobic population should be $< 10^7/g$, fecal streptococci (enterococci) or coliforms $< 10^3/g$, *Staphylococcus aureus* $< 10^2/g$, sulphite reducing clostridia $< 10/g$ (Kunz 1988).

fructose
Fruit sugar, levulose. A ketohexose commonly found in plants; fermentable by most MO.

fruit
The pH value of most fruit ranges around 4.5, and acts as important protective factor against most bacteria (exceptions, e.g., LAB and acetic acid bacteria), but not against moulds (Table 18). Most spoilage MO can only penetrate after damage of the skin (e.g., by insects, birds, hail, harvesting, etc.) (Beuchat 1987; Dittrich 1987; Müller 1983a). *Pen. expansum* may produce → patulin. Controlled atmosphere (→ CA storage) may protect the product against microbial spoilage

Table 18 pH value and common post-harvest diseases of fruit.

Fruit	pH	Parasite	Comment
Kernel fruit (apples, pears, quince)	3.1 – 4.6	P. expansum* Gloeosporium sp. Alternaria sp.*	brown rot
Stonefruit (cherries, apricots, peaches, plums)	3.2 – 4.4	P. expansum* Botrytis cinerea Alternaria sp.* Cladosporium sp.	brown rot grey mould
Berry fruit (black currants, gooseberries blueberries, etc.)	2.5 – 4.2	Botrytis cinerea Gloeosporium sp. Cladosporium sp. Alternaria sp.*	grey mould
Strawberries	3.0 – 3.9	Botrytis cinerea Rhizopus spp.	grey rot wet rot
Grapes	3.0 – 4.0	Botrytis cinerea P. expansum* Aureobasidium pullulans	noble rot blight
Citrus fruit	2.2 – 4.0	P. italicum P. digitatum	blue rot green rot

* Potential toxin producers.

(Nicolaisen-Scupin 1985). Other measures for preventing microbial spoilage during storage include optimum temperature and relative humidity, and the timely removal of the products for distribution.

fruit juice

Fruits are ground/pulped and the cells separated by use of → pectinase preparations. The cell juice is obtained under high mechanical pressure. Spoilage association: *Pediococcus, Lactobacillus, Leuconostoc,* acetic acid bacteria, yeasts and moulds [for microbiological examinations see Hatcher et al. (1984)]. Shelf life without refrigeration achieved by pasteurisation; spoilage possible as a result of heat resistant spores of → *Byssochlamys nivea* or *By. fulva.* Slow growth possible at low Eh (low O_2 and high CO_2 partial pressure) (Beuchat 1987).

fruit juice concentrates

Costs of storage and transport can be reduced by concentration of the fruit juice to a_w values between 0.73 and 0.94, with pH values ranging from <2.0 (lemon juice) to 4.5. Spoilage association mainly moulds (*A. glaucus* group) and → osmotolerant yeasts (*Hansenula anomala, Zygosaccharomyces rouxii, Z. bailii, Schizosaccharomyces pombe*) [for detection procedures see Baumgart (1990) or Speck (1984)]. Sugar addition may improve stability. – In apple juice concentrates the possible presence of → patulin should be taken into consideration.

fruit products

The relatively low pH values of fruit protect the products against bacterial spoilage. The a_w value can be reduced by addition of sugars, e.g., in marmalades, jams and jellies, and microbial activities are limited to osmophilic and osmotolerant yeasts and moulds. The keeping quality of "canned" fruit relies on sufficient heat treatment and air-tight closing of the container. Spoilage may be promoted by the formation of condensate on the lid. Filling at temp. > 80 °C ensures the destruction of possible contamination from the air; for home canning lids should be boiled directly prior to closing the container.

fruit storage
Metabolic reactions are slowed down and the number of deleterious MO is decreased when the temperature is lowered. Further protection against and inhibition of→ parasites and→ opportunistic parasites can be accomplished by decreasing the oxygen- and increasing the CO_2-concentrations in the atmosphere of the store room (→CA storage). Important parameters are: perfect timing of harvesting, the correct temperature for the specific variety, sorting out of damaged and spoiled fruit, and marketing at the correct time (Bünemann and Hansen 1973; Henze and Hansen 1988).

fruit syrup
Fruit juice and sugar are boiled together to a final sugar concentration of ca. 65%. A cold osmotic extraction procedure may also be used. Spoilage association→ fruit juice concentrates (Kunz 1988; Müller 1983a).

fruit wine
Juices of fruit or berries are fermented after addition of 1 – 5% of cultivated yeast, and typically also 0.4 g/l of ammonium sulphate, phosphate or chloride as N-source. Spoilage association: LAB (*Leuconostoc, Pediococcus, Lactobacillus*); may produce slime especially in low acid wines (Dittrich 1987).

fruiting body
Structures formed by moulds from densely grouped hyphae bearing sexually produced spores. Mushrooms. French: champignon; German: speisepilz (champignon).

fumaric acid
$HOOC-CH=CH-COOH$; *trans*-ethylene dicarboxylic acid. Colourless solid at ambient temp.; sublimes at 200 °C. Intermediate in the→tricarboxylic acid cycle (Krebs cycle). Distribution in nature: *Fumaria officinalis, Cetrara islandica,* mushrooms, meat extract, etc. – Characteristic of several genera of the Mucorales. – Production primarily synthetical or with *Rhizopus* sp. Precursor for production of→malic acid. Preliminary EC no. 297.

fumonisins
Fumonisin B, and B_2 are produced by growth of *Fusarium moniliforme* on maize. Chemically not related to other *Fusarium* toxins (Table 19). Soluble in water and methanol. Weight increase in rats stopped within a few days after oral dosage; carcinogenous changes of the liver within 4 weeks after administering 3 times [Gelderblom et al.: *Appl. Environ. Microbiol.* 54 (1988) 1806 – 1811].

fungal (population) count→mould
count.

fungi→moulds.

Fungi Imperfecti→Deuteromycetes.

fungicide
Chemical substances with antifungal action; some produced by MO. Lethal (irreversible) effect on fungi and their spores. Fungicides completely non-toxic to other organisms not known.→Preservatives→pimaricin.

fungistatic
Chemical substance that inhibits the ger-

Table 19 *Fusarium* toxins, producer organisms and effect of oral ingestion.

Mould	Action	Symptoms	Toxicity
Zearalenone			
F. graminearum	oestrogenic, hyper-oestrogenism, acute:		
F. culmorum	uterotropic	vulva and teats	1 – 5 mg/animal/d
F. avenaceum		reddened and enlarged	chronic:
F. moniliforme		uterus enlarged, fruit	0.12 mg/kg feed
F. equiseti		mummified, testis atrophy, fertility disturbances	
Trichothecene			
Fusarium spp.	dermatotoxic	skin irritation, vomiting	acute: 0.3 mg/kg feed
Myrothecium spp.	neurotoxic	diarrhoea, feed refusal	(deoxynivalenol)
Trichoderma spp.	haemorrhagic	haemorrhages	2 mg/kg feed
Cephalosporium spp.	teratogenic	necroses, declining	(other trichothecenes)
Trichothecium spp.	immunosuppressive	response, thin-shelled	chronic: 0.4 mg/kg
Stachybotrys spp.		egg	feed
Moniliformin			
F. moniliforme?	necrotic	degenerations and	LD_{50} 4.5 mg/kg BW
F. moniliforme var. subglutinans		necroses of myocard, lesions of different	(chickens, p.o.) LD_{50} 21 – 29 mg/kg BW
F. avenaceum		organs	(mouse, p.o.)
Fusarin C			
F. culmorum	mutagenic,	liver cancer	carcinogenic dose
F. graminearum	probably	papillomes	1 × 50 mg/kg BW
F. poae	carcinogenic	rat, p.o.	
F. moniliforme			
F. sporotrichioides			
F. tricinctum			
F. sambucinum			
F. avenaceum			
Fumonisin B_1 and B_2			
F. moniliforme	carcinogenic	leucoencephalomalacy of horses	

mination, growth and increase of moulds, without necessarily killing them. Sublethal conc. of→fungicides may act fungistatically.

fusarenon X
→Mycotoxin, produced by different→*Fusarium* spp. during growth on foods. Belongs to the→trichothecenes.

fusarin C
→ Mycotoxin. May be produced by *Fusarium culmorum, F. graminearum, F. poae, F. crookwellense, F. sporotrichioides, F. tricinctum, F. sambucinum* and *F. avenaceum* during growth on foods. Strong mutagenic and probably carcinogenic to man; rapidly destroyed by UV light [Steyn and Vleggaar (1986) p. 305; Thrane, U.: *Mycotoxin Research* 4 (1988) 2 – 10].

fusaritoxicosis
Mycotoxicosis caused by toxins produced by *Fusarium* spp., e.g., *F. sporotrichioides* and *F. poae* on cereals, or found in bread, from such grains; especially found in Eastern Europe. Symptoms in domestic animals: tiredness, disorientation, sometimes paralysis of hindquarters, refusal of feed and often death. For man "alimentary toxic aleucia" (ATA) is reported, which is observed especially after consumption of cereals that have overwintered on the field. Largest outbreaks reported for the region Orenburg (former Soviet Union) during 1942 and 1947, with > 10% of the population suffering and with several deaths. Causative substances produced by fusaria and other genera (Table 19), and identified as 12-, 13-epoxytrichothecene (Gedek 1980).

Fusarium
Mould genus belonging to the → Deuteromycetes (class Hyphomycetes). Septate mycelium and macroconidia produced (in some spp. also chlamydospores). Taxonomy contradictory; ca. 250 taxa described. *Fusarium roseum* frequently reported in earlier literature, which, however, is non-valid since several are coloured rose/pink → teleomorphs as far as known mainly *Nectria* and *Gibberella*. Conidia crescent shaped; microconidia single-celled and spherical; macroconidia hyaline, multiseptate with a characteristi-

cally shaped foot cell and produced from phialides (Figure 7, *Mucor*). Vegetative hyphae may be transformed terminally or intercalary into → chlamydospores (Figure 11, see page 174, *Mucor*). – Most spp. are soil inhabitants; distribution worldwide. Decaying cellulosic and other organic materials. Spoilage of stored fruit and vegetables. Several are plant parasites and specifically adapted to plant spp. or types. Up to 70% (av. 50%) of the cereals produced in Germany are infected. – During growth on raw materials → *Fusarium* toxins are regularly produced. *F. graminearum* is used for the production of single cell protein (mycoprotein) in GB. – Cultivation in laboratory in nutrient-rich media: wort agar, potato-dextrose agar, oatmeal agar, maize meal agar, rice porridge, etc. Conidia production promoted by short-wave UV light (in day-night fluctuation) (Cerny and Hoffmann 1987; Domsch et al. 1980; Gerlach 1980).

Fusarium toxins
Mycotoxins produced by *Fusarium* spp. Five groups, showing different chemical structures, presently distinguished: (1) trichothecenes, (2) zearalenone, (3) malformin, (4) fusarin C, and (5) fumonisins (Table 19). Deoxynivalenol, nivalenol and

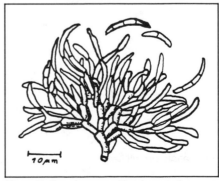

Figure 7 *Fusarium* sp. with micro- and macroconidia.

zearalenone have most commonly been found in food and feed within a range of up to 0.2 mg/kg. Greater significance in animal nutrition than for humans. Conjugate formation (glucosides, etc.) in the host plant possible; products toxicologically active but not detectable with routine methods [Engelhardt, G. et al.: *Naturwiss.* 75 (1988) 309 – 310]. – Zearalenone has oestrogenic activity and can cause hyper oestrogenism in sows. May be reduced in the organism or host plant to zearalenol, which is at least 4 times more oetrogenic than zearalenone, and may be partly excreted in the milk (→carry over). Little known about its health significance, especially in infant nutrition. The chemical structure (lactone) suggests a carcinogenic activity. It has anabolic properties and may be responsible for reduction in fertility of cattle and swine. Zero tolerance mandatory for B. – Even low amounts of "deoxynivalenol" (DON) may result in feed refusal; induces nausea and also called vomitoxin; leads to retarded growth. Neither carcinogenic nor terato-

genic. In Canada a tolerable daily intake of 0.003 mg/kg is suggested for adults. – "T-2 Toxin" is carried over in the milk in swine, and in the musculature and eggs of poultry. – "Moniliformin" has primarily been found in maize in ZA (Reiss 1981, 1986; Thalmann 1986). – "Fusarin C" has been discovered relatively recently. Strongly mutagenic, and may possibly be carcinogenic to man (Steyn and Vleggaar 1986). – "Fumonisins": fumonisin B_1 is formed during growth of *Fusarium moniliforme* on maize. Induces carcinogenic changes in the liver of rats after p.o. administration [Gelderblom et al.: *Appl. Environ. Microbiol.* 54 (1988) 1806 – 1811].

fusel oils
Undesired by-products of alcoholic fermentation. Should not be carried over with distillate under optimal conditions. Mixture of higher alcohols (propanol, isobutanol, amylalcohol, isoamylalcohol, etc.) formed by deamination, decarboxylation and reduction of amino acids.

G

galactose
Aldohexose. Occurs naturally in D and L forms. Found in several di- and oligosaccharides, e.g., → lactose (β-glycosidic bond with glucose). Metabolised by a large number of MO; released from → lactose by → β-galactosidase. Mixture of galactose + glucose is sweeter than lactose; → yoghurt.

α-galactosidase
Commercially produced with yeasts, moulds and bacteria; e.g.: *Aspergillus* spp., *Mortierella vinacea, Saccharomyces uvarum* (*S. carlsbergensis*), *Kluyveromyces* spp., *E. coli.* Trivial name: melibiase. Systematic name: α-D-galactoside galactohydrolase. E.C. 3.2.1.122. Catalysed reactions: exohydrolytic cleavage of terminal non-reducing α-D-galactose, galacto-mannans, galacto-lipids. Applications: removal of raffinose from molasses; improvement of sugar crystallisation.

β-galactosidase
Produced with the use of different MO: *A. niger, A. oryzae, Kluyveromyces fragilis, K. lactis, K. marxianus, Saccharomyces* sp., *E. coli.* – Trivial name: lactase. Systematic name: β-D-galactoside galactohydrolase. E.C. 3.2.1.23. Catalyses the hydrolysis of terminal, non-reducing β-D-galactose in galactosides. Application: to hydrolyse lactose in milk and milk products (e.g., to improve acceptability for lactose-intolerant consumers, and to increase the sweetness).

gallates
Anti-oxidants (E 310 – E 312). Esters of gallous acid (3,4,5-trihydroxy-benzoic acid) found in vegetables and other plants.

gas production
During metabolism some MO may produce gaseous end products. Especially CO_2 has significance for food, e.g., for leavening of the dough or in the production of → sparkling wines it is desired; detrimental, e.g., in the → early blowing of cheese (in combination with H_2). → Biogas is a mixture of CO_2 and CH_4, and is produced during the anaerobic digestion of sludge. CO_2 and H_2 may be produced by clostridia in canned products, and cause → swell (blown can). → Hydrogen sulphide and/or → ammonia are produced by some bacteria as a result of protein degradation → sensory properties; → grouping of MO.

gastroenteritis
Inflammatory change of the lining of stomach and intestine, usually caused by MO ingested with the food or water. → Food-borne infections; → zoonoses. Causative agents may be a number of MO that may cause diarrhoea, fever, vomiting, etc. Ca. 1 billion people are affected annually by acute infectious diarrhoea, with an estimated 5 mio deaths (mainly infants in developing countries). MO involved:
(1) Bacteria – salmonellae: *Salmonella typhi* (typhus abdominalis), *S. paratyphi* (paratyphus A and B), *Shigella* spp. (dysentery), enteropathogenic *E. coli* (enteritidis mainly of infants), *Vibrio cholerae* biotype *cholerae* and *V. cholerae* biotype *El Tor* (cholera), *V. parahaemolyticus* (Kanagawa effect), *Yersinia enterocolitica* and *Y. pseudotuberculosis* (enteritis/enterocolitis)
(2) Protozoa – *Entamoeba coli, E. histolytica* (amoeboid dysentery), *Giardia lamblia* (giardiasis, lambliasis)
(3) Viruses – rotaviruses (duoviruses), viruses of the parvo or picorna groups, coronaviruses; Norwalk agent (gastroenteritis, especially in children)

(4) Fungi – opportunistic colonisation of immuno compromised, e.g., following broad spectrum antibiotic treatment, cytostatics treatment, in diabetics, etc.; *Candida albicans, C. tropicalis, Cryptococcus neoformans, Torulopsis glabrate, Aspergillus* spp., Mucoraceae; changes in the gastro-intestinal tract are unspecific and diagnosis is difficult (Knothe et al. 1978).

gelatine
For consumption, etc. Obtained from animal cartilaginous tissue, skin and bones, e.g., by hydrolysis or boiling (heat-denaturing) of collagen (protein). Commercially available in different forms: slices (leaves), powder, granulate. For food purposes maximal aerobic microbial numbers should not exceed 10^4/g; <50/g for Enterobacteriaceae; *E. coli* absent in 1 g; salmonellae absent in 25 g; <10 anaerobic endospores per g. Used to test gelatinase activity of MO (ability to liquefy gels of 4 – 12%); indication of proteolytic activity, mainly by Gram– MO. – Has been commonly used during the era of Robert Koch for the solidification of nutrient media.

gelatine liquefication
Ability of MO to liquefy gelatine can be used as phenotypic criterion in the taxonomy of MO; → differentiation. Determination of plate counts with gelatine as solidifying agent: liquefying strains form small, "wet" zones around the colony. – Carbon-gelatine is mixed with the culture in a test tube: liquefying is followed by sedimentation of the C to the bottom of the test tube. Determination of low bacterial numbers and gelatine liquefying strains in water "DEV gelatine agar" (e.g., Merck 10685) may be used. The surface is covered with a solution of ammonium sulphate; around liquefying colonies clear zones will appear.

gene modification
Change (modification) of the genetic code of MO with the aim at their improvement. Application of→ genetically modified MO in manufacture of a specific product.

gene mutation→ AMES test.

generation time
Of bacteria: time interval (lapse) required for the doubling of the cell number ("doubling time"), given in min or h. Dependent on sp. and environmental factors (temp., pH, nutrient availability, etc.). Examples: *B. subtilis:* 26 min at 40°C; *E. coli:* 21 min at 40°C; *B. stearothermophilus:* 11 min at 60°C → multiplication.

genes
Biological units (sequence of nucleotides) of the DNA (genetic nucleic acids). Localisation in the chromosome(s); codes for a functional peptide chain or RNA molecule.

genetically manipulated MO→ genetically modified MO.

genetically modified MO
Starter cultures with an improved capability with respect to specific desired properties (may be transferred from one sp. or type of MO to another; recombinant DNA). Examples: genes for the hydrolysis of starch or cellulose may be transferred into a culture yeast, with application for the direct conversion of high polymeric substrates (without intermediate enzyme treatment) to ethanol. Cloning of the "chymosin" gene into *Kluyveromyces lactis* will enable the yeast to produce chymosin identical to natural chymosin (→ rennet). – Care should be taken that no undesired properties are accidentally cloned into the recipient strain, e.g., resistance to therapeutically valuable antibiotics, or pathogenicity, or infectivity, etc. – Release of genetically modified MO as (e.g.) starter cultures or their products

(e.g., enzymes) should be dependent on prior toxicological safety assessment.

genus

Taxonomic unit or rank; first name in species nomenclature, e.g., *Staphylococcus aureus* → systematics.

Geotrichum candidum

(*Oidium lactis*). Class Hyphomycetes. Anamorphs to *Dipodascus geotrichus*. "Milk mould" (German) or machinery mould. The teleomorph (*Dipodascus*) belongs to the division Ascomycota and the class Endomycetes. The asexually reproducing anamorph produces thalloconidia (arthrospores) by septation of the hyphae, that become rounded (oidia) after release of

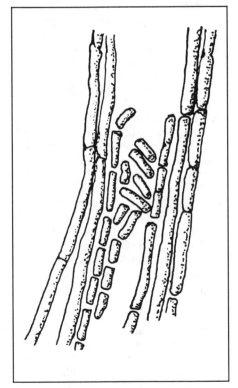

Figure 8 Talloconidia (arthrospores) of *Geotrichum candidum.*

the cylindric units; microscopical characteristics typical of this mould (see Figure 8). – Growth on yoghurt, fermented milk products, fresh cheese and sour milk cheese, soft cheese with mould, compressed yeast, etc. Especially lactic acid is decomposed, resulting in an increase in surface pH and eventual growth and increase of acid sensitive microbes. – Fragments of the mould may be removed from surfaces with sticky tape for microscopical examination, and serves as indication of insufficient hygiene during the processing of fruit and vegetables. – Damage of bakery products and on → bakery yeasts. – *Geotrichum candidum* is a desired MO in the fermentation of cocoa since it produces → pectinases (Beuchat 1987; Müller 1983a; Pitt and Hocking 1985).

Gibberella → *Fusarium.*

gibberellins

Secondary metabolites produced by some spp. of the genus *Gibberella* or its → anamorph *Fusarium*. Gibberellic acid belongs to a group of phytohormones that regulate stem elongation and seed germination, and find application (e.g.) in the induction of rapid barley germination during malting, and by improving fruit setting. Biotechnical production from *G. fujikuroi.*

Gloeosporium

→ Anamorph of *Pezicula* (Pezizales, Ascomycotina). Causes damage of stonefruit during storage (bitter rot). On apples especially *G. album* (*P. alba*) and *G. perennans* (*P. malicorticis*); the infected part dries out and hardens. None of these two spp. develops during → CA storage. *G. musarum* (presently *Colletotrichum musae*) causes crown rot of bananas; blueberries are spoiled by *G. fructigenum*. No information yet on possible mycotoxin production.

Glomerella

Sphaeriales, Ascomycetes; teleomorph of → *Colletotrichum*. *Gl. cingulata* causes "bitter rot" of stonefruit. Mummification of infected fruit; also referred to as "dry rot" (Müller 1983a).

GLP → good laboratory practice.

β-glucanases

Produced from various MO. Trivial name: laminarinase. Systematic name: 1,3(1,3; 1,4)-β-D-glucan 3(4)-glucanohydrolase. E.C. nos. 3.2.1.39 and 3.2.1.6. Catalyses endophydrolytic cleavage of 1,3- or 1,4- bonds in β-D-glucans. Applications: prevention of filtration problems, and of precipitation of β-D-glucans. MO used for commercial production: *A. niger, Trichoderma harzianum, B. subtilis* (GDCh 1983).

gluconic acid

Dextronic acid, maltonic acid, glycogenic acid. Organic acid widely distributed in nature. – → Glucoseoxydase is excreted by *Aspergillus* and *Penicillium* spp. into the substrate and transforms glucose in presence of dissolved O_2, via gluconolactone, into gluconic acid. The simultaneous formation of H_2O_2 may deleteriously influence food quality because of strong oxidative action (e.g., in delicatessen foods, mayonnaise, etc.). Oxygen-sensitive products may be protected against oxidative effect of H_2O_2 by addition of → catalase. Produced synthetically or from *A. niger*. Approval as food additive not necessary in all countries; in the EC preliminary catalogued as no. 574; application in foods to be regulated in future.

Gluconobacter

(*Acetomonas*). Genus of the Acetobacteraceae; Gram –, obligatory aerobic, non-motile or lophotrichously flagellated rods, catalase +. Oxidises → ethanol to → acetic acid, and glucose to → gluconic acid with the production of H_2O_2; → acetic acid bacteria.

glucose

Dextrose. Monosaccharide commonly found in nature; component of most polysaccharides, e.g., starch, cellulose, etc. Most important C and energy source for practically all MO; → metabolism. SO_4 reducing MO are not able to metabolise glucose.

glucose isomerase

D-xylose-ketolisomerase; xylose isomerase. E.C. 5.3.1.5. Produced from different bacteria, e.g., *Actinoplanes missouriensis, Arthrobacter* spp., *Bacillus coagulans, Streptomyces albus, S. olivaceus, S. olivochromogenes, S. rubiginosus, S. violaceoniger*. Converts D-xylose to D-xylulose and vice versa; D-glucose to D-fructose; induced in the presence of xylose. Application: production of fructose syrup reducing the glucose content; isomerisation of glucose (GDCh 1983).

glucose oxidase

Trivial names: glucose-aerohydrogenase, aero-glucosehydrogenase, glucose oxyhydrogenase, notatin, etc. Systematic name: β-D-glucose:oxygen 1-oxidoreductase. E.C. 1.1.3.4. Catalyses oxidation of glucosepyranose in presence of oxygen to D-glucono-δ-lactone and H_2O_2. Applications in food industry: for removal of glucose from foods (e.g., eggs) to prevent Maillard reactions during processing; for removing oxygen from beverages, etc., → gluconic acid. – MO for production: *A. niger, Penicillium* spp. (GDCh 1983).

α-glucosidase

Trivial names: maltase, glucosesucrase. Systematic name: α-1,4-D-glucosido glucohydrolase. E.C. 3.2.1.20. Catalyses hydrolysis of terminal non-reducing α-1,4- glucosides to α-glucose. Application: for saccharification of starches in combination with→amyloglucosidases; malt products. MO for production: *Saccharomyces* spp., *A. niger, A. oryzae, Rhizopus oryzae, Trichoderma reesei* (GDCh 1983).

β-glucosidase
1,3-β-D-glucan glucohydrolase. E.C. 3.2.1.58. Catalyses exohydrolytic cleavage of glucose units from non-reducing chain termina in 1,3-β-D-glucans. Produced from *Trichoderma harzianum*. Applications: in wine production to prevent precipitation of β-glucans causing filtration problems.

glycerol
Propanetriol. $CH_2OH.CHOH.CH_2OH$. Syrupy colourless liquid with sweet taste; hygroscopic. Alcoholic component of lipids, released by product specific or microbial→lipases. Formed also during fermentation of grape must, especially produced by wild yeasts and during development of noble rot (→*Botrytis cinerea*) on grapes. Glycerol content of superior wines may amount up to 30 g/l (Dittrich 1987).

glycolysis
Emden-Meyerhof-Parnas (EMP) pathway for glucose metabolism in cells of a wide range of organisms;→pyruvate and energy (2 molecules of ATP per molecule of glucose) yielded under both aerobic and anaerobic conditions.

GMP guidelines
Revised guidelines on "Good Manufacturing Practice" for the production of pharmaceuticals published by the WHO in 1975. GMP guidelines are applied with the aim of improving the microbiological quality and safety of the food supply; key manufacturing steps/conditions (pH, temp., a_w, application of strict raw material specifications, sampling plans, analytical methods, etc.) are stringently controlled. Approach towards application of→

HACCP concept rather than end product examination. The guidelines are related to all facets of food production, including staff, production rooms, criteria for hygiene, manufacture, testing, approval, storage, characterisation and documentation.

good laboratory practice
Principles for the examination of raw materials for the cultivation of→starter cultures and food manufactured with microbial cultures. Defined, e.g., by OECD. These principles are presently relevant to chemicals, pharmaceuticals and pesticides, but should also be made applicable to the food area.

Gouda
Refers to semi-hard cheese produced with→starter culture of *Lactococcus* spp. During maturing of ca. 5 weeks→diacetyl is formed.

grain storage
"Ripe" grains (→cereals) with < 13% moisture have a long storage life due to low respiration rate and the inability of moulds to grow. Mould and yeast growth may initiate from ≥ 16% moisture, and the→ respiration heat of the stored product and the MO may cause a strong increase in temperature. Sufficient cooling (< 14°C) and ventilation are therefore of vital importance. Spoilage of grains with moisture ≥ 13% may be prevented only for limited periods by application of fungistatics (propionic acid) or gassing with CO_2, N_2 or methylbromide. Insects may promote mould growth; xerotolerant moulds may germinate in the microclimate of their intestinal tract. Respiration moisture contributes to a favourable microenvironment in which abundant mould growth can occur.

GRAM staining
Bacterial staining procedure discovered

Table 20   Classification (grouping) of MO according to their ability to bring about changes in foods.

1. Metabolic end products	*Dematiaceae* (dark brown to black)
1.1 Acid producers	Moulds (all colours except blue)
1.1.1 Lactic acid (→homofermentative	1.6 Smell/flavour production
→heterofermentative)	*Brevibacterium linens*
coccus-shaped rod-shaped	*Pseudomonas fragi*
Streptococcus *Lactobacillus*	*Bacillus*
Leuconostoc *Microbacterium*	*Clostridium* (H_2S)
Pediococcus	several other bacteria and moulds
1.1.2 Acetic acid	causing spoilage
Acetobacter	1.7 Toxin production
Gluconobacter (Acetomonas)	*Staphylococcus aureus*
1.1.3 Butyric acid	*Bacillus cereus*
Clostridium	*Clostridium botulinum*
1.1.4 Propionic acid	*Claviceps purpurea* (ergot)
Propionibacterium	ca. 100 different mould spp.
1.2 Alcohols and ketones	some Cyanophyceae
heterofermentative LAB	some algae
Saccharomyces cerevisiae	(→paralytic shellfish poisoning)
(anaerobic) and several	
other yeast spp.;	2. Extracellular enzymes
several moulds at <0.1% O_2,	2.1 Proteolytes
ketones prod. by several moulds	*Bacillus*
aerobically	*Clostridium*
1.3 Gas production	*Pseudomonas*
Leuconostoc	*Proteus*
heterofermentative	*Rhizomucor miehei, R. pusillus*
Lactobacillus spp.	(rennet substitute)
Propionibacterium	several mould spp.
Escherichia coli	2.2 Lipolytes
Klebsiella	*Pseudomonas*
Proteus vulgaris	*Achromobacter*
Bacillus	*Alcaligenes*
Clostridium	*Serratia*
1.4 Slime production	*Micrococcus*
Alcaligenes viscolactis	*Penicillium roqueforti*
Streptococcus (some strains)	several other moulds
Lactobacillus (some strains)	2.3 Amylolytes
Leuconostoc mesenteroides	*Bacillus subtilis*
Bacillus subtilis (ropiness of bread)	*Clostridium butyricum*
Aureobasidium pullulans	*Aspergillus niger, A. oryzae*
1.5 Colour (pigment) production	seveal other moulds
Serratia marcescens (red)	2.4 Pectolytes
Brevibacterium linens (orange)	*Erwinia*
Flavobacterium (yellow)	*Bacillus*
Pseudomonas spp. (blue-green to	*Clostridium*
yellow-green)	*Aspergillus* spp.
Rhodotorula and other yeasts	*Trichoderma reesei*
(orange to red)	*Mucor* spp.

Table 20 (continued).

3.	Environmental dependence		Brevibacterium linens
3.1	Temperature		Halobacterium
3.1.1	Thermophiles (opt. >45 °C)		Sarcina
	Bacillus stearothermophilus		Micrococcus
	Clostridium thermosaccharolyticum		Pseudomonas
	Rhizomucor pusillus, R. miehei		Vibrio
	Talaromyces emersonii		Pediococcus
3.1.2	Psychrotrophs (near 0 °C up to opt.	3.3	Osmotolerance (>30% sucrose)
	bei 20 °C)		some strains of Leuconostoc
	Pseudomonas		Zygosaccharomyces rouxii
	Achromobacter		A. glaucus-group
	Flavobacterium		Xeromonas bisporus
	Alcaligenes		
	Micrococcus		
	Sarcina	4.	Food-borne infections
	Serratia		causative agents of zoonoses
	Lactobacillus		(→food poisoning)
	several moulds		Polio viruses
3.2	Salt tolerance (>7% table salt)		Hepatitis A viruses

in 1884 by the Danish physician Christian Gram, for differentiation of two main bacterial groups: Gram-positive (blue) (e.g., lactic acid bacteria) and Gram-negative (red) (e.g., salmonellae). After heat-fixing and consecutively staining with certain basic dyes (aniline or triphenylmethane dyes, e.g., crystal violet, gentian violet) and Lugol's iodine, Gram-negative cells are decolourised with organic solvents (ethanol, acetone) but not Gram+ cells. The colourless Gram-negative bacteria are then counterstained with carbolfuchsin or safranine. The mechanism of this phenomenon is not fully understood, but is related to the difference in cell wall structure and the higher murein (peptidoglycan) content of Gram-positive bacteria (30–50% of the DM as compared to ca. 10% for Gram – bacteria). Taxonomically valuable criterion in the differentiation of bacteria.

gray (Gy)→radiation treatment.

green mould
General (trivial) designation of→penicillia and other fungi producing green conidia.

green olives→olives.

green rot
General designation for fungal rots of stonefruit, berries, subtropical fruit; mainly caused by Penicillium spp. and other fungi with green conidia. May be indication of→patulin formation.

grey rot
Storage rot of fruit, vegetables and lettuce caused by→Botrytis cinerea; may cause considerable damage during storage and distribution. Its development on grapes before harvest→noble rot.

griseofulvin
Antimycotic produced by Penicillium griseofulvum. Inhibitory against actively growing dermatophytic fungi; adminis-

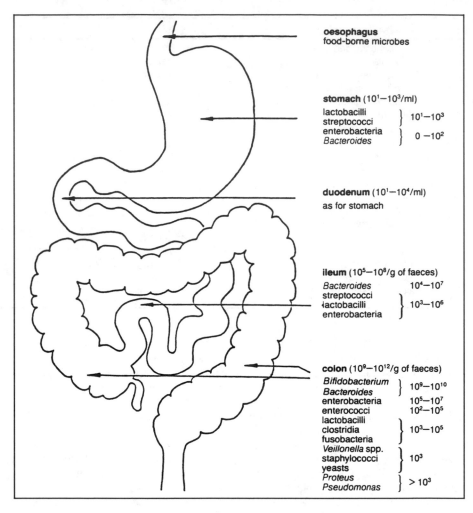

Figure 9 Scheme representing the colonisation of the intestinal tract of man and monogastric animals (source: Gedek, B.: *Aktuelle Themen der Tierernährung*. Cuxhaven: Lohmann, 1986).

tered orally and accumulates in keratin. Sometimes found in mouldy meat products (Reiss 1981). Interferes with pattern of glycoprotein and glucan deposition in the fungal cell wall, and causes curling of hyphae tips in low concentrations; lethal in higher concentrations.

grouping of microorganisms

Based on practical, food microbiological considerations, MO can be grouped in view of their role/importance in food processing and manufacture, transport and storage (in the negative and positive sense;→ sensory changes or spoilage) (see Table 20). Further criteria for grouping may include: metabolic end products, extracellular enzymes, temperature dependence, toxin production, colour (pigment), etc. It should be noted that several genera may be involved in different ways in bringing about changes in a commodity; → systematics.

growth

Expression used in general for the increase in biomass of individual organisms. Often incorrectly used in the case of bacteria – also in this book – when it concerns the → multiplication of single cells apart from the real growth, to reach the size of the mother cell after division. – In the case of fungi there can correctly be spoken of growth when implying either the increase of the colony's diameter after point inoculation under specific conditions (nutrient medium, temperature) or the mg DW/volume unit in liquid media. The → trophophase can be distinguished by the rapid assimilation of nutrients and formation of substrate mycelia at the initiation of colony formation, or at the edge of a growing colony. The → idiophase with the development of air mycelia and secondary metabolites (e.g., antibiotics, mycotoxins) follows. For colonies on foodstuffs or nutrient media in the laboratory, the central part is in the idiophase and the "young" growing colony edge in the trophophase.

gut infection → gastroenteritis.

gut microbes

Gastro-intestinal (GI) or "intestinal" microbes. Initially the GI tract of the neonate is free from any MO. Colonisation of the oral cavity and intestines within a few days. Gradual stabilisation of a "constant" population only after weaning, both qualitatively and quantitatively. In the jejunum (upper duodenum) the total population approximates 10^4/ml, in the ileum (lower duodenum) $10^5 - 10^8$/ml. In the colon the population may reach up to 10^{10}/ml, and up to 10^{11}/ml in the rectum. Mainly → anaerobic MO (Menge, H.: in Knothe et al. 1978). Qualitative distribution of the genera is shown in Figure 9. Important role of the gut MO in the activation and stabilisation of the immune system. Representatives of the oral and faecal MO are of importance as food contaminants. However, the value of → coliforms as → indicator organisms is doubted.

H

H milk→UHT milk.

H-milk products
Especially cocoa drink, but also some special products that are produced in the same way as→UHT milk. Strict criteria necessary for the ingredients; of special concern are heat resistant endospores. For the export of H-milk (UHT) products in tropical countries, internal process control is necessary by incubation of samples at 55 °C for ≥ 3 days in view of *Bacillus stearothermophilus*, before release of product for marketing (Teuser 1987).

H-milk system→aseptic system.

HACCP
"Hazard Analysis Critical Control Point" concept, representing a comprehensive systematic approach with the aim at minimising hazards (reducing the risk of harmful changes/effects) during a food processing operation.

haemolysis
The destruction or lysis of erythrocytes (red blood cells) by metabolites or excretions (e.g., nitrite) of pathogenic or potentially pathogenic bacteria. Haemolysins are excreted by the cells. Detection by, e.g.,→pour plate method on blood agar (rabbit, sheep, human). Distinguished between α-haemolysis resulting in a greenish opaque zone around a colony (e.g., by *Str. mitis* and strains of *Str. agalactiae*) and β-haemolysis characterised by a clear zone surrounding the colony (some strains of *Staph. aureus* and *Enterococcus faecalis*). Also observed by *Pseudomonas aeruginosa, Vibrio parahaemolyticus* and some strains of *E. coli*.

Hafnia
Genus of the→Enterobacteriaceae; typically found in the GI tract of animals, and also in soil and wastewater. May be of significance as→opportunists in→gastroenteritis. May be transmitted by foods (Krämer 1987; Sinell 1985). Destroyed by pasteurisation.

Halobacterium
Genus of Gram − bacteria (Halobacteriaceae), catalase-positive, cytochrome-oxidase-positive; pleomorphic rods or filaments; lophotrichously flagellated or non-motile. Dark yellow to red pigmented. Obligatory halophilic or osmophilic. Typically found in salt lakes (> 12% salt); may cause damage (spoil-age) of salted fish, intestines (natural casings) and in→brines. Carotinoids responsible for typical colour are fat soluble.

halogens
Disinfectants; antimicrobial agents by strong oxidative action on vital cell components; lethal damage (destructive action). Especially Cl and J have practical importance. Br as bromium water or as organic compound finds application in laboratory for the surface disinfection of seeds. Chlorine highly and non-selectively reactive with organic matter, by which its antimicrobial action may be decreased. Compounds releasing active chlorine ("available chlorine") find general application, e.g., for reduction of microbial populations and removal of offensive aromas in drinking water; decontamination of recycled bottles, utensils and instruments, etc., at opt. pH 6 − 8.→Disinfectants containing active Cl can easily be rinsed from surfaces, whilst chloride residues are physiologically harmless. Gaseous chlorine (Cl_2) is chemically highly aggressive and irritating to the skin and mucous membranes; corrosive towards chromium-nickel steel alloys and alu-

minium (MAK value 0.5 ppm = 1.5 mg/m^3 of air). All MO including endospores are lethally damaged; resistance not observed, although selection is possible at sublethal concentrations. Organic chloramines are less toxic and less irritating and can be used as antiseptics and disinfectants; slower release of chlorine. – Iodine is microbicidal against a wide range of MO. Application in medicine as tincture of iodine or Lugol's solution (antiseptic) for disinfection of skin surface. Iodophores find application in the food industry as disinfectants and detergents (surface active properties); complexes formed with organic matter; pH opt. 6 – 8. Lower toxicity than Cl, although→allergies (sensitivity) may be found [Kästner, W.: *Arch. Lm. Hyg.* 32 (1981) 97 – 140]. Corrosive at lower pH as for Cl. Protein acts as detoxicant. All MO including endospores are sensitive; resistance not observed (Wallhäusser 1988).

halophiles
Bacteria and some yeasts that grow optimally in presence of salt (NaCl) and other electrolytes. "Natural" MO of marine waters (2 – 3% salt), and of salt lakes and → brines. → Classification as "salt tolerant" MO.

Hanseniaspora
"Perfect" yeast genus; imperfect form *Kloeckera*. Cells lemon-shaped to oval. *H. uvarum* found in fruit juices and grape juice and possesses strong fermentative ability. *H. valbyensis* involved in spoilage of dried fruit when a$_w$ exceeds 0.72 (Kunz 1988).

Hansenula
Yeast genus of importance in the fermentation industry. Utilises nitrate and degrades starch. Type sp.: *H. anomala*. Belongs to typical surface microbial population on grapes. Involved in the manufacture of sherries and port wines because of high alcohol and osmotic tolerance. Moderate alcohol production of 4 – 5%. Several spp. are surface yeasts. Aerobic oxidation of ethanol and glycerol and utilisation of sugar; inhibition of *H.* as wine spoilage MO by exclusion of O$_2$ (Dittrich 1987). – *Hansenula* spp. are partly involved in the manufacture of→East Asiatic specialities, and play some role in the fermentation of→cocoa. May cause → blowing of fruit syrups and raw materials. – *H. anomala* is used for the production of→single cell protein from CH.

healthy unobjectionable
For the official approval of→food additives, → starter cultures containing genetically modifed MO, or enzymes produced with genetically modified MO, it is necessary that they are being proved safe for human consumption and "technologically beneficial" by the producer. → Toxicological testing. By means of extensive, long-term biological and biochemical test procedures it is to be assessed whether the utilisation of this substance or MO will induce any physical or behavioural changes in experimental animals. Any noticeable damage to organs or their functions are carefully monitored. By means of long-term investigations it is to be determined whether genetic or carcinogenic changes are effected. If no effects have been detected the daily dose will be increased up to a level at which a definitive relationship between dose and effect is observed. – Results from these toxicological tests are used to determine the daily no-adverse-effect-level. This will form the basis for estimating the life-long "acceptable daily intake" (→ ADI value) level.

heat resistance
Given for bacteria in terms of→D value or→F value relating to the respective spp. (see Table 29, page 193, for pasteurised milk). Information for moulds summarised

Table 21   Destruction of moulds in moist heat (modified acc. to Bröker, U.: Dissertation, Univ. Giessen, 19. 12. 1985).

Species	Temp. (°C)	Time (Min)	Occurrence
Alternaria solani (h)*	60	5	potatoes, vegetables
Aspergillus flavus (c)*	52	32	cereals, maize, nuts
	54	10	
Aspergillus fumigatus (c)	80	>60	soil
Aspergillus niger (h)	60	15	ubiquitary
Byssochlamys fulva (s)*	96	10	fruit juice, jams
Colletotrichum lini (h)	55	240	linseed, vegetables
Fusarium oxysporum (h, c)*	55	5	path. to domestic plants
Geotrichum candidum (h)	60	30	dairy prod., milk mach.
Neurospora sitophila (c)	72	5	bread, flour, cereals
Paecilomyces varioti (c)*	80	>60	vegetables, compost
Phialophora mustea (c)	85	90	fruit juice
Penicillium brevicaule (h)	60	5	soil
Penicillium citrinum (c)*	60	38	bread, flour, cereals
Penicillium expansum (c)*	60	58	fruit, vegetables
Penicillium roqueforti (c)*	58	33	household, air
Penicillium vermiculatum (c)	80	>60	soil, vegetables
Rhizopus nigricans (h)	60	5	ubiquitary
Rhizopus stolonifer (s)	100	10	canned fruit
Thermoascus aurantiacum (s)	88	>60	compost, air
Trichosporon variable (c)	90	>60	bread, flour, cereals
	100	>15	
Verticillium alboatrum (h, c)	47	10	path. to many plants
(sc)	47	50	

*Potential toxin producers.
(h) = hyphae; (c) = conidia; (s) = spores; (sc) = sclerotia.

in Table 21. – Numerical information can only be approximate values for bacteria and moulds, and depends on considerable fluctuations. Differences possible between strains of one sp., and a.o. determined by their previous history (composition of nutrient medium, incubation temperature, age of culture, spores, conidia, etc.) and the type of the substrate in which the thermal treatment occurs (CH, lipids, protein, NaCl, preservatives, pH, etc.) (Ayres et al. 1980; Krämer 1987; Mitscherlich and Marth 1984; Müller, G. 1983).

heat sensitivity of toxins
→ Bacillus cereus

→ biogenic amines
→ cyanobacteria
→ Cl. botulinum
→ toxic moulds
→ mycotoxins
→ Staphylococcus aureus

heat sterilisation
Commonly practised method for the destruction (killing) of all viable MO. Typically performed in closed systems, e.g., cans, → UHT systems. Products are being "cooked" simultaneously. Since the initial microbial population and the exact composition are not exactly known, the F_0 concept (F value) generally forms the basis for

calculating the efficiency in practice. The→core temperature of a product serves as decisive parameter of heat penetration during the heating period; it is reached more rapidly in motile systems (rotation autoclave, turbulent flowing liquids).

heavy metals
Metals with a specific gravity of >5; Pb, Fe, Cu, Zn, Ni, Cr, Wo, Mo, Cd, Co, U, V, Ag, Au and Pt, as well as semi-metals As, Sb and Te. They are found in extremely minute conc. in all foods and as such do not exert any inhibitory action against MO. Fe, Cu, Zn and Mo are required as trace elements for the growth or multiplication of MO.

"Hefeweizen"
Wheat beer brewed mainly with top-fermenting yeast.

Heliobacter pylori
Gram-negative slender curved rods, motile with 4 – 6 polar flagella, oxidase + , catalase + , urease + ; opt. growth temp. 37 °C. It is suggested that this organism is the causative agent of surface gastritis and peptic ulcers in the duodenum, and is probably transmitted by food, although no final proof of incriminated foods has been given yet [V. Wulffn, H.: *Immunität und Infektion* 16 (1988) 49 – 55].

Helminthosporium
→ Black moulds. Anamorph: *Cochliobolus* (Ascomycetes). Typically found on woody plants; often accompanied by *Alternaria* and *Cladosporium*. Opt. growth at ca. 22 °C and 90 – 100% rH (Rhodes 1979). Stress metabolites (→ phytoalexins) induced by *H. carbonum* on potatoes. – Involved in the degradation of plant materials; some spp. pathogenic to plants. Destroys materials. Produces → cytochalasins.

hemicellulase
Systematic name: 1,4-β-D-mannan mannanohydrolase. E.C. 3.2.1.78. Catalyses the reaction: hydrolytic cleavage of β-1,4-D-mannosidic bonds in mannans, galactomannans and glucomannans. Application: degradation of galactomannan and glucomannan for reducing the viscosity. – Industrial production mainly with moulds, e.g., *A. niger, A. oryzae, Rhizopus delemar, R. oryzae, Sporotrichum dimorphosporum, Trichoderma reesei;* and also *Bacillus subtilis* (GDCh 1983).

hemicelluloses
Non-cellulosic polysaccharides present in the matrix of plant cell walls. Include mannans, galactomannans and glucomannans; contain mannose, xylose, galactose and other sugars. Enzymatic degradation is often necessary prior to filtration. Extracellular degradation by numerous MO for utilisation.

hepatitis → virus hepatitis A.

heterofermentative lactic acid bacteria
→ Lactic acid bacteria fermenting sugars to a number of products (e.g., → acetic acid or ethanol + CO_2, or → acetoin) in addition to lactic acid. Phosphoketolase is a key enzyme in each of two possible pathways. Most common is probably the 6-phosphogluconate pathway of → *Leuconostoc* spp. and heterofermentative lactobacilli (see Table 23, page 154). In the bifidus pathway (→ *Bifidobacterium*) glucose is metabolised to lactic acid and acetic acid in a molar ratio of 2:3.

heterokaryosis
Condition in moulds when genetically different nuclei are present in the same cell or mycelium. Forms the basis of the ability for → adaptation of MO. Typically results from multinucleic conidia and/or → anas-

tomoses between individuals of the same sp. Heterokaryosis is responsible for the loss of specific properties (e.g., antibiotic production), during prolonged cultivation of mould strains in culture collections under maintenance of constant conditions (nutrient medium, temperature, etc.). May also explain the simultaneous occurrence of toxinogenic and non-toxinogenic strains of a sp. in its natural habitat, e.g., in soil (Müller and Loeffler 1982).

heterotrophy
The use of organic compounds for practically all carbon and energy requirements of an organism; all animals and numerous MO. Macromolecules are degraded extracellularly and the components transported into the cell (uptake). Most heterotrophic MO are→ saprophytes and few obligatory→ parasites.

hexamethylenetetramine
Hexamine; methenamine; urotropin. $C_6H_2N_4$. Condensation product of ammonia and formaldehyde; colourless crystals with sweetish taste. – Has long been used as preservative and for disinfection of urogenital tract. Slow cleavage at pH 6 – 8, more rapid at reduced pH values. Active against Gram + and Gram – bacteria, yeasts and moulds. Main applications: preservative for fish and delicatessen. – Because of possible health risk approval for use as food preservative withdrawn in several countries. Exception: prevention of→ late blowing in two Italian cheese types (Provolone, Grana Padano) and in real caviar (Classen et al. 1987).

high temperature processing
Process for continuous→ pasteurisation of milk at 85 – 90 °C for 2 – 4 s. At least 98 to >99% of the raw milk population is destroyed, including all pathogens (e.g., *Mycobacterium tuberculosis*) (Teuber 1987).

histamine
→ Biogenic amine, formed enzymatically by a number of bacterial spp. by decarboxylation of histidine. Histamine is associated with some cases of food poisoning, related to the consumption of certain seafoods (tuna).

HIV virus→ AIDS.

hole formation
Formation of gas holes (by CO_2 and H_2) during the ripening process of cheese. Abnormal gas formation or "gassiness" → early blowing,→ late blowing. More or less large round holes are caused by propionic acid bacteria in hard cheeses. Numerous pea-size holes in Edamer and Gouda are formed by gas-producing bacteria of the *Str.* and *Lb.* genera and they are also responsible for the aroma. For Tilsiter the curd is not pressed and the holes or slits are as large as barley seeds (Teuber, 1987).

homofermentative lactic acid bacteria
→ LAB that ferment glucose via the fructose-diphosphate pathway to either D(–), L(+) or DL lactic acid (see Tables 23 and 24, pages 154 and 155). Several representatives find application as→ starter cultures or for the biotechnical production of lactic acid.

homogenising equipment
Ultra Turrax, Waring Blender and mixing utensils used in the kitchen, with a sterilisable container. A preweighed sample, e.g., 10 g, is mechanically homogenised with 90 ml of→ diluting buffer at ca. 20,000 rpm for 1 min for the first dilution step. Sharp knives are especially important for the determination of→ mould counts (Baumgart 1990).

hospitalism
Refers to illnesses contracted during hos-

pitalisation (→ *Staphylococcus* poisoning; → penicillin). In this context especially referring to food-borne infections and intoxications resulting from management failures in the kitchen or hygienic failures in the tea kitchens. – In addition, mastitis and enterocolitis in maternity wards; → enteropathogenic *E. coli.*

hot filling
Practised for fruit juices. Following pasteurisation treatment at ca. 90 °C the hot juice is directly filled into non-sterile bottles; the vegetative microbes are killed by the combined effect of heat and the low pH. The caps should be sterile (Kunz 1988).

hot holding
Critical phase in the processing and preparation of foodstuffs in large kitchens, restaurants, canteens and the transport in→ thermophores. The period of time in which the unsterile food is kept at 15 – 65 °C should be short in order to minimise the reproduction of undesired organisms (Krämer 1987; Sinell 1985). Foods may be kept warm in, e.g., steam tables, a bain marie, hot-air cabinets or under infrared lamps (ICMSF 1980).

hot smoke → smoking.

HOWARD mould count
Microscopic counting of fungal (hyphal) fragments, especially in tomato products, etc.; in the minced/pulped raw materials and as end product control. Control of → GMP to assess whether mouldy raw materials (fruit, vegetables) have been used for processing. – Presently mainly used for internal processing control; unreliable as official control method (Beuchat 1987; Rhodes 1979).

humulon
Represents with lupulon the bitter tasting inhibitory substances extracted from the fruit flower of hops (*Humulus lupulus*). → Bacteriostatic to Gram + bacteria. Important taste component for→ beer. For→ wort agar only unhopped wort should be used because of inhibitory action of hops.

hurdle concept
Combination of different preservation parameters. May enable the reduction of one single measure, e.g., temperature for heat labile products (liquid egg, etc.), that allows some protection of the inherent product qualities. Every single factor (temperature, a_w, pH, Eh, preservative, salt, spices) can be considered to some extent as stress or inhibitory factor towards MO. Careful introduction of a combination of some of these factors in food processing may result in the optimal inhibition or destruction of MO, simultaneously keeping each "hurdle" at a minimal level and thereby preventing or reducing possible deleterious processing effects on the product (Krämer 1987; Sinell 1985).

hydrocarbons
Alkanes, paraffins, cyclic or aromatic hydrocarbons. Compounds containing only carbon and hydrogen. Bacteria present in soil and several yeasts can utilise the fossil hydrocarbons in petroleum, coal and earth gas under aerobic conditions. These processes are utilised for the production of→ single cell protein. Only a few species produce CH_4 as fermentation product (→ biogas) under anaerobic conditions. These species occur in swamps, rice fields, wastewater treatment plants and in the rumen of herbivores (Schlegel 1985).

hydrocolloids
Gums. Groups of products widely distributed in nature. Important food processing aids; provide gelling, suspending, emulsifying and stabilising properties. Such "bound" water, however, is also

available for microbial metabolism. Mainly → polysaccharides, but also→ proteins. Also provides protection against dehydration. – Several MO possess such "moisture regulatory substances", e.g.,→ capsules that surround the cell and may be the cause of slimy consistency of liquids or slime defects on meat surfaces. – Important hydrocolloids for microbiological laboratory:→ agar agar, and→ gelatin.

hydrogen peroxide

H_2O_2. Hydrogen superoxide. Commercially available as a 30 – 35% aqueous solution. Strong oxidising agent killing MO by inactivating their enzymes. Produced during metabolic activities of numerous aerobic and anaerobic bacteria and is immediately inactivated intracellularly by→ catalase present in aerobic and facultative anaerobic cells. Also produced in the enzymatic elimination of glucose by→ glucose oxidase. H_2O_2 is not officially approved as preservative in foodstuffs, since it may destroy several nutritionally valuable components. It is used in the→ sterilisation of→ packaging material in→ aseptic processing plants. MAK value 13 mg/m^3 air. Hydrogen peroxide is used as stabiliser of per-acetic acid in the formulation of→ disinfectants (Classen et al. 1987; Krämer 1987).

hydrogen sulphide

H_2S is produced by a number of bacteria during growth in presence of protein (SH groups) or thiosulphate. It is a poisonous gas, the smell of which reminds of rotten eggs, observable even in extremely low conc. (1 ng/m^3 of air). Rapidly reacts with Fe-III-ions with the production of black iron sulphide. This phenomenon is used in→ differential media, e.g., for the→ identification of→ Enterobacteriaceae or→ *Cl. perfringens*. "MAK" value 15 mg/m^3 of breathing air.

hygiene

Maintenance of food and plant hygiene is aimed in the first instance at consumer protection (→ food poisoning; minimal→ health risk). It also concerns measures for assuring the shelf life extension and prevention of spoilage. Spoiled products should not be distributed but do not necessarily imply a health risk; most spoilage MO can be considered as "harmless" with reference to human health. In addition, deleterious substances, residues and contaminations may however have hygienic consequences over longer periods of time (Berg et al. 1978; Classen et al. 1987; Sinell 1985).

hygiene regulations

Food acts or regulations dealing with hygiene; "hygienic practices"; sanitary standards. Differ from country to country; food manufacturer must comply with such legislation, typically dealing with health and cleanliness aspects of hygiene. Refers to specific product categories (meat products, or products from animal origin, bakery and confectionery products, drinking water, etc.); utensils, processing equipment and may generally refer to→ GMP, leaving details on fulfillment of the requirements in a plant to the processor. Codes of Practice are prepared, e.g., by the Federal Register Current Good Manufacturing Practice in manufacturing, packaging or holding human food (CFR 1987). In addition, the industry may also draw up codes or guidelines describing how good hygiene may be achieved in a plant (factory). Moreover, the UK Food Act of 1984 (HMSO 1984) states the responsibility of both employees and employers concerning "notifiable" or specified diseases. As in the USA (CFR 1987) employees with open cuts or sores are not permitted to handle foods in Britain (HMSO 1974).

hyphae

Filamentous (branched or unbranched) vegetative multinucleic organs typical of→fungi (moulds), but also of some bacteria (→Actinomycetales). Diameter varies from sp. to sp. between 5 and 70 μm. Tubular cell wall contains polysaccharides and chitin. The growth zone lies directly behind the distal end, at which point the branching also occurs, resulting in a dense→ mycelium. Most fungi possess septate hyphae (i.e., with a cross wall). The Zygomycetes are aseptate. Most fungi produce colourless hyphae; exception: the Dematiaceae (a group that is not precisely definable), however, deposit dark pigments, melanins, in the cell wall. These→"black fungi" have significance as spoilage microbes of stored fruit and vegetables, and also as→"wall moulds" and allergens. – Between hyphae of individual moulds cross connections,→anastomoses, may be formed that allow the exchange of nuclei, resulting in mycelia with varied genetic information (→heterokaryosis) (Müller and Loeffler 1982).

I

ice cream
Sensitive product prepared from milk, cream, sugar, egg, plant fats, whilst ingredients such as coffee, cocoa, aroma components, fruit products, etc., may be added as well. May easily spoil if not frozen. The basic mixture is pasteurised. Secondary contamination with *Staph. aureus, Salmonella*, etc., has been reported frequently and results from poor sanitary and hygienic practices. The hygienic regulations of a state or province are to be observed strictly (Krämer 1987). In D reclaims of defective ready packaged products are rare; ca. 4% for gastronomy products, and ca. 9% for retail producers. The portioning water, used during sale of ice cream, may contain from 10^3 to 10^8 MO/ml.

identification
Laboratory procedures to determine the family, genus or species relationship of a MO. The first step is to obtain a → pure culture, i.e., free from other MO. Subsequently, different characteristics are determined, e.g., the morphological form and cell arrangement, motility, capsule formation, sporulation, colony form, -colour, -surface, -edge, staining reaction (Gram + or −), enzyme secretion, O_2 requirements, nitrite formation, etc. To determine a number of characteristics, differential media are used and with the aid of specific identification keys the genus can be determined. The identification of a species requires the determination of extraordinary features and a collection of reference strains is required if the genus is not well described. Keys for bacteria by Baumgart (1990), p. 120, for yeasts, p. 159, and fungi, p. 179. Test systems designed for identification or differentiation in medical microbiology are only partially applicable to food microbiology since these organisms are often psychrotrophic or psychrotolerant and could be falsely identified.

idiophase
Fungal developmental stage following the trophophase characterised by the formation of aerial hyphae and reproduction organs responsible for the colony's species specific appearance. Secondary metabolites, e.g., antibiotics and mycotoxins are produced and excreted (Müller and Loeffler 1982).

imitations → substitutes.

immobilised cells
Microbial cells, dead or alive, contained in a reactor, physically or chemically attached either to each other or to a carrier, or they may be matrix encapsulated or enclosed in a membrane. Conventional methods are, e.g., beechwood shavings in the production of vinegar, and trickling filters for wastewater treatment where coke and slag are used as carriers of MO. Viable immobilised cells are investigated for application in the beverage industry. Dead immobilised cells are matted together by glutaraldehyde and thereby preserve the intracellular enzyme, e.g., *B. coagulans* in the production of fructose-containing syrups. The advantages are continuous treatment in a flow-through process and re-utilisation.

immobilised enzymes
Chemically bound to carrier material or physically encapsulated enzymes are used in the continuous production of, e.g., L-amino acids and milk with reduced lactose content. They must be protected against microbial activities and destruction by using, e.g., sterile production conditions, specific pH- and/or temperature conditions or by disinfection.

immunity

Antibodies in the blood are responsible for the acquired resistance against specific pathogenic agents. They are antimicrobial and present in fresh milk and eggs. They are rapidly inactivated by storage and do not affect shelf life.

immunofluorescence technique

Fluorescent antibody technique. Serological test with the aid of antibodies cojugated with a fluorochrome. Applied in food microbiology and medical microbiology, e.g., for rapid diagnosis of infectious disease (Albertus et al. 1986).

IMVEC tests

Abbreviation for 5 biochemical tests used for the→identification of faecal E. coli. I = indole production, 44°C, +; M = methyl red, +; V = Voges-Proskauer, −; E = Eijkman-test, lactose 37 and 44°C, +; C = citrate utilization, − (Baumgart 1990, p. 132).

IMVIC tests

Abbreviation for 4 biochemical tests used for the→differentiation between E. coli and Enterobacter aerogenes. I = indole production; M = methyl red; V = Voges-Proskauer test; C = citrate utilization (Baumgart 1990, p. 79).

in-process control

Monitoring of the microbiological condition of a product, from the raw material through the complete process up to packaging. Decisions on the sampling plan (site and time) should take into account the processing steps and the ecological requirements of undesired MO.→Process control.

in vitro

Latin. Process taking place in a test tube under laboratory conditions.

in vivo

Latin. Process taking place in a living organism. An inevitable procedure in toxicology for detection, e.g., of botulinum toxin.

incubation period

The time interval between exposure to a pathogen or toxin and the appearance of the first symptoms. It differs considerably between food-borne infections and intoxication (Table 22).

incubator

Standard equipment for microbiological laboratories; for the cultivation of MO at constant temperature, e.g., for use in the plate count determination. Insulated cabinets with temperature control, are equipped either with a heating device only, or with heating and cooling mechanism. The latter is especially important for maintaining a constant temperature around or below ambient (room) temperature. The former can only be used reliably if the adjusted temperature exceeds 5°C above ambient. Special airtight incubators may be filled with specific gases (e.g., N_2) or mixtures of it, for cultivation of anaerobic organisms.

index organism

Bacterial group or species that serves as indication of the success or failure of a certain production step, e.g.,→coliform or→Listeria must be absent in milk after pasteurisation. It is indicative of recontamination when found at a later stage in the production process.

indicator organism

All foodstuffs contain more or less characteristic MO. These MO are indicative of contamination that may constitute a health risk. Rapid and simple tests are used to determine these indicators. The determination of pathogens is difficult and tedious. Intestinal tract MO are suitable in-

Table 22 Incubation period and period of disease of food-borne intoxications and infections (Berg et al. 1978, supplement).

Microorganism	Incubation Period	Period of Illness
Staph. aureus (toxin in food)	2 – 6 h	1 – 3 d
Cl. botulinum (toxin in food)	1 – 3 d	6 – 8 months or death after 1 – 8 d
B. cereus (toxin in food)	1 – 12 h	½ – 1 d
Cl. perfringens (toxin production in the intestine)	8 – 24 h	12 – 24 h
Campylobacter jejuni (infection)	1 – 7 d	3 – 5 d
Enteropathogenic E. coli (infection)	2 – 6 d	1 – 3 d
Listeria monocytogenes (infection)	2 – 8 weeks	different
Salmonella spp. (infection)	6 – 40 h	1 – 7 d
S. typhi and S. paratyphi (infection)	1 – 3 weeks	3 – 5 weeks
Shigella spp. (infection)	12 h – 7 d	4 – 6 d
Vibrio parahaemolyticus (infection)	3 – 36 h	2 – 5 d
V. cholerae (infection)	2 – 5 d	5 – 7 d
Yersinia enterocolitica (infection)	3 – 7 d	different

dicators of → faecal contamination. Other entities with the same function are, e.g., aerobic organisms in shredded → salad mixes in plastic bags, yeast in yoghurt, cottage cheese, delicatessen salads, endospores in conserves, lactic acid bacteria in sausages, fungi and bacterial spores on vegetable packaging material (→ HOWARD mould count, → filth test). They are indicators of the quality of food processing conditions (→ GMP rules) or give an idea of the expected shelf life.

infant botulism
It had been accepted for several years that "sudden infant death syndrome" (SIDS) during the first year of life, and especially of infants between 1 and 6 months of age, was caused by the ingestion of → Clostridium botulinum. It was suggested that the endospores (ingested, e.g., with honey) germinate and continuously produce botulinum toxin in the gut [Arnon, S. S.: Ann. Rev. Med. 31 (1980) 541 – 560]. Recent observations do not support this hypothesis.

infant enteritis → enteropathogenic E. coli.

infant nutrition → official methods.

infection
Entry and colonisation of an infectious MO

in a living macroorganism (host). Disease does not necessarily develop, e.g. (→intestinal MO and→skin microbes), but a negative influence is also possible, e.g. (→salmonellosis, etc.). Bearing the definition in mind it is clear that non-living objects cannot be infected but→contaminated. The terminology is often misused.

infection routes → contamination routes.

infectivity
The capability of a MO to be transferred from one organism or carrier (food, water or utensils) to another, to be established in or on the host and to multiply or grow. Infectivity does not only depend on the MO but also on the host's state of immunity. The characteristic infectivity of a MO is one of the qualities that should not be transferred to another species through genetic modification.

inheritance characteristics→characteristics.

inhibition zone
Clear zones around microbial growth (e.g., on agar medium) resulting from inhibitory action of antimicrobial substances (e.g., antibiotics, → bacteriocins, → disinfectants, etc.). Samples with inhibitory substances examined by means of→ disc diffusion test, hole test, cylinder test, etc. Diameter is dependent on concentration, and can be used for quantification. No clear distinction between lethal and inhibitory action.

inhibitory disinfection methods
Processes which inhibit bacterial reproduction and/or growth. By contrast→destructive methods. These processes are physical, e.g., cooling or drying, or chemical, e.g., in order of efficiency: tensides, preservatives, selective microbistatics, chemotherapy (antibiotics and sulfon-

amide) [Mrozek, H.: *Arch. Lm. Hyg.* 31 (1980) 91 – 99].→Disinfectant.

inhibitory substances
Chemical compounds of different structure, produced either biologically or synthetically, that inhibit the growth, multiplication/proliferation or outgrowth (germination) of survival organs or microbial spores.→Bacteriostatics,→ fungistatics, → disinfection, → preservatives, → spices.

inoculation
A term often used in biotechnology where different inoculation systems are described and optimal inoculation concentrations are recommended for different purposes (Kunz 1980).

inoculation needle
A 3 cm long platinum-iridium wire inserted into a holder or fused to a glass rod used for inoculation of MO on or in a culture medium. A wire with a loop is more efficient for use in liquid media. Before use, the needle or loop must be annealed.

insects
Important vehicles for transfer of MO of all species→Arthropoda.

intermediate
Designation used occasionally when referring to intermediate products of cell metabolism.

intermediate moisture foods
(IMF). Foodstuffs with a water content between 15 and 50%, which are microbiologically stable and hygienically safe by means of specific measures. In traditional products of this nature, e.g., honey, fruit syrup, marmalade and jam, sweetened condensed milk, → dry and semi-dry sausages, etc., microbial development is inhibited due to a high osmotic pressure or brine content. New formulations for fruit cakes, animal feed and meat products are

developed by combining inhibitory factors (→ hurdle concept) (Classen et al. 1989).

intrinsic parameters
Inherent physico-chemical characteristics of food that influence microbiological growth and spoilage, e.g., → moisture content, → osmotic pressure, → pH, → oxidation reduction potential. They have a selective influence on the microorganisms. Suitable technological processes can extend the shelf life. By contrast → extrinsic parameters.

invert sugar
Immobilised invertase produced by fungi is used in the conversion of → sucrose to → glucose and fructose. The lowering of the a_w value of this mixture of liquid sugar is greater than that of sucrose and therefore increases the preservative effect (see Table 5, page 38, under a_w value).

invertase
Produced by different MO. Common names: saccharase, sucrase, invertin, β-D-fructosidase; systematic names: β-D-fructofuranoside fructohydrolase; E.C. 3.2.1.26. Catalysed reaction: β-D-fructofuranose produced by hydrolytic cleavage of β-fructofuranoside. Application: hydrolytic conversion of sucrose to fructose for the production of artificial honey, noncrystallising creams, liquid fondant centres, etc. MO used in the production are A. niger, Kluyveromyces fragilis, Saccharomyces carlsbergensis, S. cerevisiae, B. subtilis (GDCh 1983).

iodophores → halogen compounds.

ion exchanger
Apparatus for the removal of anions and cations from water, drinking water in particular. Deionised water is a substitute for distilled water in practise and particularly in the laboratory. It contains a large number of bacteria ($10^3 - 10^5$/ml) which colonise the resin and are only partially removed or killed in the recycling process. It can be used in the preparation of media only if it is sterilised shortly after the media have been weighed.

ionising radiation
High energy electromagnetic waves removing electrons from their orbits around the atomic nucleus, releasing ions, similarly to salt dissolving in water. Preservation of food by ionising radiation requires wavelengths of 10^3 and 10^{-1} nm with corresponding energies of 10^2 and 10^6 eV. MO are killed as it causes damage to the DNA of a living organism (→ D value) (Classen et al. 1987; Diehl 1989).

irradiation → radiation treatment.

Islandic toxin → mycotoxin
Produced by the growth of P. islandicum on food especially rice. It is hepatotoxic and carcinogenic. Chlorine containing cyclic peptide → Penicillium.

isoglucose → sugar.

isolate
Designation of a MO strain obtained from a mixed population in food for the strain collection or for differentiation by subculturing.

isolation
The separation of a MO from a mixed population to → identify or obtain a → pure culture. Usually the isolation is made from a plate used for plate count determination with the aid of an inoculation loop. By using a dilution streaking a new colony can be obtained that is clearly separated from the other colonies.

isomerases
A group of enzymes able to convert molecules intracellularly or when applied bio-

technologically, into their isomers, e.g., lactic isomerase produces DL-lactic acid in→lactic acid bacteria. Partial conversion of D-glucose into D-fructose leads to a mixture of "isoglucose" which is 10% sweeter than→sucrose (Bruchmann 1976; DFG 1987).

J

jams

Products made from fruit and sugar in a 1:1 ratio. In the presence of oxygen several fungal species may grow on the surface if the products contain approximately 60% sugar. Mycotoxins are not produced due to the high osmotic pressure. From 45% sugar (dietary jam) mycotoxins, especially→patulin may be excreted in the substrate.

jelly→jam.

"Jochpilze"→Zygomycetes.

K

"Karenzzeit" → waiting time.

Katadyn-filter candles
Filters used for the disinfection of drinking water, especially in warmer countries. Different sizes are available, ranging from small (1 l/min) to large technical filters (3000 l/min). The heterocapillary effect of ceramic is combined with the microbicidal, oligodynamic effect of metallic silver (Wallhäusser 1988).

Kauffmann-White classification
A scheme for the classification of the serotypes of the genus→ *Salmonella* according to their O and (where applicable) H and Vi antigens. Work of this kind can only be performed in specially equipped laboratories authorized by the local health authorities.

kefir
A caucasian fermented milk beverage initially produced from mare's milk. According to the "milk production regulation" kefir must be produced from pasteurised milk with specific cultures. The symbiotic organisms *Lb. kefir, Str. lactis, Leuconostoc* spp. and yeasts are embedded in kefiran to form granules which are used as inoculum. Apart from the microbial "community" kefiran (kefir grains) is composed of coagulated casein, and a polysaccharide from *Leuconostoc* spp. A wide range of yeasts includes a.o. *Candida kefir, Torulopsis* sp. and *Saccharomyces fragilis*. After a period of 10 – 20 h, the pH is reduced to 4.4 – 4.6 at a temperature of 22°C. The soft coagulate is stirred and the granules are separated. The slightly viscous liquid is marketed as drinking kefir and contains lactic acid 0.8 – 1.5%, ethanol 0.3 – 0.85% (up to 3%), CO_2, diacetyl, acetaldehyde and acetone. These products are not well

suited for marketing due to gas production and swelling or blowing of the container. A number of substitutes are produced with hardly any gas production and only 0.01 – 0.1% ethanol (Teuber 1987).

kieselguhr
Diatomaceous earth. Frustules of → diatoms. Material used in the manufacturing of → bacterial filters. Carrier material for silver chloride used in the clarification of wine in order to remove → H_2S and mercaptan produced by bacteria.

kiwi
The most important destructive pest of *Actinidia chinensis* after harvesting is *Botrytis cinerea*. The least damage is caused when the crop is harvested early enough and is ripened after the journey with → ethylene (Sommer, N. F.: in Kader 1985).

Klebsiella
Genus of the family Enterobacteriaceae. Non-motile, Gram – rods. Natural intestinal inhabitants. *Klebsiella pneumoniae* is pathogenic. *K. aerogenes* is responsible for → early blowing in cheese when the vat milk is contaminated (Baumgart 1990, p. 134).

Kligler's agar
Triple sugar iron (TSI) agar used as a → differential medium for Enterobacteriaceae. It contains lactose, sucrose, glucose, ferric ammonium citrate, sodium thiosulphate and phenol red. Comparable to Kligler's iron agar (KIA) or double sugar iron agar from which sucrose is omitted, and thus preventing the "masking" of lactose-negative bacteria. It turns yellow when acid is produced and red in an alkaline medium. When the thiosulphate is reduced to H_2S, it turns black, e.g., Merck no. 3915.

Koch's pour plate method

Cultural method used for→bacterial counts or to obtain a pure culture;→pour plate method (Baumgart 1990).

kojic acid

Metabolite of several *Aspergillus* spp., e.g., *A. oryzae*. It is a weak mutagen in the AMES test and has weak antibiotic activity. At present attempts are made to obtain kojic acid free species for the manufacturing of East Asian specialities (DFG 1987).

"Kölsch" →beer.

koumiss

Kumiss. An acidic, alcoholic fermented beverage from Asia usually made from mare's milk. Organisms involved are *Lb. delbrueckii* var. *bulgaricus* and *Torulopsis* spp. The ethanol content is 0.7 – 3.3%. As part of the greater demand for "bio" product imitations from cow's milk are occasionally offered in retail outlets (Krämer 1987).

KOVAC's indole reagent

Several bacteria produce indole by degradation of tryptophan which can be detected by a colour reaction. A pure culture is grown for 12 – 24 h in a culture medium. A reagent layer of ca. 0.5 cm, can be added and the reaction read after a few minutes. Indole dissolves in the butanol of the reagent and forms a cherry-red compound with the colourless 4-dimethyl-aminebenzaldehyde in acidic conditions. Tryptophan would have caused a colour reaction as well, but does not dissolve in butanol – e.g., Merck no. 9293. →IMVEC, →IMVIC,→coliforms.

Krebs cycle →tricarboxylic acid cycle.

L

laboratory safety

Pathogens are not primarily dealt with in a conventional food microbiology laboratory; however, they may have been developed as colonies among saprophytic and spoilage MO on or in nutrient media. Also, examinations for Enterobacteriaceae are regularly performed. Safety rules therefore have to be strictly observed by all personnel, even in routine laboratories. Basic rules are:
- The same protective clothing used in the laboratory should not be worn in the production plant (e.g., for sampling).
- Eating and drinking are strictly forbidden in the laboratory.
- All used glass utensils should be placed immediately in a container/basin with a disinfection solution, prior to transporting them to the washing kitchen.
- Inoculation needles are to be flamed before and after use.
- Cultures and dilutions are to be drawn only with stuffed pipettes, or pipetting aids used.
- Toxic reagents and dangerous (pathogenic) cultures should be marked clearly.
- The personnel is to be instructed regularly.
- The standard safety procedures (fire and accident) should also be observed in the microbiological laboratory, and should be practised from time to time (Baumgart 1990).

lactase

Enzyme produced by several MO; cleaves → lactose (milk sugar) into glucose and galactose → β-galactosidase.

lactate

Salt of → lactic acid.

lactic acid

CH_3-HCOH-COOH is a metabolic product of living cells formed under conditions of oxygen depletion (anaerobiosis). Two optically active stereoisomers are produced: L(+) form or meat lactic acid, and the D(−) form, as well as a racemic mixture (DL) (see Table 23). Both stereoisomers are determined quantitatively with an enzymatic method using D-lactate dehydrogenase and/or L-lactate dehydrogenase and NAD^+. – Lactic acid produced by fermentation acts as → preservative in the manufacture of → sauerkraut, → sour milk products, → cheese, → silage feed, etc. Lactic acid is metabolised preferentially by (a.o.) → surface yeasts, → milk moulds and several fungi resulting in possible spoilage of the products mentioned. Bacteria, especially Gram −, are sensitive to lactic acid. Concentrations required to kill some bacteria within 24 h at 22°C are as follows: 40 mg/l for *Staph. aureus*, 1000 mg/l for *E. coli*, 300 mg/l for *Pseudomonas*. – Neonates and weanlings up to 3 months of age should not receive D(−) lactic acid since the enzyme systems required for its degradation are available. – In Europe lactic acid is technically produced by fermentation with *Lb.* spp. (20,000 t/a). The greater part is used in the food industry and another part [the D(−) lactic acid] is used in the synthesis of plant treatment agents. According to German Food Regulations it is not listed as food additive. EC no. E 270, but does not need approval for use in food. Listed under FDA regulation §121.101 as GRAS, miscellaneous and/or general purpose food additive (FEMA no. 2611).

lactic acid bacteria

Gram +, catalase −, non-motile, micro-aerophilic to facultatively aerobic bacteria playing an important role in the manufac-

Table 23 Configuration of lactic acid produced by particular spp., genera and groups.

D(−)	L(+)	DL
Leuconostoc	Bifidobacterium	Lb. acidophilus
Lb. coryniformis ssp.	Carnobacterium	Lb. casei ssp.
torquens	Lb. alimentarius	pseudoplantarum
Lb. delbrueckii ssp.	Lb. amylophilus	Lb. coryniformis ssp.
bulgaricus	Lb. animalis	coryniformis
Lb. delbrueckii ssp.	Lb. bavaricus	Lb. curvatus
delbrueckii	Lb. casei ssp. casei	Lb. helveticus
Lb. delbrueckii ssp.	Lb. casei ssp.	Lb. plantarum
lactis	rhamnosus	Lb. sake
Lb. delbrueckii ssp.	Lb. casei ssp. tolerans	
leichmannii	Lb. ruminis	
Lb. homohiochii	Lb. salivarius	
Lb. jensenii	Lb. sharpeae	
Lb. vitulinus	Lb. xylosus	
	Lb. yamanashiensis	
	Listeria	
	Streptococcus	
	some yeasts	
	many moulds	

ture of fermented products from plants and animals; → starter culture, → sourdough, → sauerkraut, etc. L(+) or D(−) lactic acid is produced from pyruvate according to different metabolic pathways (see Figure 19, page 235). Some species possess lactate racemase and produce the optically inactive DL lactic acid mixture from the L(+) isomer (Table 23 → lactic acid). Lactic acid is the only end metabolite of many lactic acid bacteria (→ homofermentative). Other species also produce acetate and CO_2 in addition (Table 24 → heterofermentative). The genera Str., Lactococcus, Pediococcus and Leuconostoc are spherical and Lb. and Bifidobacterium are rod shaped.

lactic fermented milk products
"Sour milk" products. Produced typically from pasteurised (homogenised) milk with standardised fat content. These include cultured buttermilk, sour cream ("creme fraiche") (mesophilic → starter cultures), yoghurt (thermophilic cultures) and kefir, koumiss, Biogurt® and acidophilus milk (mesophilic or thermophilic special starter cultures) (Teuber 1987).

lactic fermented vegetables
Sour (acid) vegetables. Practically all vegetable types are suitable for lactic acid fermentation, especially green tomatoes, carrots, cauliflower and young maize cobs. Green beans should be → blanched prior to fermentation in order to inactivate the toxic haemagglutinins. The vegetables are cleaned thoroughly and peeled if necessary, and then mixed with herbs and spices. The added → brine contains sucrose and lactose to promote the rapid development of LAB; in addition → starter cultures, either single or (more commonly) multiple strains (Lb. plantarum, Lb. brevis,

Lb. bavaricus, Leuconostoc mesenteroides, Pediococcus damnosus, Str. durans, etc.) (Müller 1983a).

Lactobacillus

Genus of the family Lactobacillaceae (→ lactic acid bacteria) (LAB). Gram +, straight to irregular rods, often chain formation, in general non-motile (exceptions, e.g., *Lb. agilis, Lb. curvatus, Lb. plantarum* under certain cultural conditions and only some strains), facultatively anaerobic to microaerophilic, catalase –, nitrite – (exceptions: some strains of, e.g., *Lb. pentosus* and *Lb. plantarum*).→ Homofermentative and→heterofermentative (Bergey 1986). Killed by pasteurisation. – Several species are important in the food industry either as→ starter cultures or by causing deleterious effects (spoilage, production of→biogenic amines). Fastidious, but grow well in substrates such as milk, fruit juices, vegetable juices, seasonings, etc. Produce relatively large amounts of→lactic acid from different carbohydrates.

Lactobacillus acidophilus

Fresh cultures Gram +, old cultures Gram –. Homofermentative, DL lactic acid, microaerophilic, optimum 37 °C, minimum 23 °C, maximum 43 – 48 °C. Habitat: the intestine and faeces of infants and commonly of adults. Component of starter cultures for different kinds of sour milk and often present in kefir cultures.

Lactobacillus bavaricus

Slightly curved rods, single or in chains. Homofermentative, L(+) lactic acid. Growth at 2 – 37 °C. Habitat:→ "sauerkraut" and fermented vegetables. Used as starter culture.

Lactobacillus casei ssp. casei

(*Lactobacillus casei*). Microaerophilic, coagulates milk slowly (3 – 5 d) and produces more or less 1.5% L(+) lactic acid homofermentatively. Opt. 30 °C, min. 10 °C, max. 40 – 45 °C. Habitat: milk, cheese. Component of starter cultures for some cheeses.

Lactobacillus delbrueckii ssp. bulgaricus

(*Lb. bulgaricus*). Aerobic to anaerobic, homofermentative, D(–) lactic acid. Optimum temperature 40 – 50 °C, min. 22 °C, max. 62 °C. Symbiotic with *Str. salivarius* ssp. *thermophilus* in yoghurt

Table 24 Genera of lactic acid bacteria, their type of fermentation and fermentation products [acc. to Kandler, O.: Antonie van Leeuwenhoek 49 (1983) 209 – 224].

Genus (Subgenus)	Fermentation Type	Main Metabolic Product	Configuration
Streptococcus	A	lactate	L(+)
Pediococcus	A	lactate	DL, L(+)
Lactobacillus	A	lactate	
Thermobacterium	A	lactate	D(–), L(+), DL
Streptobacterium	A	lactate	D(–), L(+), DL
	B[1]	lactate:acetate = 1:1	D(–), L(+), DL
Betabacterium	B	lactate:acetate:CO$_2$ = 1:1:1	DL
Leuconostoc	B	lactate:acetate:CO$_2$ = 1:1:1	D(–)
Bifidobacterium	B	lactate:acetate = 2:3	L(+)

[1]For pentose-fermentation; A = homofermentative; B = heterofermentative.

and also a component of some starter cultures for cheese.

Lactobacillus delbrueckii ssp. lactis

(*Lb. lactis*). Facultatively anaerobic, homofermentative, 1.7% D(−) lactic acid. Optimum temperature 40°C, min. 18 – 22°C, max. 50°C. Present in starter cultures for several cheeses.

Lactobacillus helveticus

Microaerophilic, homofermentative, DL lactic acid. Opt. 40 – 42°C, min. 20 – 22°C, max. 50°C. Widely used in milk products. Component in starter cultures for some hard cheeses.

Lactobacillus plantarum

Microaerophilic, homofermentative (facultatively heterofermentative), DL lactic acid, acetic acid and CO_2. Opt. 30°C, min. 10°C, max. 40 – 45°C. Multiplies in media with a NaCl-concentration of up to 5.5%. Habitat: on plants as part of the normal surface population and in soil. Important in the production of sour fermented vegetables, → "sauerkraut", → pickles, etc.

Lactococcus

Lactic streptococci of the → Lancefield group N. → *Streptococcus cremoris*, → *Str. lactis*, → synonym (Bergey 1986).

lactophenol blue solution

A stain and dye used for the microscopical investigation of bacteria and fungi; → "Zupfpräparat". 20 g phenol, 20 g lactic acid, 40 ml glycerine, 20 ml water and 0.05 g methylene blue. The protoplasm is stained and a colour contour develops between the solution and fungal fragments due to refraction differences. The preparations can be kept for approximately 14 days.

lactose

Milk sugar. Disaccharide from galactose and glucose in the α- and β-form. 4 – 6% in cow's milk, 4.3 – 9.5% in mother's milk, 4.3 – 5.2% in sheep's milk and 4.5 – 4.8% in goat's milk. Hydrolysed by → β-1,4-galactosidase into both monosaccharides; important for lactose-intolerant people. A large number of MO utilise lactose (DFG 1987).

lag phase → multiplication.

lager yeasts

Yeast strains that settle on the bottom of the fermenter during beer brewing (sedimentation); typical of lager type beers → beer yeasts.

Lancefield groups

Classification of the → Streptococci in serological groups. The grouping determined by the surface CH of the cell wall (Sinell 1985). Most spp. important for the milk industry belong to group N.

late blowing

Defects of hard and semi-hard cheeses, occurring towards the end of the ripening period, and characterised by gas production (CO_2 + H_2) caused by *Cl. tyrobutyricum*. The endospores are especially found in → silage, and may contaminate raw milk in the barn; they survive pasteurisation. Cows fed on silage excrete up to 10^7 endospores/g of faeces. Ca. 200 ml of milk are sufficient to cause late blowing. In areas where milk is produced for hard cheese manufacture, silage should not be used as feed. – Up to 20 g nitrate/100 l is added as precaution against this defect. During ripening the nitrate is reduced to → nitrite which prevents the germination and thus the growth and gas production of these clostridia (Teuber 1987).

late fermentation

Feared defect in Emmentaler and similar

cheeses. Enterococci (*Str. durans, Enterococcus faecium*) and/or other propionic acid bacteria, that may be present as a result of high counts in raw milk and unhygienic brine baths, are responsible for undesirable cracks due to additional CO_2 production, subsequent to formation of desired holes. The taste is hardly affected, but the price can be (Teuber 1987).

latent phase→multiplication.

LD$_{50}$
Lethal dose of a given toxic (poisonous) agent which, when administered to experimental animals, kills 50% within a period of 7 days.→*Aspergillus*, →penicillin, →*Fusarium*.

leakage
Containers or cans not hermetically sealed, leading to spoilage of the contents. The ends may become distended due to internal pressure by gas production (blower or blown can). The causes are: microscopically small channels in the seam or edge of the tin or aluminium cans; incorrectly closed lid edges (seams) due to too fast closing procedures (6000 units/min); contaminated contact areas in flexible packaging material. A negative pressure or vacuum develops when the rigid container cools down and the cooling water (with bacteria) or air may be drawn in (Heiss and Eichner 1984).

leavening agents
"Lockerungsmittel" (Germ.). Gas production in bread and other baked products promotes a loose texture, e.g., by→sourdough and/or→baker's yeast (Spicher and Stephan 1987).

lemons→citrus fruit.

Lentinus→cultivated mushrooms.

Leptospira
Genus of Spirochaetaceae. Parasitic infection of small mammals, e.g., guinea pigs, hamsters. The parasites are then spread to water or foodstuffs where they cannot reproduce, but can be transferred to man. The incubation period is 7 – 10 days. Notifiable disease (Krämer 1987).

lethal doses→LD$_{50}$.

Leuconostoc
(*Betacoccus*). Fam. Streptococcaceae. Gram +, non-motile, facultatively anaerobic, cocci in pairs or chains, catalase –, nitrite –, indole –.→Heterofermentative lactic acid bacteria able to produce D(–) lactic acid, ethanol (acetic acid) and CO_2. Several species produce→dextran. Complex media containing vitamins and amino acids are necessary for cultivation.

Leuconostoc citrovorum
Presently classified as *Lc. lactis*. Component of starter cultures for cultured creamery butter. Produces 1.5 – 2 mg/kg diacetyl during moulding and packaging. During the next 7 days these values may increase up to 2.2 – 2.5 mg/kg resulting in the development of the full aroma (Teuber 1987).

Leuconostoc mesenteroides ssp. cremoris
(*L. cremoris*). Component of the starter culture used in the production of cultured cream, cheese, kefir, butter milk and cottage cheese. It produces→diacetyl from acetaldehyde (Teuber 1987).

Leuconostoc oenos
Occurs on wine leaves and grapes. It is also used in the production of wine and is responsible for the bacterial conversion of L-malic acid to lactic acid and CO_2 by the malolactic fermentation. Malic acid is a strong acid in comparison to the weaker

lactic acid (Dittrich 1987). The addition of *Lc. oenos* to wine has not yet been legally approved throughout the EC (DFG 1987).

levan
Polysaccharide consisting of fructose; found in the→capsules of several streptococci, *B. mesentericus* and other bacteria. It causes filtration difficulties in the fruit juice industry.

levulose
Old term for→fructose.

limulus test
LAL test. A rapid method (1 – 2 h) for the detection of viable or dead Gram – bacteria, especially in the manufacture of→ UHT milk. The→lipopolysaccharides in the cell walls not destroyed by the heating process react with the blood cells of the horseshoe crab, *Limulus polyphemus,* with gel formation. Also used to detect microbial contamination of raw materials and intermediate products in the food industry or pharmaceutical industry for the absence of pyrogens in injection solutions. Commercial test kits are available, e.g., from Biomerieux, Malinckrodt, Diagnostica and Concepts Gmbh (Baumgart 1990).

lipase
Extracted from pancreas or produced with fungi or yeasts. Common name: triacylglycerollipase, steapsin, tibutyrase, triglyceride-dilipase. Systematic name: triacylglycerol acylhydrolase; E.C. 3.1.1.3. Catalytic reaction: hydrolysis of triglycerides into diglycerides and fatty acids. Applications: to hydrolise fat, e.g., to obtain the desired aroma in certain cheeses and for defatting egg albumen. MO used in the production of lipase are: *A. flavus*→aflatoxins, *A. niger, A. oryzae, Mucor javanicus, Rhizomucor pussilus, Rh. arrhizus, Rhizopus delemar, R. nigricans, R.*

niveus, Candida lipolytica (GDCh 1983).

lipids
→Fats and oils. Occurs in the cells of prokaryotes and eukaryotes. Reservoirs of carbon or energy. It can be the cause of rancid spoilage in→single cell protein,→ biomass, → compressed yeast, → algae powder, etc.

lipoid
Fat like cell components which are→hydrophobic as are the→lipids, waxes, phospholipids, fat-soluble pigments such as carotenoids and chlorophylls and steroids (e.g.,→ergosterol).

lipolytic organisms
A collective name for MO able to split fats into glycerol and fatty acids with the aid of excreted→lipases. Classification/→identification. *P. roqueforti* is the most undesired spoilage organism, but is important for the development of aroma in blue veined cheese. – Detection: bacteria can be grown in surface culture (→surface plating) on nutrient agar containing an emulsion of tributyrin, or on meat extract-yeast extract-peptone-tributyrin agar at 20 °C. Clear zones develop around colonies with lipolytic activity. The International Dairy Federation standardised the method [Milchwiss: 23 (1986) 298 – 299]. – Fungi are grown at opt. temp. in shake culture. The cell free filtrate is poured into a test tube containing tributyrin buffer agar at pH 7. The depth of the clear zone underneath the filtrate is measured after 24 h.

lipopolysaccharide
Heat resistant components of the cell wall of Gram – bacteria. They are species specific in structure and form part of the surface- or O-antigens, and together with the H-antigens they form the basis, e.g., for →*Salmonella,* for the serological identification of the serotypes. They are

identical to the endotoxins released when Gram – cells die, and are able to cause diarrhoea, fever and hypotension even in minute concentrations. Detection with the→limulus test.

liquid culture
Growth, e.g., of bacteria in liquid media (broth):→enrichment,→coli titre,→BRILA broth,→nitrate broth, etc.

liquid smoke→ smoke aroma.

Listeria monocytogenes
Gram +, peritrichously flagellated rods, psychrotrophic, produces acid pH 5 [L(+) lactic acid]. Occurrence: soil, animal food, organs and on the skins of infected pets. Food on which *L. m.* was detected thus far in order of decreasing occurrence: minced meat, dry and semi-dry sausages, cured ham, sliced cold meats, mixed salads, prepared (ready-to-eat) salads, seafoods, soft cheeses, milk, ice cream, creams and cakes. Transfer to humans by means of foodstuffs; may cause→listeriosis. – Growth occurs at temperatures >8°C in insufficiently fermented (acidified) silage at pH > 5.6, in mixed salads, milk and soft cheeses near the skin or surface. Survives in natural substrates up to 6 a; 22% NaCl up to 46 days (surface liquid for soft cheeses); pH 4.8 at 30 °C for 14 days, at 5°C for 49 days. Thermal death time during pasteurisation < 10 seconds (Mitscherlich and Marth 1985). The detection is difficult and unsatisfactory. Rapid test: *Listeria*-selective agar basis +, *L.* selective supplement from Oxoid. Permission required for the test (BSG). More satisfactory selective detection with PALCAM *Listeria* Selective Agar Base acc. to Van Netten et al. (Merck no.11755).

listeriosis
Rarely occurring disease (3 – 10 persons/million/annum), but the mortality rate

is up to 30%. Humans are affected by→ *Listeria monocytogenes* serotype 4. Neonates are infected transplacentally and the consequences are stillbirths or prematurity or intrapartum due to the mother's vaginal MO. Healthy children and adults infected by contaminated food develop angina-like symptoms. Immunosuppressed individuals (chemotherapy, AIDS) develop encephalitis, meningitis and other severe symptoms. Although rare there is an increasing tendency in recent years [Lang, B. et al.: *Immun. Infekt.* 15 (1987) 175].

liver broth
Used as enrichment medium for *Cl.* and other anaerobes obtained from meat, etc. The reducing substances present in the liver create an anaerobic environment. – Small particles of sample to be investigated are added into the hot liver broth (ca. 80°C), which is then covered with a layer of paraffin and incubated for 24 h at 37°C. The formation of gas is an indication of *Cl.* Dehydrated media from Merck no. 5464, Oxoid CM 77, etc.

log phase→ multiplication.

Long milk ("Langmilch")→ sour milk.

longterm feeding experiments
Experiments used for testing chronic toxicity are carried out on animals for at least 200 d or optimally until natural death occurs. Important especially for microbial toxins since the limit for acute toxicity is rarely reached, but, on the other hand, a longterm subacute exposure in man cannot be excluded.

Lugol's solution
Aqueous solution of iodine and potassium iodide used in→GRAM staining: 1 g Iodine and 2 g potassium iodide are dis-

solved in 100 ml of distilled water. Named after the French physician J. G. A. Lugol.

luminescent bacteria

Photobacterium and *Lucibacterium* spp. (fam. Vibrionaceae). Aerobic bacteria occurring in ocean waters and the slime layer of fishes. During storage and transport colonies are formed and a green light is radiated. The intensity of the bioluminescence of *Photobacterium phosphoreum* in 2% NaCl solution is employed in toxicity experiments for assessing the quality of waste- and drinking water.

Lupulon

Inhibitory substance; bitter tasting→phytoncide isolated from the fruit stalks of hops (*Humulus lupulus*).→Humulon.

luteoskyrin

A yellow fat-soluble dianthraquinone produced by *P. islandicumon,* e.g., rice. Carcinogenic hepatotoxin.→*Penicillium,* →

mycotoxin.

lyophilisation→freeze-drying.

lysozyme

Extracted from egg albumen. Trivial name: muramidase, mucopeptide glucohydrolase. Systematic name: mucopeptide N-acetylmuramyl hydrolase. E.C. 3.2.1.17. Catalytic reaction: hydrolysis of glycosidic bonds between the N-acetyl-muramic acid and N-acetylglucosamine in mucopolysaccharides. Use:→lysozymes (GDCh 1983).

lysozymes

Enzymes present in tears, saliva and egg-white (albumen). It cleaves the mucopolysaccharides in the cell wall of bacteria and "dissolves" the bacteria. Several MO produce lysozymes. The approved addition of lysozyme to→cheese milk in the preparation of hard cheese, can prevent the germination of spores of *Cl. tyrobutyricum* (Teuber 1987).

M

MacConkey medium

A→selective medium combined with a→differential medium for the detection of→coliforms and other enterobacteria. Moderate reliability for diluted food samples or wastewater, and satisfactory results for→seasonings and→pitching yeasts (with actidione). Gram + organisms are inhibited by the addition of bile and crystal violet. Lactose organisms are inhibited since lactose is the only source of CH. The pH indicator is neutral red. Commercially available as, e.g., Difco no. 0075 or Merck no. 5465. Broth: Difco no. 0020 or Merck no. 5396 (Baumgart 1987).

machinery mould

→ Geotrichum candida grows well in insufficiently cleaned or disinfected machines, on conveyor belts and in pipelines over weekends, in milk, fruit and vegetable processing industries. Special attention is required for prevention and control (Beuchat 1987).

machines

The main problem of processing equipment in the food industry is to reach the different parts for cleansing and disinfection. They should be corrosion proof so as to prevent the formation of niches where MO are protected against disinfectants. Other similar niches are seals, taps and valves. The machines are to be inspected and controlled (→step control) according to the specific situation, the type of industry and application. Additional tests should be carried out after longer interruption of processing operations, e.g., week-ends, in order to prevent massive contamination. →Machinery mould (Krämer 1987).

mackerel poisoning

Mackerels and tuna contain relatively high amounts of histidine (→amino acid), which can be transformed into→histamine by microbiological→decarboxylation, especially under insufficient chilling. Histamine may cause symptomatic poisoning, like other→biogenic amines (Krämer 1987; Sinell 1985).

Madeira

Wine from southern Europe with a relative high alcohol content of 17–20%. Fermentation is stopped by the addition of 76% wine alcohol (so-called fortification or avination) (Dittrich 1987).

Maillard reaction

Non-enzymatic browning reaction between sugar and amino compounds occurring especially during heat processing, with the formation of melanoidins (brown N-containing polymers). Occurs in flour and bread crust, or during drying of liquid egg (Heiss and Eichner 1984).

Malaga→ Madeira.

malates

Salt of→malic acid.

malic acid

(Monohydroxysuccinic acid). HOOC-HCOH-CH$_2$-COOH. Formed in the tricarboxylic acid cycle (Krebs cycle) from fumaric acid to oxaloacetate. Present as L(+) malic acid in varying amounts in all fruits and vegetables. The malic acid content of grapes is strongly influenced by the climate, especially in the northern part of Europe and the USA. As dicarboxylic acid it acts sensorically as "strong" acid, and its decarboxylation in wine to the milder lactic acid by→lactic acid bacteria, e.g., Leuconostoc oenos, is considered as desirable for many types of wine. Malic acid is converted by the malolactic enzyme to L(+) lactic and CO$_2$ during this "second fermentation" which takes place at some stage after the yeast fermentation of the wine. *L. oenos* is not yet approved as→ starter culture in the EC (Dittrich 1987). – Malic acid is listed by the FDA as "GRAS" substance and does not require special approval for use as general purpose food additive (Gardner 1977). In the EC it is categorised by the preliminary number 296. The commercial synthetic product contains a mixture of the L(+) and D(–) isomers, and is manufactured by the catalytic hydration of maleic and fumaric acids. Biotechnically it is produced as the L(+) isomer by MO such as *Brevibacterium* sp., *Corynebacterium* sp., *Propionibacterium brevicompactum;* or it is produced from fumaric acid with *Aspergillus* spp. (DFG 1987). – The salts are called malates.

malt extract
Syrup extracted from barley malt rich in maltose and other organic substances. Suitable to culture and detect yeasts and fungi. For counting, the agar should be treated with 10% lactic acid to obtain a pH of 4.5 or with 20% tartaric acid to obtain a pH of 3.5 subsequently to autoclaving for 10 min at 121°C, in order to inhibit the growth of bacteria (Baumgart 1990). Malt extract agar is commercially available, e.g., Merck no. 5397 or Oxoid CM 57. – In addition to Czapek-Dox and Sabouraud agar, malt extract agar may be used to detect characteristic differences in colour, surface and growth for the identification of fungi (Cerny and Hoffmann 1987, "Pilz-Monographien").

Malta fever
Human brucellosis caused by→ *Brucella melitensis* present in unpasteurised milk and milk products of sheep and goats.

maltose
Malt sugar. Disaccharide from plants, in starch and honey. Main sugar in malt for beer manufacture. Utilised by several bacteria, yeasts and fungi.

mango
Anthracnosis (*Colletotrichum gloeosporioides*) is the most frequent post-harvest disease of *Mangifera indica*. It occurs more frequently in humid than in semi-arid regions. The infection manifests when the fruit ripens, but it is not observed before harvesting. Treatment of the unharvested fruit with copper salts is an important preventative measure. Spoilage on the stalk site is caused by *Diplodia natalensis* and/or *Botryodiplodia theobromae*. Treatment with hot water as in the case of papaya is effective and even better if combined with fungicides. The allowable residue of fungicides in the different importing countries should be borne in mind (Sommer, N. F.: in Kader 1985).

manipulated MO→ genetically modified organisms.

mannitol
Sugar alcohol (6C) with a sweet taste and used by diabetics as a substitute for sugar. Utilised by enteric MO and beer yeasts. Occurs in the thickened syrup from *Fraxinus ornus* and in several other plants.

mannose
An aldohexose widely distributed in the plant kingdom and which can be utilised by a large number of MO.

marasmins
Wilting toxins produced by plant pathogens, e.g., *Fusarium* species. Causes wilting and pre-aging of the cultured plant. Lycomarasmin occurs on tomatoes. It is harmless to man but has economic implications.

margarine
Produced from plant fats and/or hardened plant oils; used for cooking or as spread.

Spoilage is caused by *Cladosporium bu-tyri* and *Candida lipolytica* by rancidity or formation of methylketone. Coconut oil as raw product is spoiled by *Margarinomyces bubaki* (Beuchat 1987).

marinades
Mostly semi-conserves or→preserved food of which the core temperature has not exceeded 65 – 75 °C due to heat sensitivity of the product. Expected shelf life of ~ 6 months at < 5 °C. Marinated (sour) fish products with→preservatives (Krämer 1987).

marker organisms
MO which may be indicative of the presence of pathogenic organisms of a particular ecological group, e.g., faecal organisms.→Indicator organisms, e.g., enterobacteria, coliforms or *E. coli*. They are only of value if the examiner knows the technological and ecological characteristics of the product so as to exclude the possible multiplication of the MO during the production procedures. Examples: certified milk or ice cream with counts of > 100/ml enterobacteria or > 1000/g *E. coli* in tartar are alarming signs. Of little significance in soft cheese (Busse, M.: in SGLH 1984).

marzipan
Product made from minced peeled almonds and sugar in a proportion 1:1. It can be spoiled by osmophilic yeasts, e.g., *Zygosaccharomyces rouxii*. Gas production and/or colour changes can be due to→black fungi (moulds).

mash
(1) Brewery: non-microbiological process in which dried malt is ground and mixed with water to produce the→wort. (2) Distilleries: a suspension of the raw material (e.g., corn, potatoes, fruit) with malt or →amylases, in which an alcoholic fermentation takes place, often after addition

of starter yeasts. (3) Wine-production: direct fermentation of pulp of red grapes. The solids are removed by pressing after allowing different contact periods (Dittrich 1987; Kunz 1988).

mastitis
Inflammation of the udder caused by *Streptococcus agalactiae, Staph. aureus,* other bacteria and yeasts; treatment with antibiotics. Milk should not be supplied to dairies for several days (→waiting time) following therapeutic treatment, because of possible consequences for consumer's health or probable inhibition of starter organisms in fermentation processes. Widely distributed mainly because of defective adjustment or insufficient disinfection of milking machines.

maximum tolerable dose (MTD)→
NOEL.

maximum tolerable numbers
Indicate a level where either spoilage is immediately likely (e.g., ca. 10^7/g or ml) or the exceeding of which may constitute a severe risk to the consumer. Indicated by M in the three-class sampling plan, with m as the "level of no concern". Normally, i.e., under GMP, the majority (if not all) of the samples will have a microbial load between M and m. By this criterion the microbiological status of a product or group of products can be assessed, e.g., dried spices. When the value is reached or exceeded a formal complaint or intervention can be launched according to the particular case and situation. No automatic complaints according to §17 (part. 1) LMBG.

mayonnaise
Emulsion of egg yolk, vinegar and oil with a pH value of 4.5 – 3.4. Easily spoiled, especially by yeasts from the genera, e.g., *Saccharomyces, Candida, Zygosaccharomyces, Debaryomyces,* etc. Salads and

other delicatessen containing mayonnaise but no preservatives are highly sensitive (Krämer 1987).

meat

Musculature of meat animals, mainly "red meat" mammals (e.g., cattle, sheep, pigs, etc.). High spoilage potential because of water (a_w ca. 0.99) and rich nutrient content. During slaughtering, evisceration and dressing (removing of the skin) contamination cannot be completely prevented; subsequently the carcass is chilled under varying conditions, in view of the desired tenderness, till *rigor mortis*, and a core temperature of $4-7°C$ is reached. Fast chilling at low temperatures and high air speed may reduce bacterial numbers, especially at low humidities, but may also result in tough meat. Meat tenderness is improved by initial chilling temperatures around $15-16°C$ after slaughtering (e.g., $12-24$ h for beef, $4-12$ h for pigs, and ca. 10 h for lambs), but will promote surface microbial growth and reduce potential shelf life of the product. A high humidity ($85-90\%$) will reduce losses by evaporation, and favourably influences the maturation process, however, again it will enhance microbial growth, especially of *Pseudomonas* spp. and other Gram− bacteria, and *Mucor racemosus* (Heiss and Eichner 1984). → Official methods.

meat lactic acid → lactic acid.

meat poisoning → botulism.

meat products → official methods.

meat spoilage

Cutting and boning increase the surface area of the meat that is exposed to contamination. Under refrigeration ($4°C$) the spoilage association is dominated by the *Pseudomonas, Acinetobacter, Moraxella, Alcaligenes, Flavobacterium* groups; at elevated temperatures, e.g., during transport and distribution, the → Enterobacteriaceae and eventually Gram+ bacteria (*Micrococcus, Staphylococcus,* LAB and bacilli) tend to dominate. Grey to grey-green discolourations may accompany these microbial successions. Off-smells may be observed when microbial numbers reach $\geq 10^7/cm^2$, whilst the surface becomes slimy at $\geq 10^8/cm^2$ (Krämer 1987; Sinell 1985; ICMSF 1980).

Megasphaera

Gram− cocci, obligately anaerobic, catalase−, gas production from glucose, glutamic acid and other amino acids. *M. elsdenii* and *M. cerevisae* can cause spoilage of beer → beer spoilage organisms (Baumgart 1990; Priest and Campbell 1987).

membrane filter culture

Preferred method for determination of the microbial count of water and easily filterable fruit juices with a low bacterial and/or yeast count. Sterile membrane filters with pore size $0.22-0.45$ μm, fittings for vacuum and pressure filters, → nutrient disc impregnated with culture media and other accessories are commercially available, e.g., Millipore and Sartorius. The advantage is that greater volumes, up to 1 l, can be easily handled and examined. The membrane is placed onto a moist nutrient disc or a culture medium in a petri dish after filtration after which it is cultured in the same way as a → plate culture.

membrane filters → bacterial filters.

mercury

Hg. Heavy metal with no effect on MO. Sublimate, $HgCl_2$ is strong microbicidal and is used as disinfectant of inanimate surfaces in the laboratory, and HgCl as fungicide in horticulture. Organic mercurials are less toxic and less irritating to man, but quite effective against MO:

phenylmercuric borate (including nitrate and acetate) are effective antiseptics (as alternatives for allergenic iodine tincture) for wound disinfection, and as preservatives for pharmaceutics.

mesophiles
MO preferring a moderate temperature. Opt. ~37 °C, min. ~15 °C and max. ~48 °C. All the bacteria and fungi pathogenic to warm blooded animals belong to this group. See Figure 23, page 266.

mesosomes
Intracellular invaginations of the cytoplasmic membrane considered as a primitive excretion organelle. It may possibly have additional functions.

metabolism
Biochemical process in all living cells. Uptake of nutrients and assimilation in the cell. Excretion of non-utilisable end products and resulting energy yield and growth or replication (Schlegel 1985).

metabolite
An intermediate or end product of microbial metabolism partially excreted in food, with a desired or undesired effect, e.g., → organic acids, CO_2 and other gases, → ethanol, → antibiotics, → mycotoxins, flavour and other substances influencing the taste.

methane
CH_4. Simplest carbohydrate produced by → Methanobacteriaceae under anaerobic conditions from CO_2; → biogas. Methylotrophic bacteria are able to utilise methane as the only source of C and energy, e.g., *Methylococcus* and *Methylosinus* with the use of molecular O_2.

Methanobacteriaceae
Methanobacterium, Methanococcus and *Methanosarcina*. Energy obtained by complete reduction of CO_2 to CH_4 with H_2

as H-donor. They are not able to metabolise carbohydrates and proteins in addition to methanol, formic acid and acetic acid. Strict anaerobic conditions are required. Symbiotic association with other organisms while using their excreted metabolites as substrate. Habitat: rumen, anaerobic digesters in → sewage works, → biogas fermenters, lake mud. Important organisms for waste treatment in the food industry.

methanol
H_3COH, methyl alcohol, "Holzgeist". Produced in the microbial decomposition of → pectin or through their enzymatic cleavage (→ pectinase) in fruit and vegetable juices. Also produced in intestinal digestion. – In fruit juices 10–130 mg/l; ~50 mg/l in wine; maximum of 100 mg/l in wine after treatment with dimethyldicarbonate (→ cold preservatives); ~1.5 mg/l in the blood of adults; >340 mg/kg BW is acutely dangerous for man (Classen et al. 1987; DFG 1987).

methyl red reaction
A pH-indicator turning from red to yellow if the neutral medium is acidified to a pH value of 4.5. Used to differentiate between coliforms and enteric bacteria. → IMVEC; → IMVIC; → MacConkey medium.

Metschnikowa pulcherrima
(*Candida pulcherrima*). → Surface yeast producing ascospores with occasional mycelial-like growth and an intensive red colour. Unlike *Rhodotorula* the pigment diffuses into the medium. Only glucose and fructose fermented, but several sugars and organic acids can be metabolised anaerobically. Habitat: grapes, grape- and fruit juices (Dittrich 1987).

MIC → minimum inhibitory concentration.

microaerophiles → oxygen requirement.

microaerotolerant → oxygen requirement.

Microbacterium
Gram+ rods, often arranged in palisades, non-motile, poor acid production [L(+) lactic acid], catalase+, thermotolerant. *M. lacticum* in milk tolerates 30 min at 72°C and is not killed by pasteurisation. See Table 29, page 193. Up to 10^5/g are found in powdered milk. *M. flavum* may also be found in pasteurised milk, cheese and other dairy products. Due to their slow reproduction both spp. do not typically cause spoilage of milk (Teuber 1987).

microbial counting
One of the most important activities of a food microbiology laboratory. The in-plant → process control is just as important as the quality control of the end product by the trade itself or official control by authorities. This has relevance to regulations governing the highest permitted (tolerable) bacterial count in several foodstuffs, e.g., milk, ice cream, etc. (→ official methods); guideline- and risk values for other products are in discussion. The first step in each case is the → sampling for which great care is required, since it will determine the reliability of the bacterial count. The procedure must be selected from the existing methods. A → dilution series is usually prepared. – Direct bacterial assay methods are usually quick but it is difficult to discriminate between viable or dead cells. → Microscopic count. The → turbidity is measured in special cases. It is tedious to culture MO, but it gives information about the ability to multiply (→ cfu) by formation of a visible colony; → pour plate method, → surface plate method, → spiral plate method, → drop plate method, → membrane filter culture. Endospores can be counted selectively, after the vegetative cells have been killed by heat ("heat shock") treatment (Baumgart 1986).

microbial titre
Method commonly applied for counting bacteria in liquid culture. It is used to determine whether a given volume (0.1, 1, 10 ml) is free from indicator organisms or → pathogens. For larger volumes the → membrane filter culture is preferred, whilst serial dilutions are prepared for smaller volumes. – The test sample or a series of decreasing volumes are pipetted into three parallel test tubes containing 10 ml nutrient broth and are then shaken and cultured. For the evaluation the count is determined from the highest dilution at which none of the three → enriched cultures shows any growth and/or signs of typical metabolites (e.g., gas produced from lactose, acids, etc.). The most probable bacterial count is then calculated according to a specific key (→ MPN). The titre method is most suitable when low numbers (e.g., <100/ml) of the species or group are expected (Baumgart 1990).

microbicidal gases
Gaseous substances used for the decontamination of dehydrated foodstuffs, e.g., spices, flours, powders, granules, stored grain crops, leguminous crops and surfaces. – β-Propiolactone: excellent microbicide, but not allowed for the treatment of food due to its mutagenic and carcinogenic effects. → Ethylene oxide: effective against all MO, but valuable food components are destroyed and, in addition, it reacts with halogen ions with the formation of ethylene chlorhydrine. Its use in D is no longer approved for stocks and foodstuffs. – Propylene oxide: poor microbicide and not a substitute for ethylene oxide due to considerable doubts about its toxicological safety. – Methyl bromide: poor bacteriocide, good fungicide. The use for protection of stored food of plant origin is disputed because of high residue levels. – → Formaldehyde: excellent microbicide. Primarily used for the treatment

of surfaces since its penetration capacity is low (Classen et al. 1987).

microbiological criteria

Aimed at protection of the consumer. Suggestions that a.o. include proposals by the CAC concerning the detection of pathogens and include end product specifications and also are aimed at control measures of hygienic precautions; quantifiable, e.g., as microbial count, indicator organisms, mould count, etc. Microbiological guidelines are aimed at→in-plant (intrinsic) process control, and consider product specific differences in contrast to the general assessment of pathogens [Sinell, H.-J.: *Fleischwirtsch.* 65 (1985) 672–678].

microbiological limits

Refer to levels or targets that are not to be exceeded, e.g., absence of MO or a specific group (e.g., *Salmonella*) in a specific weight or volume of a food (per g or ml). To meet hygienic criteria, i.e., to protect consumers against possible health risks, microbiological criteria (guidelines) may also refer to maximum numbers (limits) of particular microbial groups (e.g., *E. coli* or→coliforms) that may be tolerated (allowed) (e.g., 10/g).

microclimate

Temperature and humidity of small spaces and rooms, rendering these more favourable for microbial growth than the direct environment: e.g., sliced bread in plastic bags, cheese domes in the fridge or in the room, small insect tunnels in grain corns or nutmeg. The temperature and humidity rise due to respiration and metabolic activities accompanying microbial growth.

Micrococcus

Genus of the Micrococcaceae. Gram + cocci, single, in tetrads or in clusters; colour varies from yellow to red, obligate aerobic, catalase + , nitrite + , growth in up to 5% NaCl, between pH 5.6 – 9.1, min. a_w value 0.9, min. growth temp. 0 – 5 °C. Spoilage of→wort is caused by *M. kristinae*. *M. varians* is used as starter culture for ripening of dry and semi-dry sausages, due to its catalase activity. White rot in eggs is caused by micrococci, which may also grow on the surface of meat during cold storage (Krämer 1987; Sinell 1985).

microscopic cell count

Direct method for determining the microbial count. Yeasts can be counted in a microscopic counting chamber with a known volume (0.16 mm^2 × 0.2 mm depth) and surface marked with bigger and smaller squares. The most widely used chamber is the THOMA chamber developed for examination of blood specimens. For bacteria, that are considerably smaller, chambers with a depth of 0.2 mm are available, e.g., Bürker-Türk, Helber or Petroff-Hauser. Counting must be under the phase-contrast microscope due to the low refraction index of bacteria. It is not possible for the unexperienced eye to distinguish between viable and dead cells (Baumgart 1990).

microwave heating

Products containing moisture can be heated in a dielectric field of 2000 – 3000 MHz; in addition, portions or pieces may be thawed and warmed. The advantage is that the product is heated evenly from the core to the surface layers and, considering the short time required, it is often preferred for use in the kitchen, restaurants, aeroplanes and trains. It is used both for heating and/or cooking. MO are killed because of the increase in temperature as in conventional heating methods. It is not possible to inactivate spores in dry products with this method.

milk

Produced by cows, sheep and goats. It is

free from any organisms in a healthy udder, but is heavily contaminated in the teat canal during the milking procedure either by hand or machine. See Table 25. An excellent growth medium for most MO because of its nutrient composition: water 860 – 880 g/l; fat 30 – 50 g/l; protein (casein, etc.) 30 – 35 g/l; CH (→lactose) 45 – 50 g/l; mineral salts 8.65 g/l; vitamins A, B, C, D, E, H, etc.→Raw milk.→Drinking milk.→UHT milk (→H milk).

milk hygiene guidelines

The EC milk hygiene guidelines (85/397/E.C. of 5.8.85) for regulating legal matters concerning health and animal diseases and the inter-community trade with heat treated milk is to be adopted on a national level in two stages: the first on 1.1.89 and the second on 1.1.93. Apart from its influence on the areas of milk production (product, stable, animal and udder control), processing and distribution, it also includes standards for raw milk, and specifications for pasteurised, sterilised and ultra-high heat treated milk, see Table 25. The temperature may not exceed 8 °C at delivery by the producer during the day or 6 °C at night. During production and transport pasteurised milk should be kept at a temperature of maximally 6 °C.

milk moulds→ *Geotrichum candidum.*

Table 25   Microbiological figures for milk acc. to EC hygiene guidelines of 1985.

Product	Level 1 (1989)	Level 2 (1993)
Raw milk		
Microbial count at 30 °C/ml	<300,000	<100,000
Somatic cells/ml	<500,000	<400,000
Freezing point (°C)	– 0.520	– 0.520
Antibiotics/ml		
Penicillin	<0.004 μg	<0.004 μg
Other	not detectable	not detectable
Pasteurised milk		
Pathogens	none	none
Coliforms/ml	<5	<1
Microbial count/ml at 30 °C	<50,000	<30,000
After incubation at 6 °C for 5 d		
microbial count/ml at 21 °C	<250,000	<100,000
Phosphatase	–	–
Peroxidase	+	+
Antibiotics/ml	not detectable	not detectable
Freezing point (°C)	< – 0.520	< – 0.520
In addition, a pyruvate test is to be performed.		
Sterilised and UHT milk		
After incubation time of 15 d at 30 °C		
a) microbial count/0.1 ml at 30 °C	<10	<10
Antibiotics/ml	not detectable	not detectable
In addition, a limulus test has to be performed.		

milk powder

Dried (dehydrated) milk. Skimmed milk powder, full cream milk powder. Produced from pasteurised milk entering a spray tower with incoming air, heated to 180°C, coming from the opposite direction and leaving the tower at a temperature of 90°C. The powder itself reaches a temperature of 70°C and heat resistant MO survive. The total microbial count is about 10 times higher in the end product than in the original material. Recontamination and multiplication before drying should therefore be avoided. Enterobacteria in the end product is an indication of → recontamination. – The same applies to whey powder. – Control: → official methods (Teuber 1987).

milled cereal products

Flours, meals. Include bran, flour, grits (from maize or wheat), semolina and breakfast cereals. Microbiological contamination similar to that of → cereals out of which these are being manufactured. Milling, sifting and separation of the hulls contribute to a reduction in the total number of MO, and health hazardous components such as → ergot or the seed of weeds are removed. The outer hulls of the grains (bran) carry a higher microbial load than the inner components (flour, grits). Storage under humid conditions may cause sensory defects (acidification, bitter taste, rancidity) and spoilage by MO (Kunz 1988). Microbiological examination methods, see Hobbs and Greene (1984).

mineral water → alcohol free beverages.

mineralisation

Continuous process of → respiration and → fermentation by MO on land and in water in which organic substances (plant or animal materials, excreta, chemicals) are metabolised, with the production of CO_2 and other inorganic compounds. Nitrogen containing molecules (amino acids, purines, etc.) are oxidised to NO_3 or reduced to N_2. These processes guarantee recycling and → biological equilibrium in nature. Green plants assimilate CO_2 and NO_3, which then serve as nutrients for animals. – Additional CO_2 is produced by vulcanoes and the combustion of fossil fuels, e.g., coal, oil and methane. The consequence is a disturbance in the equilibrium resulting in a green house effect.

minimal inhibitory concentration

MIC. Important parameter for the selection and assessment of a → preservative or → disinfectant. It gives information on the lowest possible concentration of an active substance still able to completely inhibit a given pure culture. The MIC depends on the specific MO species to be eliminated. Gram + are generally more sensitive than Gram – . Fungi and yeasts are even more resistant, with the exception where acids are concerned. Another factor is the ecological conditions: temperature of application, degree and type of contamination (protein, fat, heavy metals, etc.), the nature of the surfaces (roughness, cracks) and contact time. → Cleansing (Wallhäusser 1988).

mitochondria

Organelles in eucaryotic cells, e.g., fungi and yeasts, with their own → DNA. They play an important role in → respiration and → ATP-synthesis and are therefore important in energy generation of the eukaryotic cell. They correspond to → membrane vesicles in → prokaryotes in some respects.

mitosis

Normal cellular division in which the number of chromosomes or genes remain the same. Opposite: meiosis.

mixed culture

Opposite → pure culture. Several → starter

cultures are mixed cultures of different species and types.

mixed pickles → lactic fermented vegetables.

mixed salad
"Ready-to-eat" salad mixes. Shredded salad packed in plastic bags or transparent bowls. Ingredients depend on the season: endive, iceberg lettuce, radish, raddicchio, China cabbage, cabbage, green paprika, carrots, sweet corn, etc. The raw products are cleaned, shredded and washed up to three times in precooled water, with possible weak acidification and packed, stored and delivered at 3°C. It must be kept refrigerated below 6°C in the supermarket. Exudate from the cells and sugar especially from carrots and/or sweet corn are excellent growth media. Respiration of the salad and the MO causes a rise in the temperature and of the CO_2-content, up to 30%, and a drop of up to ~1% in the O_2-concentration in the bag (optimal in polypropylene). → CA storage. The shelf life based on sensory data in the laboratory is up to 12 d at 1 °C; 8 d at 4 °C and 3 d at 12 °C. In practice it is less since it is almost impossible to strictly maintain the cold chain. The total microbial count of the product in retail outlets may range from $8 - 16 \times 10^6$/g as compared to $10^4 - 10^5$/g directly after packaging. Nitrite may be produced by bacteria through the reduction of nitrate in foodstuffs rich in nitrate, e.g., salad in winter time (Nicolaisen-Scupin 1985).

moisture
Also referred to as water content (in %) of a product. For the growth or metabolic activity of a MO only the available water is important, which is indicated by the → water activity (Kunz 1988).

moisture equilibrium → water activity.

molasses
Viscous, dark-brown liquid; by-product of sugar refining. Contains ca. 50% sugar, 19% of non-sugar substances and 23% water. Molasses is clarified for the manufacture of → baker's yeast or → ethanol. M. from sugar cane is used for the production of Arrak and rum.

Monascus purpureus
(*Xeromyce* ssp.). Ascomycete. Anamorph *Basipetospora*. In Japan and RC used in the production of a red food colour by fermenting rice for fish and other foodstuffs ("Ang-kak"). Attempts are made for its biotechnical production. Causes spoilage of dried fruit. Extremely xerotolerant: a_w 0.61 at 25°C, min. pH 3.2.

Monilia
Heliotales, Ascomycetes. Anamorph of *Neurospora*. Red moulds growing on moist bread not well baked. – In the literature "ring rot" in stonefruit (*M. fructigena, M. laxa*) is often ascribed to this genus, but is caused by → *Sclerotinia*.

moniliformin → *Fusarium.*

monosaccharide
Sugar consisting of only one component, e.g., glucose, galactose, fructose, etc. Easily metabolised by MO. Opposite: di-, tri- or polysaccharides.

monotrichous
Equipped with one flagellum. Atrichous: none. Polytrichous: several flagella. – From the Greek trichos = hair.

Moraxella
Genus of Gram– aerobic rods, oxidase +, catalase +, psychrotrophic (Bergey 1984). Together with → *Acinetobacter* mainly on fresh meat, ocean and fresh water fish. Also responsible for spoilage (Krämer 1987; Sinell 1985). High radiation resistance.

morphological forms of bacteria

Species specific morphological characteristics of bacterial cells, related to two basic forms: rods and cocci (Figure 10). For→ differentiation purposes the typical cell arrangement and cell size can be determined by light microscopical methods, using phase contrast and→ staining procedures.

morphology

Study of the form, shape and change in form (pleomorphism) of MO. Important characteristic in the→differentiation of fungi; among bacteria only cocci, rods and other basic shapes. See Figure 10.

mould count

The enumeration of the fungal population by cultural methods is extremely inaccurate and the interpretation is unreliable. Fluctuations of up to 100-fold between different laboratories can be considered as "normal", unless the procedures (from sampling to plating) are exactly specified, e.g., in the form of a protocol. Causes: (1) extremely uneven distribution in the product (amounts of mycelium and conidia in different ratios). (2) Homogenisation of the sample (Ultra Turrax, Waring Blender, etc.) may produce hyphal fragments of different sizes, some of which may be too short and thus non-viable. Using the Stomacher ensures a better distribution of conidia, but disruption of the mycelium is limited. (3) Practically all selective fungal media contain inhibitory substances against bacteria (→selective media); these, however, may exert some inhibitory action against some fungi. – The microscopical examination (→HOWARD mould count) cannot distinguish between dead and viable cells [Bandler et al.: *J. Food Prot.* 50 (1987) 28–37]. – A number of indirect methods are available: determination of CO_2 production and O_2 utilisation; amyloglucosidase activity [Moebus and Teuber: *Kieler Milchwirtsch. Forsch.-Ber.*

38 (1986) 255–264]; detection of ergosterol [Müller, H.-M. and Lehn, Christine: *Arch. Anim. Nutr., Berlin* 38/3 (1988) 1–14]; catalase activity [Frank, H. K. and Hertkorn-Obst: *Chem. Mikrobiol. Technol. Lebensm.* 6 (1980) 143–149]. – The determination of chitin has not been proved reliable, since fragments of arthropodes may contribute to an increase in false positive values. – In some cases, especially in cereals, the mould plate count should be combined with examination for→mycotoxins (Beuchat 1987; Rhodes 1979).

mould counts in household

Domestic air generally contains lower numbers of moulds than outside. Climatic factors such as temperature, humidity and light are of secondary importance in interior rooms. The air movement, e.g., caused by activities and heating, is however a significant factor (Table 26). Both in static and moving air spores and mycelium fragments (and also bacteria) may settle continuously onto uncovered foods and thus contaminate the surfaces. – In rooms with mould growth on the walls the average values are considerably higher than in Table 26.→Toxinogenic moulds are outnumbered; the presence of→allergens can however not be excluded.

Table 26   Average mould counts in the air of living rooms [Reiss, J.: *Mycosen* 30 (1987) 127].

Room	→cfu/m^3 of Room Air
Kitchen, without air movement	566
Kitchen with one person working	700
Refrigerator	483
Bath before cleaning	905
Bath after cleaning	593
Living room without air movement	858
Bedroom before making the beds	566
Bedroom after making the beds	844

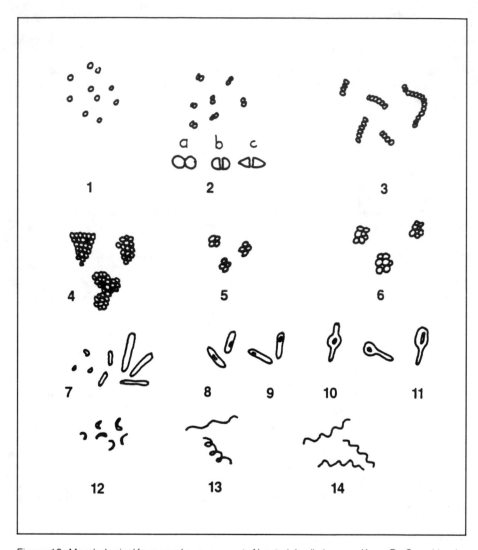

Figure 10 Morphological forms and arrangement of bacterial cells (source: Kunz, B.: *Grundriss der Lebensmittel-Biologie*. Hamburg: Behr's Verlag, 1988, S.41).

1 = monococci
2 = diplococci
3 = streptococci—chains
4 = staphylococci—clusters
5 = tetracocci
6 = sarcina
7 = rods (different forms)

8 = rods with central endospore
9 = rods with terminal endospore
10 = spindle-shaped rods with endospore
11 = drumstick-shaped rods with endospore
12 = vibrios
13 = spirillum (rigid)
14 = spirochaetes (flexible)

mould culture
Starter culture with→ P. caseicolum,→ P. roqueforti, or→ P. nalgiovensis.

mould fruiting bodies → fruiting bodies.

mould media
Fungal growth media. Weakly acid (pH 5 – 6.5) substrates are preferred by moulds, and are vitamin autotrophic as plants. A large number of food-associated moulds are relatively non-fastidious, and a medium containing a mineral salts mixture and a sugar as energy and C source may be sufficient to support growth. Examples: Czapek-Dox agar (BBL no. 11140, Merck no. 5490, Oxoid no. CM 97). Nutritiously richer substrates contain more or less undefined mixtures of organic materials originating from the food industry, e.g., malt extract, beer wort (→ beer), corn steep liquor, corn (maize) meal, etc. Examples: malt extract agar (BBL no. 11401, Difco no. 0024, Merck no. 5398), wort agar (Difco no. 0112, Merck no. 5448). Maize meal agar is especially suitable for the cultivation of→ Fusarium. For colonisers of bread "Bread Powder" agar is recommended [Reiss, J.: Chem. Mikrobiol. Technol. Lebensm. 11 (1987) 60 – 62]. Sabouraud agar (peptones and sugar) is used for the control of yeasts and moulds on packaging material, e.g., Sabouraud – 1%, glucose – 1%, maltose-agar (Merck no. 7662), but also to check for possible antimycotic substances [N.N.: Merkblatt 18, 19, 21; VerpackungsRdsch. 25 (1974), Techn. Wiss. Beilage]. Practically no→ differential media are available for moulds. The isolation of moulds from foods heavily contaminated with bacteria is enabled by the addition of an antibiotic or antibiotic mixture of copper sulphate, crystal violet, brilliant green, etc. For the differentiation of fungi defined growth media are to be used under exactly defined conditions (temperature, time, light, dark,

etc.). Compare monographies of mould genera, e.g.,→ Aspergillus, Penicillium, etc.

mould ripened salami→ fermented sausage.

mould taste
A defect in wine caused by pellice yeasts. Type-specific tastes and odours are changed as a result of ester formation.

moulds
Americ.: molds; french: moissisure; span.: moho. General term for fungi (sing.: fungus). Kingdom (regnum) of the→ eukaryotes with ca. 120,000 spp. Characteristic for all fungi is the presence of chitin as cell wall component and ergosterol in the plasma membrane (→ mould count). – From a→ spore,→ conidium or other survival body (→ sclerotium) a filamentous cell (→ hypha) develops, that may form cross walls (septa). Growth at the distal end accompanied by branching; may reach macroscopical dimensions within a few days. The visible layer of hyphae,→ mycelium, spreads over the surface of suitable substrates (foods); the surface covered as well as the→ biomass increase. The proliferation of biomass and the reactions required for energy are collectively grouped as primary metabolism, the reaction sequence of which is similar to that of all other living beings. The hyphae are in close contact with the substrate, and excrete extracellular→ enzymes that cleave macromolecules (protein, CH) or water insoluble materials (lipids), the products (amino acids, sugars, fatty acids) of which are being transported via the cell wall for assimilation within the cell. This phase of development in fungi is referred to as trophophase. Depletion of the substrate towards nm amounts around the hyphae and the accumulation of metabolites in the cells result in the reversion of the cells to secondary me-

tabolism, the "idiophase". At this stage the formation of an aerial mycelium is initiated, which gives rise to the distribution organs, and from which the colony has its soft cotton wool like appearance at the surface. The→ secondary metabolites are formed during this stage, and may in part be excreted into the environment. This can be designated as "luxury" phase, resulting from the abundant availability of nutrients, since most secondary metabolites are molecules with energy content greater than, e.g., glucose and other sugars. In the "natural" environment of the moulds, the soil, many of these substances are not produced. – → Yeasts are systematically grouped under the fungi, but are treated separately for historical and practical reasons (Müller and Loeffler 1982). – Moulds responsible for food spoilage mainly belong to the→ Deuteromycetes (Fungi Imperfecti) that either lack a sexual phase or for which such a phase is not known yet. The same applies to several spp. that are biotechnically utilised. – The→ differentiation of moulds is mainly based on microscopical and cultural criteria; the phenotypic system is based on morphological characteristics, the application of which requires extensive experience, time and relevant knowledge.

MPN counting
Most probable number. Method for counting bacteria that links up with the titer method (→microbial titre). A "yes" or "no" answer from the results in two parallel enrichment tubes enables the determination of the most probable bacterial count from tables that have been published by McCrady in 1934. – The method can also be applied for assessment of→ drop cultures (Baumgart 1990).

MTD
Maximum tolerated dose→ NOEL.

Mucor
Genus belonging to Mucorales, Zygomycota. Aseptate hyphae, occurring as plus and minus entities, able to reproduce sexually following contact with each other (heterothallic). Without contact anamorphic spherical sporangia produce a large number of sporangiospores (see Figure 21, page 245).→Chlamydospores are formed in the hyphae as a result of changes in the ecosystem (Figure 11). Commonly non-fastidious saprophytes, but in part with parasitic abilities, (→opportunists). Mycoallergoses and mycoses are known. – Under anaerobic conditions the fungi start to ferment and lactic acid or ethanol is produced. Several species change their growth pattern under such conditions (→dimorphism) and reproduce by budding, like yeasts. About 50

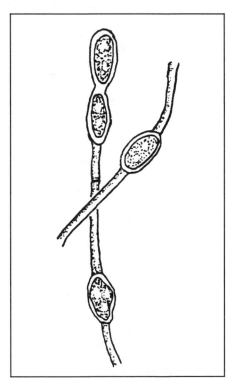

Figure 11 Chlamydospores of *Mucor* sp.

species are described; occurrence in soil and on dead plant materials. *Rhizomucor miehei* (*Mucor mihei*) and *Rh. pusillus* (*M. pusillus*) have been omitted from this genus.

Mucor hiemalis

One of the most abundant species in soil (compost). Growth conditions: opt. 25°C, min. 5°C, max. 30°C. Used in the production of sufu – cheese made from soya milk in the RC.

Mucor mucedo

Frequently on bread and sometimes also on poultry meat. Also on the walls of production areas with a high humidity. Inhalation of spores may cause an allergic reaction in sensitised persons. It is also responsible for damage to materials, e.g., paintings.

Mucor racemosus

On bread, meat, green coffee beans and cocoa. Growth conditions: opt. 25 – 28°C, min. -4 to -3°C; a_w: 0.92, opt. 0.98, spore formation 0.95. In Mexico used to produce a beverage from maize (pozol) in combination with *M. rouxianus* and several yeasts (Reiss 1986).

Muffton → cork taste in wine.

multiplication

Growth. Replication. Single species in the mixed population of MO usually found in foodstuffs, reproduce at distinctive rates. The rate depends on changes in the medium due to utilisation (e.g., O_2) and excretion of metabolites (e.g., CO_2), acids, etc., by species and/or groups rapidly multiplying at a given stage. This leads to succession by metabolic types best adapted to the prevailing conditions, e.g., as known in the production of → sauerkraut or yoghurt. – Pure cultures used for the production of → starter cultures, → single cell protein or → enzymes

follow in principle the same pattern but are more constant. Four different stages of multiplication can be distinguished when a sterile medium is inoculated with viable cells of a specific species in an enclosed system (fermentor, flask, etc.).

(1) Lag phase: adaptation to the new medium
(2) Log phase: constant → division rate following a logarithmic or exponential function
(3) Stationary phase: during which the number of viable cells remains constant
(4) The death phase: during which the number of viable cells decreases due to population pressure, the presence of toxic metabolites and depletion of nutrients

Between phases three and four the biomass in the system may still increase. To determine single parameters with the view on specific statements concerning, e.g., shelf life or warm keeping periods, etc., simple mathematical equations, valid under stable conditions, can be used:

B = initial bacterial count at t_0 (e.g., amount of organisms used for inoculation)
b = bacterial count after time, t
G = generation time (time for one division)
g = division rate (number of divisions per h)
n = number of divisions (generations)
b = B × 2^n
log 2 = 0.301
log b = log B + n × log 2

$$n = \frac{\log b - \log B}{\log 2} \quad (\log 2 = 0.301)$$

$$g = \frac{n}{t} = \frac{\log b - \log B}{0.301 \times t}$$

$$G = \frac{t}{n} = \frac{t}{g}$$

e.g., division rate of a bacterial suspension

in which the cells increase from $10^6 - 10^9$ within 5 h.

$$g = \frac{\log 10^9 - 10^6}{0.301 \times 5} \sim \frac{3}{1.5} = 2$$

This gives a generation time of:

$$G = \frac{1}{2} = 0.5\,h$$

Equivalent calculations are not possible for fungi and yeasts. → Mould count.

murein

Peptidoglycan. Characteristic glucoprotein of all bacterial → cell walls responsible for the stratified network (murein sacculus) and mechanical strength (rigidity). It is constructed by alternatingly linked units of N-acetyl glucosamine and N-acetyl muramic acid woven into each other. → Lysozyme cleaves the glycosidic bonds and dissolves (lyses) the cell wall.

muscarine

Muscarin. Toxin of the fly agaric. Tetrahydrofuran, a → biogenic amine. Occurs in *Amanita muscaria* and other species, *Clitocybe* spp., *Inocybe* spp. The following symptoms are elicited after ingestion: hypotension, bradycardia, hypersalivation, perspiration and hallucinations. Excreted in the urine (Roth et al. 1988).

mushroom

Generally refers to any edible mushroom, which may include enokitake, oyster fungus and shii-take (→ cultivated mushrooms). More specifically the edible fruiting body of the cultivated mushroom: *Agaricus brunnescens* (= *Ag. bisporus*) (brownish) and *Ag. hortensis* (whitish); related to the "wild" mushroom *Ag. campestris*. The dikaryotic → mycelium is grown, e.g., on moist grains in the laboratory. The "spawn" serves as inoculum for cultivation on a substrate prepared from composting horse manure, straw, wood chips, etc. Incubation is at 24 °C and 70 – 80% rH. The mycelium takes ca. 14 days to grow through the compost which is then covered (4 – 5 cm) with a casing layer of soil (pH ca. 8). During the following week the temp. is maintained at 24 – 25 °C and the rH at > 95% the mycelium grows through to the surface; fruiting initiation is induced by reducing the temp. to 18 °C or less and the rH to 80 – 90%. Mature fruiting bodies may be harvested after 2 – 3 weeks. – Parasites include: *Mycogone perniciosa* (wet bubble disease), *Verticillium fungicola* (dry bubble), *Hypomyces rosellus* (cobweb disease), *Pseudomonas tolaasii* (brown blotch), *Pseudomonas fluorescens* (ginger blotch); some *Erwinia* and *Bacillus* spp. may cause specific blight diseases. Larvae of some fly spp. may cause damage (Bötticher 1974; Müller and Loeffler 1982; Hayes 1985). – Recent reports suggest the presence of genotoxic substances, agaritins, in raw mushrooms. Specific knowledge on the health risk is still lacking. – Worldwide annual production has increased from ca. 250,000 tonnes in 1960 to > 1 million tonnes in 1980; consumption in Germany in 1986: fresh 0.77 kg/capita; processed 2.5 kg/capita (Hayes 1985).

must

Juice from grapes or other fruit heavily contaminated with MO originating from the surface of the fruit. Spontaneous → fermentation starts rapidly.

mutation

Random change in the inherited genotype usually influencing one gene, seldom more than one gene. It results in a new phenotype different from the wild type. Most commonly there are changes in the sequence of the nucleotides and bases of a gene. Changes due to an unknown reason are probably caused by cosmic radiation, UV radiation or → nitrite, etc., are called spontaneous mutations. Intentive induction of mutation in bacteria is possible by application of natural or artificial mutagens, e.g., high-energy radiation, ethylmethane sulphonic acid, nitrosomethylguanidine, etc. The mutation rate is defined as the frequency of spontaneous mutations. – Examples: *B. megaterium*: susceptibility to phages – 5×10^{-8}; *E. coli*: streptomycin resistance – 10^{-10}; *E. coli*: penicillin resistance – 10^{-7}. – Most mutations are lethal, thus the bacteria die before or during division. The capability for mutations to take place is a criterion of life and forms the basis for the development of the large species diversity on earth.

Mycelia sterilia
→ Deuteromycetes. No reproductive organs (conidia or spores) are formed. Extremely difficult to classify.

mycelium
Group or mass of fungal hyphae. Substrate mycelia grow on the surface of the growth media or can penetrate it, but only up to a depth of 1 mm. It is responsible for the nutrition of the fungus. The air hyphae grow away from the medium and often bear the reproductive organs which produce spores and conidia on/in them and which are distributed by air, water or insects. They contaminate the environment, e.g., in a bundle or storage room, etc. Although the air mycelia obtain their nutrients from the substrate mycelia, they can also take up water from the air if the relative humidity is >72% (Müller and Loeffler 1982).

mycetism
Mycetismus. Fungal poisoning. Disease caused by the consumption of poisonous vegetative fungal fruiting bodies (Roth et al. 1988).

Mycobacterium bovis
A responsible agent of tuberculosis in cat-

tle, humans, pigs, cats and dogs. In earlier days it occurred in raw milk, but presently it is rare. Due to proper cattle management and hygiene, and → pasteurisation of drinking milk the possibility of infection is very low. Destruction time in milk: 60 °C <10 min; 71 °C <40 s; 85 °C <8 s. It can survive >270 d in frozen meat and >126 d in smoked sausages (Mitscherlich and Marth 1984).

mycogenic allergy
Allergic diseases in sensitised persons due to fungal allergens. Most commonly it is caused by their spores or conidia, but → enzymes produced by the fungi may also elicit the allergic symptoms. Inhalation is more important than ingestion. It is suspected that the excreted enzymes are the actual allergens (Müller and Loeffler 1982; Schachta and Jorde 1989).

mycology
The study of fungi (Beuchat 1987; Dorfelt 1989; Müller and Loeffler 1982; Rhodes 1979).

mycoses
Human and animal diseases resulting from the infection with parasitic fungi (Gedek 1980).

Mycota → fungi (moulds).

mycotoxicosis
Any disease of man and animal most commonly resulting from the oral ingestion of → mycotoxins in foodstuffs. In contrast to bacterial toxins and mushroom toxins (→ poisonous moulds) which most often cause acute poisoning, mycotoxicoses result from a chronic ingestion of small amounts of the toxin over a period of time. St. Anthony's fire or ergotism has been known for a long time. It is caused by → *Claviceps purpurea* growing on grain used for bread. Toxic alimentary aleucia is caused by grain which was still

Table 27 Some mycotoxins, their producing strains and effects (acc. to Sinell 1985, modified).

Substance Class/Toxin	Producing Strains	Effects on Mammals
Cumarin derivatives		
Aflatoxins	*A. flavus, A. parasiticus*	carcinogenic, hepatotoxic
Ochratoxin A	*A. ochraceus, P. virdicatum* and many others	carcinogenic, hepato- and nephrotoxic
Sterigmatocystin	*A. versicolor, A. nidulans* a.o.	carcinogenic
Anthrachinones		
Luteoskyrin	*P. islandicum*	carcinogenic, hepatotoxic
Pyrones		
Patulin	*P. expansum, P. patulum, A. clavatus, Byssochlamys fulva, B. nivea* a.o.	general cell poison, haemorrhagic
Citreoviridin	*P. citreoviride* a.o.	nephrotoxic
Kojic acid	*A. oryzae, A. candidus, A. nidulans, A. tamarii* a.o.	neurotoxic, mutagen
Polypeptides		
Islanditoxin	*P. islandicum*	hepatotoxic
Macrolides		
Zearalenon	*Fusarium* spp.	oestrogen, infertility, abortion
Trichothecenes	*Fusarium* spp.	dermatotoxic, enterotoxic, haematotoxic
Ethanol	*Saccharomyces cerevisiae*	neurotoxic
Ergotalkaloids	*Claviceps purpurea*	neurotoxic

on the field and covered with snow during winter time. It corresponds to endemic Balcan nephropathy caused by ochratoxin A. The same toxin causes nephropathy in pigs in DK, N, S, IRL. → Zearalenone can cause vulvovaginitis in sows and infertility in cattle. Several other mycotoxins are described for animals and, in the tropic regions, for man (Gedek 1980).

mycotoxins
Natural occurring metabolites of → moulds growing on or in foodstuffs. The MW is <500, e.g., → aflatoxins are the toxins most frequently excreted by fungal cells; they are toxic to animals and partly to plants. They are also → teratogenic and/or → carcinogenic. Most of the mycotoxins are resistant to acid and heat. The distinction by definition from the → antibiotics is unclear. → Mycotoxicosis can occur in humans and pets due to commonly chronic ingestion with the diet (see Table 27).

Myrothecium
Imperfect fungal genus responsible for damage to plant products, e.g., straw, textiles and paper. *M. verrucaria* is strongly cellulytic and grows on media with 20 – 40% sucrose. Opt. 27°C, min. 12°C, max. 35°C. The species produces different → mycotoxins (verrucarine, roridine, gliotoxin, rugulosin, etc.) and antibiotics with unknown structures. Several diseases are described in domestic animals (Domsch et al. 1980; Gedek 1980; Reiss 1981, 1986).

mytilotoxicosis → mussel poisoning.

N

naringinase
Enzyme produced with *A. niger*. It cleaves "naringinin", the rhamnoglucoside, of naringenin, naringinin, and is used to remove the bitter taste in citrus juices (DFG 1987).

natamycin
Pimaricin. Antimycotic produced by *Streptomyces natalensis*. It acts like all polyene antimycotics on the cellular membrane and the presence of → ergosterol. LD_{50} for rats and mice is 1.5 – 4.5 g/kg BW p.o. The preliminary → ADI value is 0 – 0.3 mg/kg BW. It does not show any genetic and reproductive toxicity. Allowed in the treatment of the surfaces of cheese to inhibit the reproduction of moulds and the production of mycotoxins. It is not allowed in the treatment of sausage casings (Classen et al. 1987).

Natick process
Microbiological process for recycling of cellulose waste which is converted into a glucose solution. → *Trichonema viride* is cultured during the first step to produce a mixture of enzymes for the hydrolysis of the cellulose during the second step (Müller and Loeffler 1982).

natural "sour"
→ Sourdough. Made by adding water to flour and allow the "natural" MO present to multiply. → Spontaneous sourdough. Opposite: "Reinzuchtsauer" (pure culture sourdough) (Spicher and Stephan 1987).

necrobiotrophs
Perthophytes. Saprophytic fungi which destroy living plant tissues, usually after physical damage, by secreting toxic substances, e.g., organic acids and to some extent also mycotoxins and enzymes able to degrade pectin and cellulose hydrolytically.

necrosis
Localised death of plant or animal tissues, usually accompanied by an exudate. Noted as spoilage in the case of fruit and vegetables. It can be caused by microbial toxins, organic acids, etc., which destroy the cells in combination with proteases, pectinases and cellulases. The site is prone to secondary infection.

Nectria
A genus of fungi which forms part of Sphaeriales, Ascomycetes, and includes several parasitic spp.: teleomorph of some *Fusarium* spp. *N. galligena* is responsible for severe damage to stored stonefruit.

needle mount
Method often used in the microscopical examination of fungi. A small part of the mycelium is picked with two needles or a tweezer from the nutrient medium and transferred onto the slide into → lactophenol. It is carefully spliced in pieces and covered with a cover slip to avoid the inclusion of air, after which it is examined microscopically.

NEL
No effect level → NOEL.

nephrotoxin
Toxins responsible for damage to kidneys, e.g., ochratoxin A.

Neurospora
Ascomycota. Teleomorph of → *Monilia*. Fungal genus often used in different genetic and biochemical studies. Identification key by Baumgart (1990).

Neurospora sitophila
Anamorph *Monilia sitophila* or red bread mould with thermotolerant conidia and as-

cospores (5 min, 75°C). Growth: opt. 36°C, min. 4°C, max. 44°C, a_w min. 0.88 – 0.90, opt. 0.95 – 0.98, pH >3.0 to >8.0, opt. 5 – 6. Habitat: soil, meat, bread. Used in the preparation of an Indonesian food, oncon, from peanuts. Due to the high content of essential→ amino acids it is also used in the production of→ single cell protein from molasses.

nisin

A polypeptide antibiotic (lantibiotic) produced by several strains of *Lactococcus lactis* ssp. *lactis* (→*Streptococcus lactis* ssp. *lactis*). Naturally occurring in milk and dairy products in varying concentrations. Active against Gram+, but no effect against Gram− organisms and fungi. LD_{50} mouse (p.o.): ~9 g/kg BW. ADI value 0 – 33,000 IU/kg BW. It does effectively prevent→ late blowing in semi-hard cheeses. It however inhibits the germination of endospores and reduces their heat resistance. Due to this characteristic it is used in the production of→ processed cheese, especially when increased numbers of anaerobic endosporeformers (*Cl. butyricum, Cl. tyrobutyricum* or *Cl. sporogenes*) are expected in the bulk milk, e.g., as a result of silage feeding. Nisin is allowed in 32 countries for this purpose, but not in Germany (Classen et al.1987).

nitrate

NO_3. Most important N-containing compound in nature utilised as nutrient by plants. N-containing organic compounds are produced by microbial oxidation or it is used as mineral fertiliser in soil. It is a source of N for several bacteria and moulds. Several MO reduce nitrate to nitrite and eventually to ammonia. Excessive nitrite is excreted from the cells because of its toxicity. – Due to microbial and in vivo nitrite production, foodstuffs for neonates and toddlers should contain not more than 250 mg of nitrate/kg, and not more than 50 mg/l NO_3 in drinking water

(Hahn and Muermann 1987). EC no.: E 251, sodium nitrate; E 252, potassium nitrate.

nitrite

NO_2 is an active ingredient of curing salt. It is the first metabolite in dissimilatory→ nitrate reduction in plants and MO. As it is excreted by several organisms in presence of sufficient nitrate and since it is easily detected, its production is used as characteristic in the→ differentiation of bacteria. It can be detected in broth cultures with Griess-Ilosvay's reagent, by, e.g., Merck no. 9023, whereby a red diazo dye is produced in a positive reaction. Due to its toxic effect on neonates and toddlers only 0.1 mg/l is allowed in drinking water, 0.01% (100 mg/kg) in meat products, and 0.015% (150 mg/kg) in raw ham. EC no. E 250, sodium nitrate. ADI value for children older than 7 months is 0 – 0.13 mg/kg BW (Classen et al. 1987).

nitrite curing salt

Salt mixture containing nitrite. Maximal 0.4 – 0.5% $NaNO_2$.

nitrite excretion

Several Gram− bacteria and moulds growing on nitrate rich substrates excrete nitrite in the environment. It may cause a toxic risk for smaller children through, e.g., spinach, or it can exceed the allowed values in, e.g.,→ mixed salads. By including nitrate reducing bacteria (e.g., *Staph. carnosus* or *Micrococcus varians*) in starter cultures used in the production of→ dry and semi-dry sausages a desired nitrite concentration can be reached for a specific period during ripening, resulting in the inhibition of anaerobic sporeformers, e.g., *Cl. botulinum* (DFG 1987).

nitrosamine

A strong carcinogen occurring naturally in food. The average daily intake is about 1

μg per human adult. It is produced by nitrosation of amines ($\rightarrow$ biogenic amines, $\rightarrow$ amino acids), with nitrite which is produced by microbial reduction of nitrate in food ($\rightarrow$ oral MO; $\rightarrow$ nitrite excretion), or which is present as additive in meat products, $\rightarrow$ nitrite curing salt (Preussmann 1983).

nivalenol
$\rightarrow$ Mycotoxin, belonging to the trichothecene group and produced by $\rightarrow$ *Fusarium* spp. during growth on foodstuffs. It has been detected in wheat, rye, barley, maize and millet.

NIZO methods
Methods to produce sour cream butter ($\rightarrow$ butter; $\rightarrow$ cultured creamery butter), without the use of ripening cultures, developed in Ede/NL. Sour cream butter is manufactured with a moisture content of 12% which is then increased up to the allowed value of 16% by kneading, while a lactic acid bacterial and $\rightarrow$ butter aroma culture is added simultaneously. Skim milk is fermented with *Lb. helveticus* and then concentrated by evaporation. $\rightarrow$ *Str. lactis* ssp. *diacetylactis* and *Leuconostoc cremoris* are used as aroma cultures. Acid concentrate and cultures are added to acidify the butter to a pH of 5.2 – 4.9 and a $\rightarrow$ diacetyl value of 1.5 – 2 mg/kg, which increases up to 2.2 – 2.5 mg/kg within a week during storage.

noble rot $\rightarrow$ *Botrytis cinerea.*

NOEL
"No observed effect level". NEL: no effect level. MTD: maximal tolerated dose. Highest daily dose/level of a toxin or foreign substance causing no change in experimental animals, in comparison to the control group, within the duration of the experiment. The reliability of the assessment depends on the selection of intervals (daily, weekly, etc.) for multiple applications, e.g., in 90-day tests or generations experiments. $\rightarrow$ ADI value.

nomenclature
Scientific naming of organisms in biology. The binomial nomenclature of C. von Linne has been used since 1735 for the names of the genera and species. The name of the first author to describe a species is often linked to the name, e.g., *Aspergillus flavus Link* in order to prevent any confusion. – The systematic units or taxa are indicated by standardised suffixes: section of bacteria, or classes of fungi, -etes. Orders of bacteria or fungi: -ales. Families of bacteria or fungi: -aceae. Genera of bacteria or fungi: *-a, -us, -um, -er, -ix*. Species are identified by the name of the genus written with a capital letter and the name of the species with a small letter, e.g., *Staphylococcus aureus*. The abbreviation ssp. is used for the subspecies or var. often for variants of fungi, e.g., *Streptococcus salivarius* ssp. *thermophilus* or *Rhizopus microsporus* var. *oligosporus*.

non-alcoholic beverages
Soft drinks. Unsweetened carbonated ("soda") or non-carbonated ("still") water, fruit juice containing beverages (with or without CO_2), lemonades with or without caffeine, and special beverages for diabetics. $\rightarrow$ Potable water, natural mineral or spring water ("table water") constitutes the basis for these beverages; these "natural" waters may contain variable numbers of bacteria such as $\rightarrow$ *Serratia* spp., $\rightarrow$ *Pseudomonas* spp., $\rightarrow$ *Flavobacterium* spp., $\rightarrow$ *Xanthomonas,* etc. (see Table 13 under "drinking water"). – During processing the product may become contaminated by $\rightarrow$ *E. coli* and other $\rightarrow$ Enterobacteriaceae, $\rightarrow$ *Enterococcus* spp., $\rightarrow$ *Clostridium* spp., etc. Substrate properties favourable for microbial growth are enhanced by ingredients such as sugar, fruit juices, aroma components,

etc., by which additional contamination may also be introduced. → Acetic acid bacteria, → LAB and → yeast are typical of the → spoilage association of such beverages. The German "Mineral and Table Water" legislation of 1.08.1984 represents a typical example which serves as basis for the microbiological examination of these commodities. According to the recommendation 250 ml of the product should be free of → E. coli, → coliforms, → entercocci, and Pseudomonas aeruginosa; sulphite-reducing anaerobic sporeformers should be absent in 50 ml. For examination methods see Baumgart (1986), p. 365ff (also: Console and Crowman 1984; Korab and Dobbs 1984; Krämer 1987; Kunz 1988).

non-thermic decontaminant
Substances used in fruit beverages, wine and beer to minimise the bacterial count and eventually degrade into harmless components. Diethylpyrocarbonate (Baycovin®) was used till 1979. It was withdrawn on account of the possible production of ethylcarbonate besides ethanol and CO_2. The half life is 40 min at 10 °C and 15 min at 20 °C. The by-product methylcarbamate is harmless. The application concentration is up to 200 mg/l at room temperature (Classen et al. 1987).

normal yeasts → baker's yeast.

Norwalk viruses → viral gastritis.

notifiable disease
The frequency of notifiable diseases transmitted by food reached a plateau during 1987 in Germany (see Table 28), but has increased since then dramatically (compare Figure 18). Enteritis infections ("enteritis infectiosa") are caused by → Campylobacter, → Yersinia, → enteropathogenic E. coli, → Cl. perfringens. The number of registered cases in Germany and the USA may represent about 10% or less of the actual number (Sinell 1985).

nucleic acids
Polynucleotides. Carriers of the genetic information of cells. Highly polymerised consisting of deoxyribose (→ DNA) or ribose (→ RNA), phosphoric acid and N-containing bases: adenine, guanine, cytosine and thymine (Alberts et al. 1986). The nitrogen bases are often only partly decomposed to uric acid which is blamed for causing gout. – Offall and MO contain 6% nucleic acids; it should be decreased to 2% in → single cell protein for human consumption (DFG 1987).

Table 28 Notifiable diseases transferred by foods or associated with foods ("Stat. Bundesamt", Subject Series 12, Part 2, 1988).

	Cl. botulinum		Salm. typhi Cases	Salm. paratyphi Cases	Salm. (Other than typhi and paratyphi)		Other Forms of Enteritis Infectiosa		Shigella Cases
Year	Cases	Deaths	Cases	Cases	Cases	Deaths	Cases	Deaths	Cases
1982	34	1	231	196	40,977	88	5,393	1	1466
1983	30	1	212	165	34,989	81	10,067	3	1611
1984	32	0	205	134	31,701	71	15,313	1	1637
1985	26	1	227	167	30,566	89	15,690	5	1599
1986	30	1	230	153	33,271	34	19,508	7	1635
1987	18	0	228	149	39,342	35	22,932	3	1871

nucleus
Cellular nucleus. → Prokaryotes (bacteria and blue-green bacteria or cyanobacteria) possess only a closed circular strand of nucleic acids (bacterial chromosome or genophore) which is not enclosed by a nuclear membrane. Eukaryotic cells (fungi and algae) possess specific numbers of single chromosomes enclosed by a nuclear membrane.

nutrient broth
A liquid substrate containing nutrients for the culture of MO. Originally referring to a solution of meat extract, "broth" in present-day sense implies any liquid nutrient providing growth factors required by MO.

nutrient discs
Paper disc with the size of a membrane filter used for the determination of → bacterial counts in water and other liquids. The sterile disc is saturated with medium and dried. When moistened it serves as carrier and supplier of nutrients to the MO retained on the filter. Nutrient discs for the determination of the bacterial count and → differentiation of MO in water are available from, e.g., Sartorius (Göttingen).

nutrient medium
Sterile substrate solidified with agar or gelatin, but also potato slices or rice, etc., used for the cultivation of MO in the laboratory. → Bacterial counts. Agar melts at 95°C and solidifies at 40°C.

nuts
Occasionally also named shell fruit since several of these products are not nuts in a botanical sense, e.g., walnuts, brazil nuts, etc. This group of products is prone to fungal infections and may be contaminated in the kernel or skin. The kernel is easily infected, with concomitant mould growth, if not sufficiently dried or if stored in a humid environment, e.g., when transported (shipped) overseas. → Mycotoxins are commonly produced. This aspect is considered in regulations on aflatoxins. Ground products are extremely at risk.

obligate
Exclusively, indispensably. The term is used to describe specific microbiological characteristics, e.g., obligate anaerobe, obligate parasite, etc. Opposite: facultative.

ocean fish
Contamination mainly from the surface slime, and decay mainly originates from the gills and intestines. The catching methods (nets, drag nets, etc.), the climate zone (tropical, subtropical, arctic waters) and the season exert a qualitative and quantitative influence. Fish from deep waters (4 – 5 °C) are contaminated mainly with → psychrophiles, showing considerable metabolic activity even during transport under ice coverage. It is estimated that at 0 °C the quality of ocean fish is retained very well for 6 d, satisfactorily for 10 d and doubtful for 15 d. The keeping quality of non-eviscerated herrings is considerably worse, ca. 2 – 3 d (Heiss and Eichner 1984). – Dominating spoilage association: spp. of *Ps., Alteromonas, Flavobacterium, Pseudomonas, Cytophaga, Moraxella* and *Acinetobacter*. Intensive smelling substances are produced such as ammonia, H_2S, methylmercaptan, dimethyl and trimethylamine, etc. Dark coloured gills and turbid surface slime are visible indications of impending spoilage. In histidine rich spp. (mackerel and tuna) → histamine is formed by action of amino acid decarboxylases, and may result in poisoning (Krämer 1987). A few bacteria may constitute direct health risks, e.g., *Cl. botulinum* type E and non-proteolytic types B and F (toxinogenic), and the pathogen *Vibrio parahaemolyticus* (ICMSF 1980).

ochratoxins
A group of mycotoxins produced by → *A.*

and → *P.* growing on foodstuffs. Ochratoxin A is the best known. It is nephrotoxic, carcinogenic and causes endemic Balkan nephritis in man and nephropathy in pigs. It is not toxic for ruminating animals, since it is inactivated in the rumen. Occurrence: e.g., wheat, rye, barley, oats, maize and their products, as well as, e.g., figs. Often found in feedstuffs from developing countries. Transferred from feed via kidneys, blood and serum of pigs to sausages. → Carry over effect. It has commonly been detected in human serum [Bauer, J. and Gareis, M.: *J. Vet. Med.* B34 (1987) 613 – 617; Scheuer, R. and Leistner, L.: in *Jahresber. der Bundesanst. f. Fleischforsch.* (1986) C21; Reiss 1981]. The effect on human health cannot be estimated yet. It has been detected in chicken livers, albumen and egg yolks due to the carry over effect [Bauer, J. et al.: in *Arch. Geflugelkunde* 52 (1988) 71 – 75]. In D a highest tolerable level for ochratoxin A in foods has not been made mandatory yet. The daily intake amounts to ca. 80 ng/adult [DFG: "Ochratoxin A". Weinheim: VCH (1990)]. The Joint FAO/WHO Expert Committee on Food Additives suggested 112 ng/kg BW as → PTWI, at its 37th meeting of 14 June 1990. – Detection: TLC, HPLC, ELISA.

Oechsle degree
°Oe expressing the specific gravity of the must used for wine making. The grape juice is spindled with the must balance at 20 °C. Not only sugar but also acids, minerals, protein, pectin, etc., contribute to the specific gravity of the grape. 1 °Oe is approximately equivalent to the amount of sugar from which ca. 1 g ethanol is formed during fermentation.

off-flavours
Variety of adverse odours and/or tastes

often caused by MO in foodstuffs, e.g., eggs infected with *Pseudomonas*.

off-odour

Adverse odours of foodstuffs caused by MO, e.g., in poultry and fresh meat stored for extended periods formed by, e.g., *Pseudomonas* or other Gram – organisms.

offensive

Difficult, sometimes subjective interpretation of consumer response to the sensory or ethical acceptability of a product. Not necessarily related to sensory observation or microbiological detection of possible food spoilage. May be sufficient justification for withdrawal of a product, e.g., handling of paper money prior to touching sliced meat or other unpackaged food commodities with bare hands.

official control → control.

official methods

Official methods for the examination of foods are being published (e.g.) by the German Ministry of Health since 1975 on a regular basis, as part of its regulatory duties according to §35 LMBG. This "Official Collection of Examination Procedures according to §35 LMBG" is extended regularly, and new methods are added that have been verified in ring tests. This collection is arranged according to the "System Product Code" in method numbers. – In addition to → sensory test procedures, chemical, physical, immunological and microbiological methods and procedures for → sampling and → sample preparation are included, several of which are in accordance with EC methods. – In the edition of May 1988 the following relevant methods are described:

- detection and determination of aflatoxin B_1, B_2, G_1, and G_2 in foods
- detection and determination of aflatoxin M_1 in weaning food

- determination of preservatives in
  - low fat foods
  - high fat foods
- determination of inhibitory substances in milk based weaning and infant foods
- histamine content of fish, fish products and crustaceae
- preparation of samples for microbiological test procedures
  - milk and liquid milk products
  - dehydrated milk products, cheese and butter
  - meat and meat products
  - sugar, ice cream and semi-cooked products
  - weaning and infant foods on milk basis
- determination of the → pyruvate content of milk
- determination (enzymatic) of the content of $L(+)$ and $D(-)$ lactic acid in milk and milk products
- determination of the microbial numbers in milk and milk products, meat, ice cream, and semi-processed foods, weaning and infant foods, and processed cheese products
- determination of the number of coliforms in milk, milk products, butter, cheese, ice cream, lactose, and weaning and infant food
  - procedures with liquid growth medium (broth)
  - procedures with solid substrate
- determination of *E. coli* in milk and milk products, lactose, weaning and infant food, cheese
  - procedures with liquid (broth) media
- determination of coagulase-positive staphylococci in milk and milk products, meat, meat products, cheese, processed cheese products
  - colony counts
  - procedures with liquid nutrient medium
- detection of salmonellae; reference procedures in milk, milk products, lac-

tose, ice cream and semi-processed products, weaning and infant food on milk basis, processed meat products and sausages
- determination of the viable number of yeasts and moulds in milk products, and weaning and infant foods on milk basis
- determination of *Clostridium perfringens* in meat and meat products
- determination of Enterobacteriaceae in meat and meat products, and sausages
  - surface plating (reference procedure)
  - drop-plate method

For the examination of water→ drinking water.

oligotrophic
Surface water containing low levels of nutrients and therefore a low MO load. Opposite: eutrophic.

olives
Green olives are treated with 0.5 – 2% NaOH under air supply to remove the bitter substance oleutropin. The largest .part of the surface MO is inactivated. The fruit is then added into→ brine containing 10 – 13% NaCl, 0.5 – 1% sucrose and a starter culture of *Lb. plantarum* (10^7/ml brine); *Lb. brevis, Leuconostoc mesenteroides* and *Pediococcus damnosus* can also develop. The pH drops to 3.8 – 4.2 within 1 – 3 d due to the production of 0.7 – 1% lactic acid. Several yeasts, e.g., *Saccharomyces lactis, Candida utilis, Hansenula anomala,* etc., also take part in the fermentation.
Black olives: the bitter taste is removed before or after fermentation. 2.5 – 10% NaCl-solution (according to size) is used to cover the olives. Most of the time no starter cultures are used. Little is known about the bacteria involved. Yeasts present in the brine in order of decreasing numbers are: *Saccharomyces oleagino-*

sus, Hansenula anomala, Torulopsis candida, Debaryomyces hansenii, Candida diddensii and *Pichia membranaefaciens.* The acid content ranges between 0.4 and 0.6%.

Oomycetes
A class of aquatic and terrestrial fungi with sexual reproduction. Also parasites of terrestrial plants, e.g., *Peronospora, Phytophthora,* etc. The cell walls mainly consist of→ cellulose. (Müller and Loeffler 1982).

opportunistic parasites
Common expression in phytopathology, referring to→ parasites which are infective only under conditions where the intrinsic protection of the plant or plant product is weakened or destroyed (e.g., by climate, pests, ageing/maturing/ripening, or damage). Parasites causing damage often as a result of wrong storage practices. – In animal and human medicine reference is made to→ opportunists.

opportunists
Bacteria occurring more or less regularly on foodstuffs and to some extent also in drinking water, that may be classified as spoilage agents but that may cause lethal alimentary infections in immune suppressed or weakened individuals. Special care is taken in hospitals. – Bacteria clearly diagnosed as opportunists with relation to practical incidents are: *Proteus, Providencia, Citrobacter, Enterobacter, Edwardsiella, Ps. aeruginosa, Aeromonas hydrophila.* – Isolation of some of these microbial spp. or genera from drinking water does not necessarily imply a health risk, since they are ubiquitous and their presence generally unavoidable. They may act as pathogens in extraordinary cases only (Sinell 1985).

oral microbes
Naturally occurring MO in the oral cavity, rapidly building up in neonates and

changing throughout life according to life-style and nutrition of an individual (vegetarian, smoker). Apart from mainly acid producing bacteria, responsible for caries, yeasts and protozoa are also to be found. The organisms are immobilised on the mucous membrane and teeth by a slime layer (→ adhesin) which acts as a bioreactor, e.g., nitrate secreted in saliva is immediately reduced to nitrite and swallowed.

oranges → citrus fruit.

order
Latin: *ordo*. Taxon in biological → systematics.

organelles
Very small particles in cells with specific functions. They can be compared to organs in highly organised organisms, e.g., plasmids, plastids, mitochondria, ribosomes, etc. (Alberts et al. 1986; Schlegel 1985).

organoleptic → sensory.

Orleans process
Oldest commercial method to produce vinegar from wine by using acetic acid bacteria forming a → skin (surface layer). → Vinegar (Rehm 1985).

orthophenyl phenol
2-Hydroxy diphenyl, E 231, and its salt, sodium phenylphenolate, E 232, are used as fungistatic → preservatives. It is added to the washing water of citrus fruit in the producing country for prevention of mould growth on the peels (blue moulds, green moulds, stalk rot, brown rot). It kills *Salmonella* and *Staph. aureus* at higher concentrations of 500 – 1250 mg/l (Wallhäusser 1987). LD_{50} rats p.o. ~3 g/kg BW; ADI value presently still 0 – 1 mg/kg BW (Classen et al. 1987).

osmophiles
More appropriate term: osmotolerant, as these MO are able to grow and reproduce faster in conditions with a low osmotic pressure or high a_w value than in extreme but tolerated conditions, e.g., yeasts and fungi able to metabolise at a_w value < 0.72. → Water activity. They colonise and cause spoilage of marzipan, honey, and confectioneries.

osmotic pressure
According to Van't Hoff the osmotic pressure of a diluted solution equals the pressure that would have been exerted by the vapour of the undissolved substance at the same concentration. For low molecular weight substances it can be derived from the resultant lowering of the freezing point or increase of the boiling point. – The osmotic value of the cellular contents is equivalent to that of a 10 – 20% sucrose solution. The amount of water taken up from less concentrated solutions in which MO are typically found, will be determined by the cell wall. If the concentration of the solution is increased above that of the cell contents, the protoplast loses water. – Growth or multiplication is reduced as the osmotic pressure of the substrate increases, whilst → secondary metabolites are not produced, but primary (CO_2) production takes place and indicates high energy input. → Jam; → marzipan; → *Zygosaccharomyces rouxii*.

overlay
To cover inoculated agar media in a petri dish with the same nutrient medium (45°C) directly after the agar has solidified; used in methods for → bacterial counts, e.g., to suppress bacterial spreaders. Overlay 1 h after incubation for → resuscitation (Baumgart 1990). A mixture of paraffin oil and paraffin can be used as overlay in test tubes for anaerobic cultivation and serves to exclude oxygen or for the detection of gas production,

where positive reactions can be recognised when the layer rises.

oxalic acid
HOOC-COOH. Saturated dicarboxylic acid. An intermediate metabolite in cells. Stored in several plants as Ca-oxalate, e.g., rhubarb. Degradation product of vit. C in organisms, and excreted by several fungi, e.g., *A. niger*.

oxidising agents
→ Disinfectants with strong oxidation potential, e.g., → ozone, → hydrogen peroxide, potassium permanganate, → halogen compounds. Commercially available in concentrations of 20 – 40%. Application ("use") concentrations 0.02 – 0.05% (Kunz 1988).

oxidising disinfectants
Destructive → disinfectants, containing, e.g., strong oxidative compounds such as H_2O_2 and/or "per-acetic acid". An aqueous solution can decompose into H_2O + O_2 or acetic acid by which residues do not cause any problems. All MO and endospores are killed. Hydrogen peroxide is used in the → sterilisation of packaging material. It should be handled with care, even the use (application) concentrations are corrosive to the skin. "MAK" value: 1 ppm = 1.4 mg/m^3 H_2O_2 (5 mg/m^3 "per-acetic acid" in the former GDR). → AMES test is negative for both substances [Kästner, W.: *Arch. Lebensm. Hyg.* 32 (1981) 117 – 124].

oxygen demand
MO may be grouped according to their oxygen demand for metabolism and growth. This approach has been found valuable for judging the microbial behaviour during processing procedures, the selection of packaging materials, etc.
(1) Aerobes require O_2 as electron acceptor during glycolysis, with the production of H_2O.

(2) Anaerobes grown only in oxygen-free medium (e.g., canned foods); oxygen is toxic although not necessarily lethal.
(3) Facultative anaerobes prefer an aerobic environment, but also grow under oxygen restrictions; facultative aerobes prefer an anaerobic environment (e.g., GI tract).
(4) Aerotolerant anaerobes multiply in presence of O_2, but cannot utilise the oxygen.
(5) Micro-aerotolerants are adapted to low conc. of O_2 (5 – 0.1%) in which their growth is optimal (e.g., several → LAB).

Moulds are generally aerobes. Many spp. involved in spoilage, however, are still able to grow at 1 – 0.1% O_2, and also produce conidia, and some also → mycotoxins. Many yeasts, moulds and bacteria use aerobic respiration when sufficient O_2 is available, but under O_2 depletion they are able to "switch" over to fermentation and produce (e.g.) ethanol, with simultaneous strong reduction in growth (→ Pasteur effect).

oyster fungus
→ Cultivated mushrooms.

ozone
O_3. Toxic gas with a characteristic smell noticed at even 1:10^6. MAK value 0.1 ml/m^3 or 0.2 mg/m^3 air (DFG 1985). Bacteria, endospores, fungi and viruses are easily destroyed in 1 – 5 mg/l water. More effective at 0 °C than at 20 °C. Approved for the treatment of water for domestic purposes. Easily inactivated by inorganic and organic substances in solution, including Cl and I. It is also more effective in the dark since dissolved O_3 is destroyed by light (Wallhäusser 1988). Considerable research efforts have been devoted to possible applications of the gas and aqueous solutions for decontamination of foodstuffs, however without applicable results.

packaging
Produced from→ packaging materials for the protection and transport of food commodities. Should protect the food against mechanical and climatic influences, and keep harmful animals and MO out. In addition its role as bearer of information and advertisement is well known.

packaging material sterilisation
Aseptic packaging systems may rely on H_2O_2 (solutions of 15 – 35%) as sterilant, by which the material is treated by means of submersion in a bath or by jet spraying. The peroxide is removed by means of a subsequent thermal treatment (hot air, infrared radiation, etc.; UV may support the effectivity). Residual H_2O_2 should not exceed 0.5 ppm in the final container as stipulated by the FDA (Shapton and Shapton 1991).

packaging materials
Manufactured from different materials or combinations of such materials; may be contaminated with MO to different degrees, depending on type of material, transport, storage, etc. Should be free from pathogens. The requirements for the packaging material will be determined by the product properties (dry, easily deteriorating, sterile, pasteurised, etc.), the length of storage and the transport conditions. – For citrus fruit the use of packaging paper impregnated with→ diphenyl is allowed as protection against moulds (Reuter 1989; Cerny 1986; Kunz 1988).

Paecilomyces
Mould genus of the Deuteromycetes. Teleomorphs of these spp. belong to→ *Byssochlamys*. *Pae. variotiia* is weakly xerophilic. Growth up to a_w 0.84; opt. 35 – 40°C, min. 5°C, max. 45 – 48°C; >0.1% O_2. Distribution: raw materials,

food; in soils, especially in warmer regions. Growth range of *Pae. fulvus* as for *Pae. variotii,* but distribution different: acid fruit juices, soil, surfaces of plants (Pitt and Hocking 1985). – Several isolates from foods may produce→ patulin and byssochlamic acid.

papaya
Carica papaya produced in humid areas; post-harvest deterioration often caused by anthracnosis (*Colletotrichum gloeosporioides*), as for→ avocado. Rots starting from the stem are caused by *Ascochyta carica-papaya, Phomopsis carica-papaya, Phytophthora nicotiana* var. *parasitica* or *Ph. palmivora, Botryodiplodia theobromae* or *B. natalensis;* all may penetrate via damage in the epidermis. The fruits are cold sensitive, and refrigeration below 12°C may provoke infection by *Alternaria alternata* and/or *Stemphylium* spp. – Hot water treatment (49°C for 10 – 20 min, or 60°C for 0.5 min) prior to packaging and transport has proved effective. Damage of the epidermis by excessive heat may result that the unripe fruit does not reach typical ripe colour (Sommer, N. F.: in Kader et al. 1985).

para-hydroxybenzoic acid→ PHB
esters.

Paracoccus denitrificans
Gram –, immotile, aerobic, oxidase +, catalase +, non-halophilic, facultatively chemolithotrophic bacterial sp. Opt. 25 – 30°C. Found in soil, surface of weeds and root vegetables. Anaerobic growth in presence of NO_3, NO_2 and NO, with the production of molecular nitrogen (Bergey 1984). Shows potential for the removal of→ nitrate from vegetable pulp and juices for infant and diet food products, since no

accumulation of nitrite (Kerner et al. 1988; ZFL 7, 564 – 570). May be used for the same purpose for purification of drinking water.

paraffins

Paraffin waxes, paraffin oils, water insoluble and chemically inert→hydrocarbons. Application in minute quantities a.o. as antifoam agents, e.g., in cooking and thickening of foaming liquids and in→fermentation. Toxicologically safe; no ADI value suggested (Classen 1987). Oxidative degradation of n-paraffins to propanol or acetate by *Mycobacterium paraffinicum, Flavobacterium* and *Nocardia* spp.

paralytic shellfish poisoning

Mytilotoxicity. PSP. Type of food poisoning. Disease following the consumption of shellfish (molluscs, crustaceae) contaminated with toxins of a variety of poisonous dinoflagellates (unicellular algae) on which these shellfish have fed. The algae reproduce in the Atlantic Ocean, North and Baltic Seas when the temperature rises. Mussels should therefore be eaten only in months with an "r" in the name! Toxins are formed by the following known species: *Goniaulax tamarentis* (northern Atlantic Ocean), *Pyrodinium phoneus* (Belgian coast) and several related species in the USA. Several thousands of poisoning cases have been reported in the last few years and 105 cases were lethal. – The toxins are only up to 70% inactivated or destroyed by cooking and 90% by heat sterilisation. One of the toxins, saxitoxin, is a purine derivative with a MW of 372. The lethal dose is estimated at 1 – 10 mg/meal. – Symptoms appear 1 h after ingestion of the toxin. It includes a tingling and burning sensation or numbness around the mouth and in the finger tips, dizziness, delirium, fatigue, sleepiness, sensation of an obstruction in the throat, inability to speak, incoherent speech exanthema, fever, respiratory

paralysis (Krämer 1987; Sinell 1985). – Mussels from coastal waters which were not heat treated before consumption may also transmit→ polio or→ hepatitis viruses.

parasitism

Life on, within, or at the expense of, another organism; the latter ("host") represents the→ecosystem for the former. However, the separation between→ symbiosis and→ pathogenicity by definition may overlap, and will depend on the reaction situation of the "partner" or the host. Most parasites of importance in this context are facultative, i.e., they are also able to live→ saprophytically, e.g., microbes of the GI tract or the→ oral microbes.

paratyphus→ *Salmonella.*

Pasteur effect

Change in the basic type of metabolism of a cell, from→ respiration (oxidative phosphorylation) in the presence of air, to → fermentation (substrate level phosphorylation) in absence of oxygen, or vice versa. Typical of a number of yeasts (e.g., *Saccharomyces cerevisiae*). For moulds (filamentous fungi) the change to fermentative metabolism may be accompanied by→ dimorphism.

pasteurisation

Thermal processing, named after Louis Pasteur (1822 – 1895), mainly applied to liquid or pasty products below 100°C, with the aim at destroying all pathogenic vegetative bacteria. Simultaneously, some of the product associated enzymes are inactivated, but the nutrition physiological and sensory properties are largely retained. For the manufacture of drinking milk 3 different time-temperature combinations are legally approved; the latter of these (long time treatment at 62 – 65°C for 30 min) is practically not applied any more. Short-time heating at 71 – 74°C for 15 – 40 s, and the high-

Table 29 Bacteria surviving the pasteurisation of milk. List in order of increasing thermoresistance.

Acinetobacter calcoaceticus
Enterococcus faecalis
Enterococcus faecium
Streptococcus bovis
Streptococcus salivarius ssp. *thermophilus*
Micrococcus luteus
Micrococcus varians
Microbacterium lacticum
Bacillus cereus
Bacillus licheniformis
Bacillus pumilus
Bacillus megaterium
Bacillus circulans
Bacillus subtilis
Clostridium spp.

temperature treatment (also called short-time or HTST treatment, at 85 – 90 °C for 2 – 4 s are generally practised. Both are continuous flow processes in → plate heat exchanger systems. Depending on the initial (raw milk) microbial population and type of thermal treatment, 95 – 99% of the vegetative microorganisms are destroyed. Practically all → endospores survive, and may cause → sweet precipitation. In addition vegetative organisms such as *Sc. thermophilus*, *Sc. durans*, *Enterococcus faecium*, *Sc. bovis* and *Microbacterium lacticum* may survive. The latter, although a common heat resistant MO is not of significance in spoilage because of its slow growth rate. The same applies also to *Alcaligenes tolerans*, *Micrococcus luteus* and *M. varians*, as well as to the → ascospores of *Byssochlamys nivea* and *Monascus purpureus* that are found rarely (Teuber 1987). See Table 29. Similar thermal processing treatments, adapted according to the type of product, are applied for the shelf life extension of → liquid eggs (64 – 65 °C, 2.5 – 3 min), → fruit juices

(85 – 87 °C) or → vegetable juices or pulps (Krämer 1987).

pasteurised milk
Misleadingly also called "fresh milk", although the product is distributed in a processed form (consumers milk, drinking milk). Free from pathogenic MO, as long as no → recontamination has occurred. → Thermoduric MO may survive → pasteurisation (Table 29), and are mainly → mesophilic. Practically no growth at 5 °C; aseptical filling (no recontamination) will therefore result in an expected shelf life of ca. 1 week. Most detrimental spoilage MO is → *Bacillus cereus*, which is practically always present in raw milk, and causes → sweet curdling. Bacteriological examination of drinking milk → official methods (Baumgart 1990; Teuber 1987).

pathogenic microorganisms
Microorganisms (bacteria, yeasts, fungi and viruses) which may infect plants, animals or man orally, inhalatively or parenterally, and which may detrimentally influence their state of health. Obligatory or facultative parasites, opportunists or toxinogens. Of special importance here are the "food poisoning" microorganisms, i.e., those causing food-borne infections and intoxications. In addition, some moulds may be of importance that could cause → "farmer lung" after inhalation via cereal dust. In Table 30, differentiation is made between obligatory infectious and intoxicatious organisms, and so-called "risk" microorganisms, the latter of which only become significant when a specific "maximum value" ("infectious dose") per meal is exceeded (Sinell 1985).

patulin
Synonyms: clavicin, clavatin, claviformin, expansin, penicidin. → Mycotoxin produced by several → penicillia (Table 31, under *Penicillium*). Formerly designated

Table 30 Pathogenic MO that may be transferred by food, food dust, or water to man [from Sinell, H.-J.: *Fleischwirtsch.* 65 (1985) 627–678; modified].

Obligatory Agents of Infections and Intoxications

Zoonoses:
B. anthracis
Mycobacterium spp.
Brucella spp.
Listeria monocytogenes
Coxiella burnetii
Salmonella spp.
Campylobacter jejuni
Yersinia enterocolitica
Contaminated foods:
Vibrio cholerae
Shigella spp.
Enteropathogenic E. coli
Cl. botulinum
Hepatitis A viruses
Polio viruses
Different other viruses

Risk Microorganisms

Staph. aureus or its toxins
Bac. cereus
Cl. perfringens
Toxinogenic moulds or mycotoxins
Toxinogenic algae or phycotoxins in molluscs

Pathogenic MO Transmitted by Food

A. fumigatus
Different Mucorales
Thermoactinomyces vulgaris
Micropolyphora faeni
All allergenic moulds

as antibiotics; could not be used therapeutically because of toxicity. General cell toxin, haemorrhagic, orally not carcinogenic. Distribution: apple juice from fruit spoiled by brown rot, in fruit juice concentrates, mouldy fruit and tomatoes, wheat. – The WHO suggests 50 µg/kg as highest tolerable dose; this has been accepted by legislation in A, N and S (Reiss 1981).

Paxillus involutus → allergy.

pectin
Methylated polysaccharide that is deposited as "connecting component" between all plant cells. Several bacteria and moulds produce → pectinases during growth on or in plant parts, causing softening or "wilting" at an early stage. During ripening of fruit pectin is decomposed slowly, thereby causing a gradual change in the consistency and preparing the release of seeds.

pectinases
Trivial name pectin depolymerase; pectin glucosidase; polygalacturonidase. Systematic name: poly(1,4-α-D-galacturonide) glucanohydrolase. E.C. 3.2.1.15. Catalyses the hydrolytic cleavage of α-1,4-D-galacturonide bonds in pectates and other polygalacturonides. – Application: degradation of pectin, pectic acid or protopectin to oligo-uronides or galacturonic acid, e.g., in the manufacture of clear or naturally turbid fruit and vegetable juices, maceration of plant tissues, etc. – MO for the production: *A. awamori, A. foetidus, A. niger, A. oryzae, P. simplicissimum, Rhizopus oryzae, Trichoderma reesei, T. viride; Bacillus* spp. (GDCh 1983). – Because of the possible presence of the enzyme pectin esterase in pectinase preparations (→ enzyme preparations), the potential formation of → methanol is to be taken into account (DFG 1987).

Pediococcus
Genus of the Streptococcaceae. Gram + cocci, single, in pairs or tetrads. → Homofermentative; produces DL or L(+) lactic acid; catalase –, nitrite –, diacetyl + (for some strains). *Ped. acidilactici* may be

Table 31 Some mycotoxins and antibiotics that may be produced by penicillia (Leistner: in Rodricks et al. 1977; Reiss 1981, a.o.).

Component Effect	Penicillium*
Citreoviridin neurotoxic; cardial beriberi (man)	P. citreo-viride; P. miczskii; P. ochrosalmoneum; P. pulvillorum
Citrinin LD_{50} 110 mg/kg BW mouse p.o. nephrotoxic carcinogenic	P. canescens; P. citreo-viride; P. citrinum; P. expansum; P. fellutanum; P. implicatum; P. lividum; P. palitans; P. purpurescens; P. roqueforti; P. spinulosum; P. steckii; P. virdicatum; P. velutinum
Cyclopiazonic acid LD_{50} 50 mg/kg BW mouse p.o. neurotoxic, milt	P. caseicolum; P. crustosum; P. cyclopium; P. patulum; P. puberulum; P. viridicatum
Emodin vomiting, diarrhoea	P. avellaneum; P. islandicum
Luteoskyrin carcinogenic	P. islandicum
Ochratoxin A LD_{50} 21 mg/kg BW rat p.o. carcinogenic	P. chrysogenum; P. commune; P. cyclopium; P. palitans; P. purpurescens; P. purpurogenum; P. variabile; P. virdicatum; P. verrucosum
Patulin LD_{50} 35 mg/kg BW mouse p.o. general cell toxin	P. claviforme; P. chrysogenum; P. cyaneo-fulvum; P. cyclopium; P. expansum; P. granulatum; P. lanosum; P. lapidosum; P. novae-zeelandiae; P. patulum; P. rivolii; P. roqueforti; P. rugulosum; P. terrestre; P. variabile
Penicillic acid LD_{50} 35 mg/kg BW mouse p.o. carcinogenic	P. aurantio-virens; P. baarense; P. canescens; P. chrysogenum; P. commune; P. cyclopium; P. expansum; P. janthinellum; P. lilacinum; P. lividum; P. martensii; P. olvino-viride; P. palitans; P. piscarium; P. puberulum; P. roqueforti; P. simplicissimum; P. stoloniferum; P. suaveolens; P. virdiactum; P. verrucosum
Penitrem A neurotoxic, tremorgen	P. canescens; P. citreo-viride; P. citrinum; P. claviforme; P. crustosum; P. implicatum; P. jenseni; P. lividum; P. notatum; P. palitans; P. purpurescens; P. spinulosum; P. steckii; P. velutinum; P. viridicatum
Rubratoxin A, B. LD_{50} 7 mg/kg BW rat p.o. carcinogenic	P. purpurogenum; P. rubrum

(continued)

Table 31 (continued).

Component Effect	*Penicillium***
Rugulosin hepatotoxic carcinogenic	*P. canescens; P. cyclopium; P. islandicum; P. rugulosum; P. tardum; P. variabile; P. wortmanni*
Griseofulvin antimycoticum	*P. albidum; P. bredfeldianum; P. decumbens; P. melinii; P. nigricans; P. patulum; P. puberulum; P. raistrickii; P. viridicatum*
Penicillin antibioticum	*P. chrysogenum; P. notatum*

*Taxonomy acc. to Raper et al. 1968.

a component of some→ starter cultures for the manufacture of fermented meat products (mainly in the USA). *Ped. inopinatus* and *Ped. dextranicus* may be found in beer and sometimes in wine. *Ped. damnosus* may detrimentally affect the flavour of beer by production of diacetyl. *Ped. halophilus* is resistant against up to 18% of NaCl and may be found in → brines. The spp. mentioned may play a role in the manufacture of→sauerkraut,→lactic fermented vegetables and→pickles (Bergey 1986; Müller 1983a; Priest and Campbell 1987).

pelleting
The forming of thoroughly mixed feed, hops, herbal tea, etc., into pellets, tablets or small units under high pressure. The short-time heating affects the reduction of the microbial population by several log units without the simultaneous loss of nutrients; no pasteurisation or sterilisation! The process is most effective when the product is immediately packed, so as to prevent recontamination and possibly rehydration; it is also necessary to store the product at <70% rH.

PEMBA-agar
Selective medium for the detection of→ *Bacillus cereus*. Polymyxin B, egg yolk,

mannitol, bromothymol blue, agar (e.g., Oxoid CM 617). Definitive conclusion is only possible after the microscopical streaking of presumptively positive colonies and→spore staining (Baumgart 1990).

penicidin→patulin.

penicillic acid
Carcinogenic→mycotoxin which may be produced by several→*Penicillium* spp. during growth on food (Table 31). Presence has been detected in maize, confectioneries and meat and meat products. Stable at pH 2 – 9 at 100 °C for more than 15 min (Reiss 1981). Penicillic acid is often accompanied by→ochratoxin A.

penicillinases
Enzymes produced by several bacterial spp. which catalyse the hydrolysis of the β-lactam ring and thereby inactivate→ penicillins. E.C. 3.5.2.6. In the detection of→inhibitory substances in milk, meat products and offal, a control can be treated with penicillinase for the verification of the presence of penicillin.

penicillins
Group of→antibiotics with the longest therapeutical history. Characterised by a

β-lactam-thiazolidine ring, with a variable acyl (R) group. Best known are benzyl-penicillin (penicillin that G) and phenoxy-methyl-penicillin are ineffective against β-lactamase producing Gram + bacteria; ineffective generally against Gram – bacteria, yeasts and fungi. Active by inhibition of cell wall (peptidoglycan) synthesis of Gram + bacteria. Degradation by→ *Staph. aureus* and other bacteria that are able to produce penicillinase; selection towards resistant pathogenic strains was the cause of the first "wave" of hospitalism in the 1950's. – Technical production with *Penicillium chrysogenum*, e.g., 6-amino-penicillanic acid as precursor for the organic synthesis of therapeutically valuable new penicillins; also with *Acremonium kiliense* (*Cephalosporium acremonium*) for penicillin N. – Application as veterinary medicine may cause severe production failures in the manufacture of yoghurt, cheese and other fermented milk products, if the→ waiting times are not observed; → LAB are relatively sensitive to penicillins. – Sensitisation in man has been described.

Penicilliopsis clavariaeformis
Monotypical sp. of the Ascomycetes that mainly occurs in the soil of tropical regions, and may spoil fruit and cause root damage of seedlings. Produces emodin, skyrin and penicilliopsin that is transformed in presence of oxygen and light to hypericin.

Penicillium
Genus of the fungal class Hyphomycetes; teleomorphs, if known, belong to *Eupenicillium* or *Talaromyces* (Ascomycetes). Depending on biography, up to 678 spp. have been described (Thom 1930; Raper and Thom 1968; Pitt 1979). – Occurrence mainly in moderate zones, as→ *Fusarium,* and a siginificant part of the soil microbes. Most important spoilage fungi; some used as→ starter cultures, and some produce→

mycotoxins and/or→ antibiotics (Table 31). Parasites rare; opportunistic parasites (storage rot) common. Some spp. used for the production of→ enzyme preparations. "*P. glaucum*", an invalid sp., is often referred to in food literature as spoilage mould; more correctly the causative spp. may be *P. expansum* or *P. cyclopium* (Cerny and Hoffmann 1987). – Septate mycelium, colourless aerial mycelium; conidiophores with metulae and phialides (Figure 12); conidia in chains typically dark green; occasionally sclerotia. Temperature: – 2°C to + 40°C; opt. ca. 22°C. Ca. 30% produce→ nitrite on nitrate containing substrates. Killed by pasteurisation. Huge potential of adaptation and variability (Domsch et al. 1980; Pitt and Hocking 1985; Reiss 1986). – All species are non-fastidious and grow on mineral salts medium with nitrate and glucose of sucrose (→ Czapek-Dox agar).

Penicillium caseicolum
(*P. candidum*); used for the production of

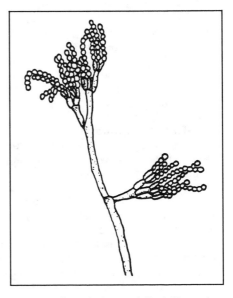

Figure 12 Conidiophore of *Penicillium chrysogenum.*

soft mould cheeses (Camembert, Brie, etc.); inoculation with conidium suspension. – Opt. 22°C; min. 4°C; max. 30°C. Proteases +, lipases +, amylases –, antibiosis –, may produce mycotoxins incl. cyclopiazonic acid; excretion of NO_2 +. Colour: white. – *P. camemberti* in the sense of Raper and Thom (1968) is practically not used any more, since growth and ripening do not conform any more to the manufacturing conditions. Acc. to Samson et al. [*A. van Leeuwenhoek* 43 (1977) 341 – 350], *P. camemberti, P. caseicolum* and *P. album* are collectively grouped in the sp. *P. camemberti.* – White mould found on other cheese types and especially in household, most probably is *P. caseicolum* and is transferred via refrigeration air, cheese plates, etc.

Penicillium digitatum
"Green mould". One of the most important spoilage organisms and colonisers of citrus fruit. By the excretion of → galacturonic acid the infection of healthy fruit is enhanced. Occasionally on bread. – Opt. 22°C, min. 10°C, max. 30°C; proteases –, lipases(+), amylases –; antibiosis –, mycotoxins; pH min. 3.0; colour yellow green. – Control: → thiabendazole (treatment of the peel), → diphenyl (impregnation of the packaging material) (Reiss 1986).

Penicillium expansum
Ubiquitous on foods. Spoilage of stone and kernel fruit (brown rot) after damage of the skin (e.g., by insects or during harvest), tomatoes and vegetables in store room. On cereals during storage at > 82% rH. Destroys leather, wall paintings, paint, etc., and is involved in the weathering of sandstone and other minerals through the production of acids. – Opt. 22°C, min. – 3°C, max. 30°C; a_w growth 0.82, conidia production 0.85, germination of conidia 0.82 – 0.85. CO_2 tolerance < 6%; proteases + + +;

lipases(+), cellulases + +, amylases +, antibiosis 0 to + + +. Mycotoxins: patulin, citrinin. NO_2 formation. – Control: → preservatives, → pimaricin. Desiccation (Reiss 1986).

Penicillium italicum
"Blue mould", common on citrus fruit; occasionally on green coffee. – Opt. 22 – 24°C, min. – 3°C, max. 33 – 35°C; a_w growth 0.87, production of conidia 0.89, opt. 0.96 – 0.98; pH 1.6 – 9.8; D_{65} value 0.6 – 0.7 min. Antibiosis –; mycotoxins? Control: → *P. digitatum* (Reiss 1986).

Penicillium nalgiovensis
"Edelschimmel Kulmbach" ("noble" mould sp. named after the town of Kulmbach Germany where the German Meat Research Centre is located); used as starter culture for the manufacture of mould ripened meat products such as mould ripened fermented sausage, South Tyrolean type of ham and "Bündner Fleisch". – Opt. 15°C, min. 4°C, max. 30°C; a_w > 0.82; proteases + +, lipases +, amylases(+); antibiosis; mycotoxins –; NO_2 production –. The use of the mould also serves as protection against possible mycotoxin production by the "natural" moulds on the meat products mentioned (Reiss 1986).

Penicillium roqueforti
Starter culture for the production of blue veined or blue mould cheeses (Roquefort, Gorgonzola, Stilton, Danish Blue, etc.). Normal coloniser of refrigerators in households. May be found on all types of food, e.g., bread, fat, crown cork (→ musty smell), barley, fruit juice, etc.; not present on living plant tissues. – Opt. 22°C, min. 4°C, max. 30°C; proteases + +, lipases + (prefers 1- and 3-position of the triglyceride), amylases –, antibiosis –. Mycotoxins: PR toxin, roquefortin A and B, mycophenolic acid; some isolates

produce patulin. – Among several investigations, roquefortin A and B and mycophenolic acid have been detected only in a few cases. The quantities, however, were so minute that in no case a health risk could be expected for the consumer. – Control: → preservatives; desiccation (drying); storage at <4°C (Reiss 1986). – The purity of starter cultures may be checked by performing a standard plate count on Czapek-Dox agar, and, parallel to it, a plate count on the same medium containing 0.5% conc. acetic acid (pH 3 – 3.5). Only *P. roqueforti* will grow on the acid medium [Engel and Teuber: *Europ. J. Microbiol. Biotechnol.* 6 (1987) 107 – 111].

Penicillium verrucosum var. *verrucosum*

(*P. viridicatum*). Common sp. found on food of animal origin, and on cereals as field fungus that develops during storage under humid conditions; may produce → ochratoxin A and → citrinin. Growth: 0 – 31°C, a_w 0.95 – 0.99, pH 3.6 – 8.4. Toxin production: 4 – 31°C, opt. 16 – 24°C, a_w 0.83 – 0.99, opt. 0.95 – 0.99. Inhibition: pH 4.5 and 0.02% of potassium sorbate or 0.2% sodium propionate.

pentoses

Sugars (aldoses) with five C atoms. Monosaccharides widely distributed in plants (especially in the middle lamellae), either free, or as pentosans, or in combination with other sugars. Examples: ribose, xylose, rhamnose, arabinose. A large number of MO are able to utilise pentoses.

peptides

Chains constituted of two or more → amino acids covalently joined by peptide bonds; formed a.o. during the hydrolytic cleavage of proteins. Peptides with specific amino acid sequences taste bitter and may cause taste defects in milk products. *B. cereus* and also the → milk mould are often the causative agents of bitter peptides. → Sweet curdling.

Peptococcus

Gram +, anaerobic cocci, single, in pairs and chains or aggregates. Normally occurs in the vagina and intestines. Destroyed by pasteurisation (Bergey 1986).

peptones

Mixture of polypeptides produced by the partial tryptic or peptic degradation of protein, either from meat or casein. Used as ingredients of growth media. Commercially available from BBL, Difco, Merck or Oxoid.

peptonisation

Hydrolytic cleavage of protein resulting in curdling. Enzymatic coagulation (clotting) of casein in a tube with ca. 10 ml of milk, is a character that can be easily determined, and may be applied in the → differentiation of bacteria and fungi. → Sweet curdling. Syneresis causes whey to separate from the casein fraction. Common for → *Bacillus* spp. and some spp. of → *Pseudomonas*.

Peptostreptococcus

Anaerobic, Gram + cocci, in pairs, chains or clumps (aggregates). Catalase –. Occurrence: skin, mucous membranes of animals, intestines; occasionally infectious. Destruction by pasteurisation (Bergey 1986).

peracetic acid

Peracids. Strong oxidative agent for surface sterilisation at room temperature. → Disinfectant; → hydrogen peroxide (Kunz 1988; Wallhäusser 1988).

perfect stage

→ Teleomorph. Growth form of a fungus (→ mould), which is able to multiply sexually. Opposite: → imperfect stage or → anamorph; → Deuteromycetes (Müller and Loeffler 1982).

peritrichous
Refers to flagellation of bacteria that is relatively uniformly distributed over the whole cell surface, e.g., for *Bacillus* or *Salmonella*.

perle wine→ sparkling wine.

permanent carriers→carriers (excreters).

permeases
Enzymes which support the transport of nutrients via the semi-permeable→cytoplasmic membrane towards the reduced concentration gradient.

Peronosporales
Order of fungi (Oomycetes) which includes the genera→*Phytophthora* (downy mildews causing potato rot), and *Plasmopara* (false mildew of grapes). Significant economic importance, mainly prior to the introduction of fungicides for plant protection (e.g., copper-based antifungals).

peroxide-catalase process
Process used back in the 1930's for reducing the number of microbes in raw milk. A conc. of 0.3–0.5% perhydrol (30% hydrogen peroxide) kills 99–99.9% of the microbial population, including endospores, by nascent oxygen within 30 min. Excess peroxide is removed with the aid of→catalase. Vitamin C is quantitatively destroyed. Only applied in tropical regions where→pasteurisation is not feasible yet. – In the 1950's a similar process had been developed for the treatment of liquid egg; with the aid of→glucoseoxidase, gluconic acid and H_2O_2 are formed from glucose. Main purpose is the elimination of glucose, so as to prevent the Maillard reaction (non-enzymatic browning) during heating or drying; at the same time, however, the microbial load is reduced.

persipan
Ground endosperm from peach kernels in 1:1 mixture with sugar. Similar to→marzipan. Reclamations common because of→aflatoxins resulting from mould contamination after harvesting and during transport. Spoilage possible by osmotolerant yeasts.

personal hygiene
Starts with the selection of employees and the provision of amply equipped sanitary rooms. Special care should be taken to observe requirements, e.g., for handwashing, wearing of protective clothing, as well as continuous education and training of personnel, which should be the direct responsibility of management. This also includes regulations restricting the handling of foods by→excreters (e.g., German no. §17 BSG) ("Bundesseuchengesetz"). Measures of this kind are not only applicable to processing and handling of foods on commercial or industrial scale, but also to commercial kitchens, restaurants, and catering in general (e.g., →airline foods) (Sinell 1985). Application is also important in private households!

perthophytes→necrobionts.

pesticide
Refers to→plant treatment agents, applied for the protection of cultivated plants against "pests" (e.g., insects).

petri dish
A round, flat-bottomed glass or plastic dish or "plate", named after the bacteriologist R. J. Petri (1852–1921), a colleague of Robert Koch. Used for the cultivation of MO. Two flat loosely fitting discs with vertical sides; the top or "lid" part is slightly larger than the bottom structure in which the molten agar→nutrient medium is poured. After solidification the dishes are incubated before determination of the→microbial count. Plastic dishes

may fit tightly on account of condensed water, and thus promote anaerobic conditions; this is prevented by "nops" in the lid that allow free exchange of air.

pH value
Measure of the "acidity" or "alkalinity" of a water containing product, given as negative logarithm of the H_3O^+ ion activity (hydrogen ion exponent). Not identical with Sr^0, ^0SH or titratable acidity! → degree of acidity. The pH value is a decisive parameter that affects the ability and rate of multiplication of MO, as well as the germination and outgrowth of → endospores. Based on their pH values, foods are generally classified into 3 groups, thereby indicating their relative spoilage potential.

(a) Low acid to neutral products, with pH 7 – 4.5: meat, poultry, fish, milk, eggs, peas, beans, carrots, potatoes
(b) Acid products, with pH 4.5 – 3.5: apples, pears, oranges, tomatoes, fruit juices, tomato juice, tomato puree
(c) High acid products, pH <3.5: lemons, rhubarb, berry fruit, several fruit juices, sauerkraut, wine, vinegar

Ranges for growth and multiplication: fungi, pH 1.5 – 9; yeasts, pH 1.5 – 8; *Bacillus* spp., pH 4.5 – 8.5; LAB (excluding *Enterococcus* and *Carnobacterium*), pH 3 – 7; acetic acid bacteria, pH 2.8 – 7.5. The type of acid is an important factor. Examples: *Staph. aureus* is inhibited at pH 4.6 by acetic, at pH 4.3 by lactic acid, at pH 4.1 by citric acid and at pH 3.9 by tartaric acid (Heiss and Eichner 1984). Through adjustment of the pH, preferably in combination with other factors (a_w, preservatives, etc.), extension of the shelf life may be attained (→ hurdle concept).

phages → bacteriophages.

PHB
Poly-β-hydroxybutyrate.

PHB esters
Approved → preservatives E 214 – E 219. The methyl, propyl and ethyl esters of para-hydroxybenzoic acid and their sodium salts. Undissociated compounds with pH independent activity, especially against yeasts, moulds and Gram+ bacteria. – Sensory changes in foods caused by concentrations ≥0.08%. – LD_{50} mouse p.o. ca. 8 g/kg BW; sodium salts 2 – 3.7 g/kg BW. → ADI value 0 – 10 mg/kg BW (Classen et al. 1987). For maximum tolerable amounts in particular foods, see German "ZZulV" – legislation – of 22.12. 1981, §3, addendum 3, list B.

phenols
Phenol (carbolic acid, hydroxybenzol; C_6H_5OH) had been one of the most important hospital → disinfectants at the beginning of this century. It served as reference substance (standard) for the development of new, more effective and less offensive smelling disinfectants (phenol coefficient). Its derivatives ("phenolics") are generally either Cl- or CH_3- substituted products, and are more effective and in general less toxic than phenol itself. Order of increasing effectivity as fungicide: phenol < cresol < thymol < chlorthymol. Generally microbistatic or microbicidal, depending on conc., temp. and other factors in the environment. Areas of application: Lysol for hand disinfection, whilst Thymol may be used as disinfectant, antiseptic (e.g., mouthwashes) and as a preservative for non-food samples (Wallhäusser 1988).

phycotoxins
Collective name for toxic metabolites of → Cyanophyceae. When ingested in large quantities by fish, the fish meat becomes toxic (→ water blooms) (Steyn and Vleggaar 1986).

phytoalexins
Chemical protective substances pro-

duced by viable plant tissues, e.g., in reaction to mould contamination, heavy metals or freeze-thaw damage. These are generally low MW antimicrobial ("antibiotic") protective compounds that are normally either absent or present only in minute quantities in the early phases of development; toxic to insect larvae, fungi and rodents, and even to man. – Under moderate to cool climatic conditions 5-methoxypsoralen (bergapten, heraclin), 8-methoxypsoralen (xanthotoxin, methoxalen) and 4,5,8-trimethylpsoralen may be present, e.g., in celery following infection with→ *Sclerotinia sclerotiorum*. These furocumarins sensitise the skin against UV light and cause reddening of the skin when the infected areas are touched. Inflammations, formation of blisters and even gangrene may follow; disease has earlier been found among agricultural labourers. – Indications of skin carcinogenicity, in combination with UV light (280 – 320 nm), resulting from the ingestion of only mg quantities per meal by fair-skinned people [Schlatter, J.: *Die toxikologische Bedeutung von Furocoumarinen in pflanzlichen Lebensmitteln. Mitt. Gebiete Lebensm. Hyg.* 79 (1988) 130 – 143]. – Carrots infected during storage with *Ceratocystis fimbriata* produce 6-methoxymellein; infection with *Sclerotinia sclerotiorum* induced the formation of the so-called psoralens. – Potatoes react on infection with *Phytophthora infestans* with accelerated production of solanidin and its glucoalkaloids α-solanin and α,-chaconin, chlorogenic acid, scopolin and rishitin. – Sweet potatoes (*Ipomoea batatas*) produce hepatotoxic ipomearone and similar sesquiterpenes upon infection by *Fusarium solani* or *Ceratocystis fimbriata*. – The same phenomenon has been observed for peas, beans, soya and clover; whereas the protective substances may differ from sp. to sp., an inhibitory action against mould infections is still exerted. – Cultiva-

tion of agricultural plants for resistance should not be practised by inducing phytoalexin formation, since a number of these substances are toxic to mammals (Rodricks 1976).

phytoncides
Antimicrobial substances produced by plants, conferring resistance against potential parasites or pathogens. Commonly volatile etheric oils, e.g., in→ spices, but also alkaloids (pepper), tannins, hydrocarbons, alcohols, ketones, terpenes, as well as sulphur containing mustard oils, partly free and partly bound glycosidically. Their action may be relatively specific, i.e., they inhibit different MO either to variable extents or not at all. Several are toxic to warm blooded animals and humans; some are genotoxic. By this standard nutmeg would presently not have been legally approved as food!→ Thymol (up to 3 g in 100 g of thyme) has the strongest antimicrobial action, being 20 times stronger than phenol. The shelf life of foods is not significantly improved by phytoncides, because of the low concentrations present. – Example: salicylic acid in grapes, tannins in unripe pears, benzoic acid in cranberries, allicin and acrolein in garlic, etc.

phytopathology
Discipline of the plant diseases, caused by MO, viruses, nematodes or insects.

Phytophthora
Genus of the fungi (class Oomycetes, order Peronosporales); highly specialised parasites of terrestrial plants. Most important is *Ph. infestans* (late blight of potato). Worst epidemics in Central and Western Europe that have decisively influenced history in Ireland during the years 1845 – 1850. Presently effective plant protection agents are commercially available. For stored potatoes the losses are presently < 1% of that caused by other

fungus storage rots. (Müller and Loeffler 1982). For apples and pears *Ph. cactorum* and *Ph. syringae* are important causative agents of storage rots; for *Citrus* spp.: *Ph. citrophthora*.

Pichia

Sporeforming yeast genus with elongated, spheroid or ellipsoid cells; vegetative reproduction by multilateral budding. *Pi. membranaefaciens* produces a skin (Germ.: "Kahmhaut") on the surface of old→ brines, and causes sensory defects in wine; production failures in→ sauerkraut by utilisation of lactic acid; spoilage of delicatessen. *Pi. farinosa* and *Pi. fermentans* are involved in the fermentation of→cocoa. Some spp. are xerotolerant (Glaubitz and Koch 1983; Beuchat 1987).

pickled cucumbers

Lactic fermented cucumbers ("gherkins"). Cucumbers are selected acc. to size, etc., washed and covered with → brine (salt + sucrose solution in water); starter culture may be added. Aerobic MO develop first until oxygen is depleted, followed by→ LAB which are responsible for the main fermentation process at room temperature for several days. The heterofermentative LAB (initially *Leuconostoc* spp. followed by *Lb. brevis*) are soon overgrown by homofermentative spp., e.g., *Lb. plantarum* and *Pediococcus pentosaceus* (*Ped. damnosus*). In addition to lactic acid, acetic acid, ethanol and an amount of CO_2 is formed. For "salt stock" pickles the initial intermediate salt level of 8 – 10% is gradually increased to 15% so as to inhibit growth of pectinolytic MO that may cause softening of the cucumbers (Müller 1983a).

pickles

Mixtures (or single products) of lactic fermented vegetables. Practically all vegetable types are suitable. Well cleaned/ trimmed raw materials are mixed with herbs and spices and→ brine containing sucrose or lactose and starter cultures added. Typically the starter may contain a mixture of two or more strains of *Lb. plantarum*, *Lb. brevis*, *Lb. bavaricus* [L(+) lactic acid!], *Leuconostoc mesenteroides*, *Pediococcus damnosus*, *Str. durans*, etc. Typically, *Lb. plantarum* is the dominating organism (Müller, G. 1983a).→ LAB.

pigments

Coloured products of→ secondary microbial metabolism. When specific MO are present in sufficiently large numbers undesired colour changes may be caused in foods (Table 33). May cause undesired or even desired changes in foods and utensils (see Table 32). Distinguished between water soluble pigments (may, e.g., form coloured zones around colonies) and fat soluble pigments that diffuse into fat layers and synthetic materials. – The colours of→ conidia, causing the characteristic appearance of fungal colonies, are in most cases probably not pigments but structural colours.

pili

Fimbriae (sing.: pilus). Filamentous proteinaceous structures extending from the cell surface of some bacteria. Diameter: 3 – 25 nm; length: 0.2 – 12 μm. Functions a.o. for nutrient uptake,→ agglutination, → plasmid transfer (conjugation); some present binding sites for certain bacteriophages (Alberts et al. 1986; Schlegel 1985).

Pilsener→ beer.

pimaricin→ natamycin.

pineapple (Ananas)

The fruit do not possess a climacteric period, and have to be harvested when practically ripe. Infestation mainly by *Thielaviopsis paradoxa;* infection through the

Table 32 Colour products (pigments) of microbial origin.

Name	Microorganisms	Colour	Soluble in	Comment
Ankaflavin, etc.	*Monascus purpureus*	yellow	water	
Anthrachinones	*Fusarium* spp.	reddish	water	
Aterrimin A + B	*B. subtilis* var. *aterrimus*	black	ethanol	
Carotinoids	yeasts, micrococci, etc.	yellow-red	fat, synthetics	
Citrinin	*P. citrinum* and other fungi	yellow	water, ethanol	mycotoxin
Emodin (acid)	*P. cyclopium*	orange	water	mycotoxin
Monascorubramin	*Monascus purpureus*	red	water	
Prodigiosin	*Serratia marcescens*	red	fat, synthetics	
Pulcherrrimin	*Candida pulcherrima*	red	insoluble	
Pyocyanin	*Ps. aeruginosa*	blue-green	alkaline liquids	
Rugulosin	*P.* spp., *Endothia parasitica*	yellow	water	mycotoxin
Sarcinaxantin	*Sarcina lutea,* micrococci	yellow	fat	
Volacein	*Chrombacterium violaceum*	purple	fat, synthetics	

cutting surface at bases. This mould grows very slowly at 11°C; rapid cooling down to 8 – 10°C and transport at a temperature just below 10°C are therefore recommended. Treatment of the cutting surface with fungicides has proved effective. – *Fusarium moniliforme* var. *subglutinans* often causes rotting of brasilian cultivars. It is, however, not known whether mycotoxins are being formed on such cutting surfaces (Sommer, N. F.: in Kader et al. 1985).

pitching sour → sourdough.

pitching yeast
Pure culture of yeast used for the inoculation of a substrate/medium during the production of baker's yeast, feed yeast and beer. As preventative measure against possible contaminations, the initial concentration in the fermenter/bioreactor should amount to at least 10^6 cells/ml.

plant diseases
Diseases of microbial origin, either latent or visible on cultivated plants. May also become visible as spoilage deterioration

("rot") during storage of vegetables and fruit, resulting in reduced shelf life. Examples: → *P. expansum,* → *Phytophthora,* → anthracnose, → *Monilia,* etc. Measures when sowing out on the field (→ plant treatment agents) or dips after harvesting (→ post harvest protection; → stock protection), may keep damage within limits. Appropriate storage methods (temperature, storage atmosphere, rH, etc.) are helpful measures (Heinze 1983; Kader et al. 1985; Weichmann 1987).

plant treatment agent
Active components or preparations thereof; applied for the protection of (1) plants against damaging organisms and, diseases (plant protection), (2) plant products against damaging organisms (stock protection). Deleterious organisms are: (a) animal parasites, (b) plant types, especially weeds, parasitical plants, and mosses and algae, (c) microbes including fungi, bacteria and viruses in all developing stages (German Plant Protection Law in the version of 22.05.1979; regulation for maximum concentrations of plant protection agents in the version of

Table 33 Some pigment producing bacterial genera and species.

Red	*Serratia marcescens* *Candida pulcherrima* *Monascus purpureus* *Rhodotorula* *Sporobolomyces salmonicolor*
Orange	*Brevibacterium linens* *Micrococcus* *Staph. aureus*
Yellow	*Flavobacterium* *Sarcina*
Blue-green	*Pseudomonas*
Violet	*Arthrobacter* *Chromobacterium violaceum* *Ps. idigofera*
Black	*B. subilis var. aterrimus* *B. subtilis var. niger* *Auerobasidium* *Dematiaceae*

18.04.1984). – Differentiated between: insecticides, acaricides (against mites), fungicides (against moulds), herbicides (against weeds), nematicides (against nematodes), molluscicides (against snails), rodenticides (against rodents), as well as growth and inhibitory agents (Berg et al. 1977). – All commercial substances are to be approved (e.g., in Germany by the Federal Centre for Biological Research in Braunschweig) for effectivity and degradability, and (e.g., by the Federal Health Authority – "Bundesgesundheitsamt" – in Berlin) for toxicity. On release the maximum tolerable concentration for foods in general, and/or for special commodities including dietetics has to be specified. – For all plant protection measures the residual concentrations in foods are to be taken into account.

plasma
Refers to the fluid part of blood; obtained, e.g., by centrifugation; defibrinated pig blood free from cell components. Up to 10% fresh or reconstituted plasma may be added to→cooked cured sausages. Feeding of mouldy grains containing→ ochratoxin A, this toxin accumulates in the blood of pigs and is transferred in this way to blood, liver and cooked sausages. About 20% of the sausage samples examined in Germany were found to contain ochratoxin A, although in small amounts. During spray drying of pork blood plasma 90% of the ochratoxin is destroyed.

plasmids
Extrachromosomal, covalently closed circular (ccc)→DNA molecules, that may a.o. transfer desired or undesired properties from one cell to another or even between spp. In this way, e.g., resistance (R) factors against antibiotics, ability to produce particular enzymes, N-fixation genes etc. are "transported". Gene technologically the plasmids are used for the construction of starter organisms with specific properties. Found in prokaryotic and eukaryotic cells, and, although dispensable, are probably responsible for maintaining the variability and adaptive ability of the organism; → selection (Alberts et al. 1986; Schlegel 1985).

Plasmopara viticola
Fungal genus of the Peronosporales (Oomycota). Parasitic, causing downy mildews and peronospora disease of grape and sunflower (*P. viticola* and *P. halstedi*). Control with Bordelaise broth, which was developed in Southern France as one of the first→ plant treatment agents.

plate→ petri dish.

plectenchyma
→ Mycology. Tissue-like structures composed of thick-walled atypical hyphae; resembling three-dimensional cell grouping. Commonly as→ survival forms

or→ fruiting bodies of fungi, or lichen thalli. Two principal types: prosenchyma (prosoplectenchyma) with loosely woven (individually distinguishable) hyphae, and pseudoparenchyma (paraplectenchyma) with tissue of closely packed and hyphae not individually distinguishable.

Pleurotus ostreatus → cultivated mushrooms.

pockets
Contact surfaces ("saddles") of→cooked sausages, etc., during→smoking. Not influenced by the smoking process, these surfaces remain more moist and thus not well preserved with a higher spoilage potential (Sinell 1985).

polar
Refers to the position of either intracellular bodies, inclusions, granules, etc., or of filaments or→flagella, at one or both poles of the cell.

poly ...
Greek: polys = many, e.g., polyethylene, polysaccharide, etc.

polyethylene glycol
PEG. Carbowax. Lubricant, precipitation agent; used, e.g., in cell fusion, as humectant and lubricant in the production of pharmaceutics, cosmetics, sanitisers and occasionally for tobacco.→*Pseudomonas aeruginosa*, commonly associated with drinking water, is able to degrade PEG enzymatically up to MW of 20,000, with the release of ethylene glycol and diethylene glycol in water solution [Haines and Alexander: *Appl. Microbiol.* 29 (1975) 261 – 263].

population
Total sum of the individual microbial cells and viable units (fragments, filaments) (colony-forming units) within an ecosystem niche or (food) sample, e.g., in ground meat, fruit juice, etc.

post harvest protection
Precautionary steps to prevent losses or spoilage of harvested crops. Measures include pre-harvest treatment with fungicides and insecticides, followed by careful handling during harvesting to prevent damage to the epidermis of fruit, seeds or nuts and by choosing the appropriate time for harvesting. Other steps determined according to the climate and product type involved are: washing procedures, cleaning, drying to a_w 0.70, rH control, temperature control in the silo or store, elimination of→ethylene from the atmosphere in the store room (→CA storage), and the addition of sprouting inhibitors to potatoes to prevent losses. – The following factors are considered to contribute decisively to worldwide post-harvest losses: rodents 10%, insects 10%, birds 5%, MO 5%, and X% due to human mistakes. By taking proper precautionary steps, part of the losses can be prevented to the benefit of the consumer, especially in developing countries (Kader et al. 1985; Weichmann 1987).

potato bacillus→ *Bacillus cereus*.

potatoes
Basic food due to nutritional physiologically valuable protein (lysin) (15 – 21%) and vitamin C. It can be stored for up to 8 months at 6 – 8 °C and 90 – 95% rH. After harvesting a period of approximately 14 days must be allowed for wound healing due to harvesting damage. During the storage period damage is caused by the following organisms: *Fusarium coeruleum* (45%); *F. sulfureum* (21%); *Colletotrichum atramentarium* (9.5%); *Alternaria alternata* (9.4%); *Fusarium* species; *Phoma* species; *Phytophora infestans* and *Rhizoctonia* species. There may be an annual shift in the frequency of a par-

ticular organism responsible for the damage (Frank: in Heinze 1983). General steps to prevent damage during storage: only mature crops should be stored; harvest in dry conditions; minimal damage to the crop; the crop must be sorted; temperature control in the store; ventilation at intervals and the crop should be marketed in time (Müller, G. 1983a).

pour plate

Plating procedure developed by Koch for counting the viable microbial population (→ plate count) in a (food) sample. A measured volume (e.g., 1 ml) is pipetted from a dilution series (in test tubes) into a petri dish and ca. 12 – 15 ml of molten (45 – 50 °C) agar medium added (poured), followed by swirling so as to disperse the "inoculum" in the medium. Transfer to a petri dish (duplicate for each dilution) is performed stepwise, starting from the tube with the highest dilution (see Figures 13 and 14). After solidification the petri dishes are incubated upside down. Single cells, chains or clumps give rise each to only one colony; the number of colonies is multiplied with the dilution factor and designated → cfu ("plate count") per g or ml.

pour plate method → pour plate.

pre-enrichment

First step in the detection of *Salmonella*. After the specimen has been mixed, 25 g is added to 225 ml buffered peptone water, homogenised and then incubated at 37 °C for 20 h (Baumgart 1990).

pre-ripening

First step in the production of → rennet cheese. The milk in the cheese vat is incubated overnight at 12 °C with ca. 1% starter culture. Care should be taken as coliforms can also multiply as a result of recontamination. The pre-ripened milk

must eventually be pasteurised or heat treated again.

predisposition

Factors which promote or enable an infection with → pathogenic or → opportunistic MO, e.g., via food or water. This situation exists, e.g., in infants, old-aged, diabetes, undernourished, patients during rehabilitation following disease and injuries, immunodeficient persons, etc. (Krämer 1987; Sinell 1985).

preservation

Various methods to extend the shelf life of food by preventing the multiplication and/or growth of the MO to a certain extent and to minimise possible chemical and sensory changes. → Dehydration; → heat sterilisation; → freezing; → radiation; → preservatives; hurdle principle.

preservatives

Antimicrobial substances (Table 34) in combination with physical methods, e.g., heat and other chemical inhibitory substances, e.g., table salt, organic acids, etc., which prevent multiplication and toxin formation of MO or destroy them when applied appropriately. Their application is regulated by the "Zusatzstoff – Verkehrsverordnung und der Zusatzstoff-Zulas-sungsverordnung (ZZulV)" (Germany), or by regulations approved by other governments (e.g., detailed in "Food Additive Orders" of the FDA in the USA), and must be declared. Their addition is limited to particular products and their components or raw materials (Table 35). – In addendum 3, list B of the "ZZulV" (Germ.) the maximum amounts are stipulated. Special regulations for maximum concentrations of combinations of preservatives – often advantageous by microbiological considerations – are presented in §3 part 3 of the same regulation. Suboptimal dosage may be dangerous as certain microorganisms could be favoured by

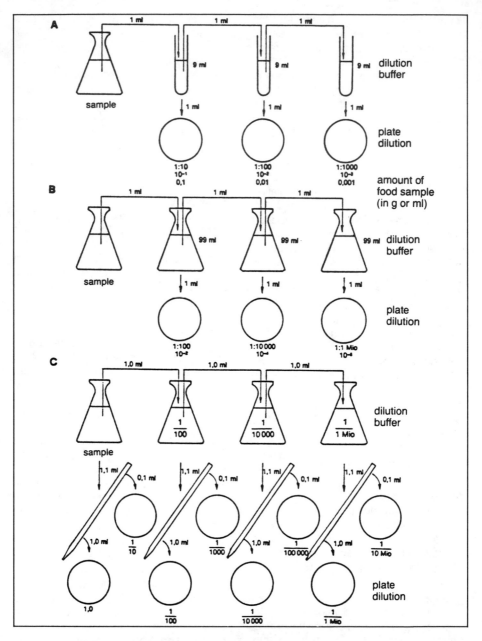

Figure 13 Schematic representation of the procedure for the pour plate method: A = dilution 1:10; B = dilution 1:100; C = dilution 1:10 using the "Demeter" pipette (source: Baumgart, J.: *Mikrobiologische Untersuchungen von Lebensmitteln*. Hamburg: Behr's Verlag, 1990).

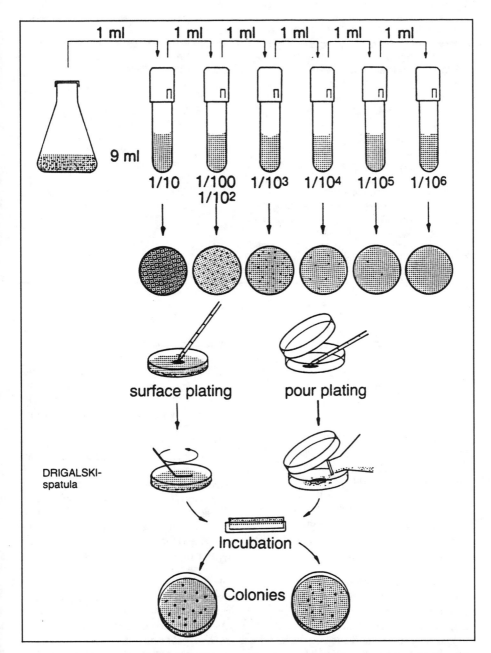

1 ml 1 ml 1 ml 1 ml 1 ml 1 ml

9 ml

1/10 1/100 1/10³ 1/10⁴ 1/10⁵ 1/10⁶
1/10²

surface plating   pour plating

DRIGALSKI-
spatula

Incubation

Colonies

Figure 14 Schematic representation of the procedure for performing the pour plate culture and the → spatula (surface plate) culture (source: Baumgart, J.: *Mikrobiologische Untersuchungen von Lebensmitteln*. Hamburg: Behr's Verlag, 1990).

Table 34 Preservatives with corresponding EC numbers.

E 200	Sorbic acid	E 230	Biphenyl-diphenyl
E 201	Sodium sorbate	E 231	Orthophenylphenol
E 202	Potassium sorbate	E 232	Sodium phenylphenolate
E 203	Calcium sorbate	E 233	Thiabendazole
E 210	Benzoic acid	E 236	Formic acid
E 211	Sodium benzoate	E 237	Sodium formiate
E 212	Potassium benzoate	E 238	Calcium formiate
E 213	Calcium benzoate	E 280	Propionic acid
E 214	p-Hydroxybenzoic acid ethyl ester	E 281	Sodium propionate
E 215	p-Hydroxybenzoic acid ethyl ester, sodium compounds	E 282	Calcium propionate
		E 283	Potassium propionate
E 216	p-Hydroxybenzoic acid-n-propyl ester		
E 217	p-Hydroxybenzoic acid-n-propyl ester, sodium compounds		
E 218	p-Hydroxybenzoic acid methyl ester		
E 219	p-Hydroxybenzoic acid methyl ester, sodium compounds		

the selective effect, and may (a.o.) result in spoilage, whilst the risk of food poisoning is drastically increased.

Preventol® → orthophenylphenol.

primary contamination

Unavoidable initial contamination during the harvesting and collection of raw materials. These may be→ pathogens such as causative agents of→ zoonoses, salmonellae, polio or hepatitis viruses, or spoilage organisms, transferred by air or water, or by contact with soil, harvesting equipment or hands (Sinell 1985).

primary metabolites

Molecules excreted during→ respiration or fermentation of microbial cells for energy yield. These substances may affect the sensory quality and shelf life of a product. Examples: CO_2, H_2S, CH_4, H_2; organic acids; ethanol, etc. Contrary to:→ secondary metabolites. → Sensorics.

process control

Microbiological and chemical control and monitoring of the manufacturing process of a particular product, from raw material through the different processing steps up to the end product, including the final packaging [Bryan: *Food Technol.* 35 (1981) 78 – 87]. Sampling plan, procedures for determination of viable population (→ plate count), control of the R&D measures on conclusion of the working day, and prior to continuation of processing on the following day, hygienic control of the water quality, sanitary facilities and the staff, are dependent on the type of product, the turn-over, the season, the weekdays and other parameters specific to the production plant. Technical and psychological training of the staff responsible for sampling is of prime importance. Only by means of a complete and protocolised process control is it possible to relate defects in the end product to processing failures and their causes. In-plant process control is more important for

Table 35 Foodstuffs allowed to contain preservatives.

1. Fish or shellfish marinade or mussels including dressings and dips
2. Fried, boiled and picked fish, including dressings and dips
3. Fish pies containing less than 10% salt
4. Salted herring in oil
5. Salmon products in oil
6. Roe products, except smoked roe
7. Anchovies including dressings and dips
8. Preparations from crab, unsterilised, except crab soup powder
9. Shrimp products, unsterilised
10. Liquid full egg, liquid egg yolk
11. Mayonnaise and mayonnaise-like products
12. Seasoning and salad dressings
13. Seasoning from lemon juice
14. Meat salad, aspic, vegetable salad and potato salad
15. Edible crust for meat products containing gelatine
16. Margarine with a moisture content > 15%, half fat margarine, skim milk products
17. Fruit pulps and concentrates, and fruit used for processing in the confectionery and beverage industries
18. Fruit juices and concentrated fruit juices with a specific gravity of up to 1.33 used for further commercial processing, except those concentrated fruit juices and fruit "nectars" for direct supply to the consumer
19. Bases and basic ingredients for alcohol free beverages containing fruit juices, e.g., lemonade, sparkling lemonade, and artificial hot and cold beverages
20. Boiled fruit including rhubarb and pumpkin, except products in closed containers heat treated for shelf life extension
21. Fruit and nut preparations for the production of fruit and nut yoghurts and other milk products
22. Hip pulp for further processing
23. Dried prunes and figs with a water content of more than 20%
24. Pectin solutions used for the treatment of dried fruits and grapes
25. Slices of citrus peel
26. All types of acidic conserve except "sauerkraut" and ready-to-use mushrooms
27. Onions, horseradish mince and paprika pulp
28. Preserved olives
29. Mustard
30. Marzipan and marzipan-like products manufactured from other oil seeds, macaroons, nut macaroons and macaroon substitutes, with added milk, fruit and other ingredients containing water and fat; fillings for confectionery, chocolate and durable bakery products including other sorts of baked products
31. Packed sliced bread, low energy bread
32. Pastry products with a moisture content of more than 22%, low kilojoule pastry products, biscuits with a moist filling
33. Semi-moist instant pastry
34. Separated emulsions
35. Watery flavours with less than 12% alcohol
36. Liquid enzyme preparations
37. Low kilojoule jams and similar products
38. Citrus fruit
39. Dried citrus peel for the production of candied lemon and orange peel
40. Bananas
41. Baking and biscuit creams used for dressings

food safety assurance than official control and inspection (Krämer 1987; Sinell 1985).

process types

Differentiation in biotechnology of processes that are either performed on the bases of living systems (MO or higher organisms) and non-living biogenic systems (enzyme preparations, immobilised enzymes) (Kunz 1988). → Lactic fermented milk products; → cheese; → fermented sausages, etc.

processed cheese and products
→ official methods.

prokaryotes

Cellular organisms (MO) with a cell wall containing murein (peptidoglycan) or pseudo-murein, or, in rare cases, consists of an S layer. The chromosome ("nucleus") is not separated from the cytoplasm by a unit membrane, and the cytoplasmic membrane is devoid of steroids; mitochondria and chloroplasts absent. This group also includes the cyanobacteria (see Figure 15). The development of the prokaryotes goes back much further, phylogenetically, than the → eukaryotes.

prophylaxis

Prevention. Measures, typically used in medicine, taken to prevent contamination or infection (disease). In food microbiology it refers to measures for the prevention of spoilage or quality loss, e.g., by precautions against contamination and/or proliferation of food-associated MO.

Propionibacterium

Gram +, non-motile, irregularly shaped (pleomorphic) bacteria, anaerobic to aerotolerant, catalase +. Produce propionic acid, acetic acid, valeric acid, formic acid and CO_2. → Propionic acid fermentation. Opt. 30 – 37 °C; pH ca. 7.

Generally slow growth. Distribution: in the rumen and intestines of herbivores, milk and dairy products, rennet from calves, soil. Of decisive importance for gas production (and hole formation) in some hard cheeses. If pasteurised milk is used under maintenance of good hygienic practices, starter cultures (*Pr. shermanii* and *Pr. petersonii*) are to be added as → starter cultures to the cheese milk. – *Pr. freudenreichii* and *Pr. shermanii* are used for the biotechnical production of cobalamin (vit. B_{12}) (DFG 1987). Enumeration: Baumgart (1987); Gilliland et al. (1984).

propionic acid

CH_3-CH_2-COOH. Irritative smelling, colourless liquid. Simple fatty acid. Salts: propionates. Natural occurrence: in sweat, bile, different plants, especially in etheric oils. Produced during → propionic acid fermentation. LD_{50} (rat) p.o. 2.6 – 4.2 g/kg BW. Chronic toxicity: no deleterious effects in rats by 3.75% in feed. ADI value unlimited. (Classen et al. 1987); "MAK" value 10 ml/m^3 or 30/m^3 of air (Roth 1989). Approved for use as → preservative against yeasts and moulds. Optimum activity at pH 3.5 – 4.5 (Wallhäusser 1987). Application especially for the prevention of defective fermentations in silage. EC no. E 280; sodium propionate E 281; calcium propionate E 282; potassium propionate E 283.

propionic acid fermentation

Some bacterial spp. produce propionic and acetic acids as main end products from glucose or lactate. Succinate is formed via → pyruvate, oxaloacetic acid, malic acid, fumaric acid, and is finally decarboxylated to propionic acid. This relatively uncommon biosynthetic pathway (methyl-malonyl-CoA-pathway) is typical of the genera → *Propionibacterium, Veillonella* and *Selenomonas*. Propionic acid is synthesized via a simpler pathway

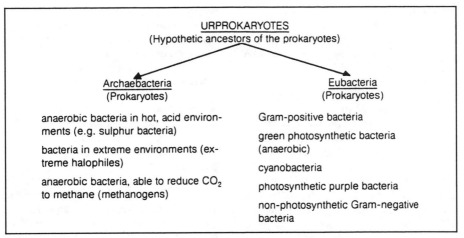

Figure 15 Possible development and relationships among the prokaryotes.

from lactate (via acrylic acid) by representatives of the obligately anaerobic genera → Clostridium, → Bacteroides and → Peptostreptococcus, associated with the GI tract. → Propionic acid is produced during the ripening of → hard cheeses and in the intestines and rumen.

proteases
May be produced from animal organs, plants, or with MO. – Trivial names: proteinase, peptidase, collagenase, keratinase, rennin; → rennet; → rennet substitute (chymosin). Systematic name: peptide hydrolase. EC numbers: 3.4 group. Catalyses the hydrolysis of peptide bonds (in proteins and peptides), and to some extent of amino acid esters and amides. – Application: hydrolysis of proteins and peptides, e.g., in the processing of cereals and meats, in the beverage industry, etc. – MO for production: A. melleus, A. niger, A. oryzae, Endothia parasitica, Rhizomucor miehei, Rh. pusillus, B. cereus, B. licheniformis (GDCh 1983).

protective colloids
High molecular compounds (protein, CH,

and to some extent also lipids) referred to in food technology, that reduce the destruction rate of MO during heat processing (→ D value; → z value). The mechanism of this phenomenon is not fully understood.

protective gas
Protective gas (→CA; →MA). Increase of the N_2 or CO_2 content of the atmosphere and simultaneous reduction of the O_2 concentration reduces or inhibits the metabolic activity of aerobic MO; however, anaerobes and micro-aerophiles are beneficially influenced. The use of >50% CO_2 has proved effective for use in air-tight packaging of sliced bread. Meat and processed meats produce CO_2 in such packages and utilise the residual oxygen. This principle is applied in→CA storage, resulting in a reduction of metabolic activity of stored products and inhibition of MO, and thus an increase in shelf life (Bünemann and Hansen 1973; Krämer 1987; Nicolaisen-Scupin 1985). A desirable additional effect is the protection against oxidation by atmospheric oxygen.

protein

Major component of the dry material of all living cells. Basic component, in addition to CH and lipids, of all foods. High MW polymer of→amino acids linked in different possible sequences by peptide bonds, typical of sp. (type) and functional specificity. Tens of thousands of protein types, differing in their primary and tertiary structures and functionality. Their biosynthesis is genetically encoded and controlled; all may however be degraded by extracellular→proteases of MO, and, in part, also by animal (e.g., pepsin, trypsin) or plant (e.g., papain) enzymes. Type specific proteins are the most important→allergens for sensitised persons.

protein yeast→single cell protein.

proteolysis

Hydrolytic cleavage of protein to→peptides and/or→amino acids by→proteases. Important step in the degradation of organic residues in nature; also in the manufacture and ripening of cheese, in the maturing of meat, the production of→peptone, and in the purification of wastewater.

proteolytes

Refers to MO which are able to degrade (decompose) proteins by extracellular enzymes.→Classification. Many cause food spoilage; some are applied as→starter cultures (→Brevibacterium;→P. roqueforti); some are important components in wastewater treatment. Detection: gelatine liquefication; on skim milk agar plates positive (proteolytic) colonies are surrounded by clear zones.

Proteus

Genus of the→Enterobacteriaceae (Bergey 1984). Gram −, motile, straight rods without capsules; facultatively anaerobic, lactose −, catalase +, nitrite +, H₂S +, urease +. Distribution: soil, intestinal tract, wastewater. Typical spoilage association of protein rich foods. Common spp.: *Pr. mirabilis, Pr. vulgaris, Pr. myxofaciens*. May complicate→plate counting by "swarming" on moist agar medium surfaces, as a result of its motility.

protoplast

Cell membrane plus cell contents. Osmotically sensitive structure that remains after the complete removal of the cell wall, e.g., by→lysozyme for bacteria. So-called L-forms may result from the inhibition of cell wall biosynthesis by→penicillin. These forms may survive considerable time in isotonic liquids, e.g., blood or serum.

prototrophy

Nutrition of a large number of MO which do not require special growth factors (e.g., particular amino acids, vitamins, etc.); their nutritional requirements are similar to those of the wild-type strain. Opposite: →autotrophy.→AMES test.

protozoa

Group of eukaryotic unicellular MO, classified as the subkingdom Protozoa within the kingdom Animalia. Ca. 25,000 spp. are described; about 20% are parasites, of which ca. 30 spp. may be associated with man. Transmission by foods possible of: *Entamoeba histolytica* (amoeboid dysentery);→*Toxoplasma gondii* (toxoplasmosis) and *Isospora hominis* (coccidiosis). Transmission typically by raw meat and faecally contaminated vegetables, salads and fruit.

Pseudomonas

Important genus of Gram −, aerobic, rod-shaped bacteria (Bergey 1984). Motile polarly flagellated (mono- or polytrichous), catalase +, nitrite +, or −, proteases +, lipases +, often psychrotrophic. Pseudomonads in ground water often show→denitrification ability. Several spp.

are phytopathogens, some animal pathogens. Some spp. produce yellow-green or blue-green pigments; fluorescent under UV. Relatively resistant to→disinfectants. Distribution: surface, terrestrial and drinking waters, washed salads, salad mixes, refrigerated protein rich foods (especially ground meat). Detection: Baumgart (1990), p. 140. Regularly accompanied by representatives of the *Acinetobacter-Moraxella* group on meat.

Pseudomonas aeruginosa

(*Ps. pyocyanea*). Infectious to man. Degradation of cosmetics, pharmaceutics, cryolubricants, kerosine, etc. Growth: $2-42$°C; pH $5.6-8$; a_w (min.) 0.97; good growth at 20°C in drinking water; colonises ion exchangers and plastic utensils causing blue-green colouration (pyocyanin), tooth brushes, mouth sprays, etc. Grows in mineral salts solution without added C sources. Survival: on porcelain (china) and glass at 11% rH >7 d; mineral water (soda water) at 20°C >28 d; epiphytic on vegetables and tomatoes 10 d; drinking water at $7-37$°C 300 d (Mitscherlich and Marth 1984). Distribution: water, milk, ready-to-eat foods, salads, refrigerated fresh meat, etc. Some strains produce an enteropathogenic toxin that may cause→food-borne infections;→opportunists (Sinell 1985; Stiles 1989). Problematic organism in hospitals because of perseverance, resistance and (opportunistic) infectivity. Detection and differentiation: Baumgart 1990; Wallhäusser 1988.

Pseudomonas fluorescens

Distribution: water, soil, faeces, egg shell; intestines and faeces of moths, flies and other insects; surface of refrigerated meat and on Crustaceae. Strongly lipolytic and causes rancidity in cream and butter; "green rot" of eggs; spoilage of fish by reduction of trimethylamine-N-oxide to trimethylamine (fish smell).

Pseudomonas fragi

Distribution: surface of meat milk products, egg shell; in soil, water. Spoilage of cream, butter, fresh cheese and quarg resulting from colouration, rancidity and fruity smell.

pseudomycelium→sprout mycelium.

pseudoparenchyma→plectenchyma.

psoralens→phytoalexins.

psychrophiles

MO growing optimally at $10-15$°C with a max. temp. around 20°C; lower limit ca. 0 to -5°C. In nature associated with deep waters (oceans and lakes), glacier streams and in the tundra. Importance as spoilage association (e.g., some spp. of *Pseudomonas* and *Vibrio*) only of ocean fish during transport and distribution; see Figure 23, page 266.

psychrotrophs

Cold tolerant (psychrotolerant) MO, able to grow at low temperatures to 0°C; opt. ca. 24°C; max. 35°C. Most spoilage MO, including the fungi, are psychrotrophic. See Table 20; Figure 23, see page 266.

PTWI

Provisional Tolerable Weekly Intake. Preliminary tolerable weekly intake by the consumer. Compare→ADI value.

pullulanase

Produced with bacteria. A debranching enzyme. Trivial name: R-enzyme, limit dextrinase. Systematic name: pullulan 6-glucanhydrolase. E.C. 3.2.1.41. Catalyses endohydrolytic cleavage of 1,6-α-D-glucosidic bonds in pullulan, amylopectin and glycogen. Application: degradation of starch, e.g., in the manufacture of glucose syrup and maltose syrup. MO for commercial production: *Klebsiella aero-*

genes and *Bacillus cereus* var. *mycoides*
(*B. acidopullulyticus*) (GDCh 1983).

Pullularia→ *Aureobasidium.*

pure culture

Culture in which all the cells are of the
same systematic unit (strain, variety, subspecies). It can be called a clone if all the
cells originate from a single cell. To obtain
a pure culture requires special care, since
it is a vital prerequisite for the→ differentiation and identification of MO. Methods,
e.g., described by Baumgart (1990) and
Smith (1981). Bacteria: typically follows a
bacterial plate count. One colony of interest (often one of the most frequent types)
is picked from a plate where colonies are
lying separately. A colony is picked up
with the edge of a loop or inoculation
needle. A→ dilution streak is made. The
surface of the agar used for the streak
method, should be dry. A single colony
developing during incubation is then
transferred to an agar slant or stabbed into
the butt of an agar culture tube. This
serves as culture for further examinations.
Yeasts: the drop method combined with
microscopical selection of a colony on a
concave microscope slide is suitable for
samples free from bacteria (to be determined beforehand). A→ dilution series is
prepared from a single colony and
streaked out onto a yeast extract glucose
chloramphenicol agar plate. A colony is
picked from a plate with 50 or less colonies
and then inoculated onto an agar slant.
There is a 98.5% probability that the obtained culture is free from foreign MO.
Fungi: the fungus of interest is picked up
with a moist inoculation needle under a
stereomicroscope. A three point inoculation is performed with the needle in an
upright position onto a plate turned upside
down. Conidia or fragments (hyphal parts)
of the fungus are taken after the incubation
and a dilution series is made. It is streaked
out and "stars" (small visible colonies) are
transferred to fresh plates with a sterile
loop after outgrowth.

pure culture facilities

Rooms, containers, pipelines, valves,
pumps, etc., used for the propagation of
starter cultures in larger quantities as required in the process. These facilities may
be located within the plant or outside.
"High security area" with all the precautions and measures taken, also with reference to personnel, to prevent contamination with foreign MO.

purines

Heterocyclic compounds with condensed
pyrimidine-imidazole ring systems. The
derivatives adenine and guanine are components of the→ nucleus and the→ plasmids, form→ nucleic acids (RNA and
DNA) with the→ pyrimidines.→ Single cell
protein.

putrefaction

Microbial degradation, mainly by Gram−
bacteria, of proteinaceous and fatty products, accompanied by the production of
off-odours: e.g.,→ deamination   of→
amino acids with release of→ ammonia;
degradation of sulphur-containing amino
acids (cysteine, methionine) produces
H_2S in addition.→ Decarboxylation of
amino acids (e.g., ornithine and lysine)
results in the formation of→ biogenic
amines (e.g., putrescine and cadaverine).→ Lipases release fatty acids and
glycerol from fats; especially butyric acid,
caproic, capronic and caprylic acids have
an intensive taste and smell (rancidity).
Such spoilage processes may occur
under both anaerobic and aerobic conditions (Krämer 1987; Kunz 1988).

pyocyanin

Pyocyanine. Water- and chloroform-sol-
uble, blue-green, non-fluorescent phena-
zine pigment of→ *Pseudomonas aeruginosa.* Coloured red in acid matrix. May

cause colour defects on plastics that cannot be removed by cleaning procedures.

pyrimidines
Bases: uracil, thymine and cytosine; nucleotides: uridine, thymidine, cytidine. Components of the→ nucleus and the → plasmids; form→ nucleic acids with the → purines.→ Single cell protein.

pyruvate content of milk
Criterion for the bacteriological quality of refrigerated→ raw milk, applied in some provinces in Germany, as alternative to plate count methods. Pyruvate (salt of pyruvic acid) is an intermediate metabolite of all microbial cells during glycolysis, and is partly excreted by some→ psychrotrophic bacteria. Advantage that the sample can be chemically preserved upon collection, especially prior to long transportation. Official analytical procedures according to "§35 LMBG" L 01.00 – 19 – 1988. Up to 300,000 psychrotrophs/ml correspond to ca. 1.4 mg pyruvate/kg; up to 3 mio = ca. 2.5 mg/kg (Teuber 1987). – From 1.1.1989 to 31.12.1992 the pyruvate test (according to the third rev. regulation of the milk quality legislation of 21.07.1988 – in agreement with the EC hygiene guideline of 1985) will be mandatory for some German provinces ("Länder").→ Milk hygiene guideline and Table 25. From 1993 a viable→ plate count determination is to be performed.

pyruvates
Salts of→ pyruvic acid.

pyruvic acid
CH_3-CO-COOH. Important metabolic intermediate in aerobic and anaerobic metabolism, especially in glycolysis. Salt (anion): pyruvate. – Pyruvate detection in raw milk quality estimation may be a substitute for plate count methods.

Pythium
Mould genus of the Oomycota (order Peronosporales); includes plant parasites. *Py. debaryanum* causes root blight of sugar cane and different vegetables, whilst *Py. ultimum* is associated with watery wound rot of potatoes during storage (Kunz 1988).

Q_{10}

Temperature coefficient ($\emptyset^{10}$) for chemical or biochemical reactions. According to the Q_{10} rule of Van't Hoff the reaction rate constant (death rate) increases by a factor of $2-4$ for every 10°C temperature increase. The metabolic changes in the microbial cells follow this rule within the range between opt. and min. growth temperatures. Different Q_{10} values of individual metabolic steps in the cell may result in blocking or undersupply, resulting in the death of the cell near the min. or max. growth temperatures.

quarg → fresh cheese.

quaternary ammonium compounds

QAC's or quats. Group of → inhibitory and partly → destructive cationic surfactants which are commonly used for → disinfection, or as antiseptics or preservatives, either singly or in combination with other compounds. Action by disruption of plasma membrane and denaturing of proteins. Gram + bacteria more sensitive than Gram − , bacteriostatic at low conc. and bactericidal at high conc.; some pseudomonads and Enterobacteriaceae are relatively resistant, as are endospores and some viruses. Selectively allows survival of, e.g., some Enterobacteriaceae; results in higher contamination risk in processing environments. Activity reduced by protein and Fe^{+++}; thorough cleaning prior to application is important! Opt. pH $7-8$. Residue formation on food contact surfaces is to be avoided. Effectively reduces the total microbial population on surfaces in production rooms. Non-toxic in the normal conc. of application; non-irritating to the skin and neutral smell; non-mutagenic. Caution should be exercised when used on contact surfaces in the dairy and beer brewing industries (Wallhäusser 1988).

R

racemases

Enzymes able to change a stereospecific compound into the other isomer: e.g., lactate racemase catalyses the formation of DL-lactate from L(+)-lactate (Bruchmann 1976).

radappertisation

Radiation sterilisation. The destruction of MO on or in a product with the aid of ionising radiation. It is only used for medical disposable articles or aids, e.g., suturing material, swabs, tampons, bandages and syringes.

radiation pasteurisation→radurisation.

radiation sterilisation→radappertisation.

radiation treatment

Procedure for the decontamination of food, water and air. Ionising radiation effectively destroys MO without affecting the substrate to any severe extent. Of practical importance are: electromagnetic waves with wavelength of 10^{-6} cm (UV light), 10^{-9} cm (X-rays) and 10^{-11} (gamma rays). The radiation dose, comparable to temperature-time-product in heat treatment, determines the effectivity of the treatment. It is expressed in Gray (Gy) as absorbed energy per mass unit of the irradiated food. 1 Gy = 1 J/kg; 1000 Gy = 1 kGy. Formerly the unit "rad" (radiation absorbed dose) was used; 100 rad = 1 Gy. – It is advantageous that radiation can be applied at room temperature as terminal processing step after final packaging of the product. Radiation doses necessary for pasteurisation treatment (→radurisation) do not constitute any health hazard with relation to a specific food. No other process (including heating,

chemical preservation) has been subjected worldwide to so much investigation as radiation treatment. In 1981 the Joint Expert Committee of the WHO/FAO/IAEA came to the following conclusion and recommendation: " . . . The irradiation of any food up to an average dose of 10 kGy constitutes no toxicological risk. Further toxicological examinations of such treated foods are therefore not required. . . ." (wholesomeness of irradiated food; WHO Technical Report Series 659, Geneva 1981). – The radiation treatment of particular food commodities is approved in ca. 20 countries; the most important of these applications concern the decontamination of spices and other dehydrated plant products, as well as the destruction of→salmonellae in poultry (Diehl 1989, 1990; Mossel 1977). In D only UV treatment of drinking water, the surface of fruit and vegetable products, hard cheeses and packaging materials for→aseptic systems is allowed, including air decontamination. – It would especially be feasible to approve the radiation treatment for the decontamination of spices (e.g., for processed cheeses) and dehydrated vegetables (e.g., for instant products), since treatment with ethylene oxide is not practised any more (→spices). For regulatory bodies radiation processing does not represent any health problem, but it has become an emotional and therefore a political issue (Krämer 1987; Sinell 1985).

radicals

Highly reactive chemical compounds with a short life span, e.g., O^-, O_2^-, O_2^{--}, CH_3^-, $C_2H_5^-$, etc. They may be produced during several technological processes and can destroy labile ingredients in foodstuffs, e.g., the loss of carotene and vit. A or E; autoxidative spoilage of fat. Radicals are produced by peroxidases, lipoxigenase, → glucose oxidase and other enzymes of microbiological origin, or are produced by UV light and/or ionising radiation as well as by mechanical processes, e.g., mincing or blending. The radicals react with the macromolecules especially → DNA and → proteins and cause damage resulting in the destruction of MO (Classen et al. 1978).

radicidation
Ionising radiation to destroy pathogens. Used to eliminate, e.g., → Salmonella in chicken and approved in several countries, but not in Germany. 2.5 – 3 kGy is usually applied for a number of frozen foodstuffs.

radurisation
Radiation pasteurisation. Decontamination (reduction in the viable bacterial count) of a product with the use of ionising radiation. It is suitable for treatment of heavily contaminated ingredients of feeds, spices for meat products, cheese and instant soups. This process is allowed in several countries, but not in Germany or Denmark yet. A dosage rate of 0.5 – 5 kGy is applied, depending on the product type and contamination load.

rancidity
Defect in taste and odour of fat containing products. The fats are hydrolysed by microbial or product → specific lipases to glycerine and fatty acids. Free fatty acids (up to C_{10}) are mainly responsible for the sensory detection of defects. Lipoxigenase commonly contributes to this effect.

raw sausage → fermented sausage.

reactors → bioreactors.

recontamination
Most important error/failure in the preparation and processing of food. Renewed → contamination with "domestic" MO after the first processing step. It has a negative influence on the hygienic status. Special care should be taken in the case of milk and meat products and in canteens and large kitchens. → Airline foods.

recycling
Methods and technologies used to reduce environmental contamination and pollution, e.g., by microbiological methods; → biological sewage treatment, anaerobic digesters, soil filters, composting, but also the use of by-products or waste as animal feed may contribute significantly.

redox potential
Measure of the degree of oxidation in foodstuffs. It characterises the tendency of a product to take up electrons (reduction) or to lose them (oxidation). The difference in potential is measured in mVolt against a reference electrode. The Eh value is an important indication of the expected growth of aerobic, microaerophilic or anaerobic MO, i.e., which kind of selection is to be expected. This value can be modified technologically by, e.g., the addition of reducing substances such as ascorbic acid or by O_2 dense ("air-tight") packaging (possibly following heat treatment).

refrigerator
Most important household utensil for the storage of perishable food. The temperature should be adjusted at 5°C so as to ensure that → food poisoning organisms do not reproduce or produce toxins. It should be cleaned with a detergent and disinfected with a vinegar solution.

reinfection → recontamination.

rennet substitute
Enzymes as substitutes for rennin produced by certain MO, e.g., *Rhizomucor miehei, Rh. pusillus, Endothia parasitica* or *B. subtilus*. The characteristics are

more or less the same as for → rennin and they are approved (e.g.) for Germany (DFG 1987; Teuber 1987).

rennin
Rennet. Chymosin. → Proteases derived from the abomasum of calves. The optimal pH range is 6 – 7 and it is used to form a curd of milk in cheese-making. Commercial preparations often contain pepsin as associated enzyme which may cause off-tastes by cleaving bitter peptides from the casein. A rennin-unit is defined as the amount of rennin able to coagulate milk under standardised conditions at a temperature of 35°C. Up to 12 mg of preservatives may be added per kg of rennin before any declaration is required on the cheese. – Natural identical rennin, chymosin from MO, has recently appeared on the market, but is not allowed in Germany yet. It is produced by the genetically modified yeast *Kluveromyces lactis* or *E. coli* (DFG 1987).

rennin cheese
General designation for cheese made by coagulating milk with the aid of → rennet or rennet substitute before acidification. After the curd has been sliced (cut), whey separates by the process of → syneresis. Hard, semihard and soft cheeses. By contrast: sour milk cheese, → cottage cheese, cooked cheese (Teuber 1987).

reserve materials
Accumulate in microbial cells, depending on the presence of the particular precursor components, and may be induced by inhibitors or the absence of essential nutrients by which growth or multiplication is limited. Examples: starch, glycogen, fat granules or drops, polyphosphate (volutin) granula or elemental sulphur in sulphur bacteria. All reserve materials are water insoluble and have no osmotic effect (Schlegel 1985).

respiration
Oxidative metabolic process; provides energy in the form of ATP to cells. An exogenous electron acceptor is involved in the oxidation of an energy rich substrate. Energy yield is greater than that obtained by fermentation of a substrate. Aerobic respiration involves molecular oxygen as inorganic electron acceptor, as compared to anaerobic respiration where

Table 36 Respiration energy of some vegetables (acc. to Hansen, H.: *DKV-Arbeitsblatt*. March 1967, 8 – 22, abbreviated and modified).

Type	Respiration Energy in MJ/t x 24 h at			
	2°C	5°C	10°C	20°C
Cauliflower	3.0 – 6.0	4.6 – 6.6	10.6 – 12.0	26.4 – 34.7
Endive salad	11.5 – 13.4	15.8 – 17.6	21.5 – 24.0	44.4 – 47.7
Cucumbers	1.0 – 2.1	2.1 – 2.9	4.4 – 5.2	13.2 – 15.1
Potatoes	0.9 – 2.6	1.0 – 1.6	1.4 – 2.0	2.1 – 3.7
Carrots	1.9 – 2.9	2.4 – 3.3	2.7 – 3.8	7.7 – 11.7
Brussels sprouts	4.8 – 6.7	8.4 – 11.7	14.4 – 19.7	42.3 – 44.8
Spinach	6.7 – 7.1	11.0 – 17.1	17.9 – 27.0	54.4 – 77.4

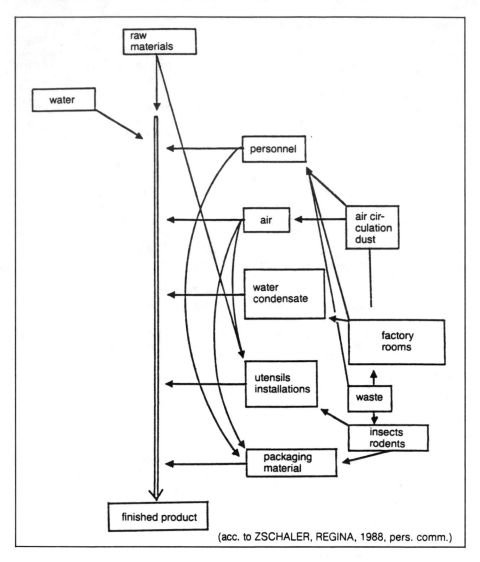

Figure 16 Contamination chains in the processing and handling of foods (source: Zschaler, R.: *Einführung des Begriffs GMP; Personalhygiene.* Behr's Seminar, March 1988, 8 – 9, Hamburg).

either an inorganic (e.g., nitrate) or organic (e.g., fumarate) agent serves as electron acceptor. In addition to the formation of energy rich→ ATP a small portion of the energy is lost as heat and plays a part in the storage of respiring vegetable matter (→ respiratory heat). For some biotechnical processes this heat has to be removed continually, e.g., in the production of → baker's yeast where a temperature of 26 – 28°C should not be exceeded.

respiration energy
During the storage of living plant matter, e.g., fruit, vegetables and potatoes, respiratory heat is generated. The amount of heat energy produced increases with increasing storage temperature (Table 36); it is released from the stored matter into the environment, concomitantly with a loss of valuable nutrients. These values may be much higher for damaged or mechanically stressed products, e.g., → salad mixes. An increase in the temperature of the stored matter enhances the metabolic activities of contaminating MO, i.e., spoilage is accelerated. A similar situation is given when the water content of cereals exceeds 14%.

ripened cream
Sweet pasteurised cream is homogenised and cultured at 18 – 22°C with a → butter starter culture (→ cultured creamery butter) until a pH of 4.9 – 5.1 is reached. Sour cream is stored and distributed at 5°C.

ripening rooms
Rooms (chambers) with heating/cooling, humidity and ventilation control for enhancement of microbial processes in the ripening of → rennet cheese, → fermented (dry/semi-dry) sausages and other meat products. Special attention must be paid to the filtration of the → supplied air in order to avoid → contamination from outside or the same plant. The air may be treated with UV light, but care must be taken that neither the products, nor the personnel are subjected to radiation (§13 LMBG).

room air
Most important vehicle of contaminants in processing plants. It is of utmost importance to separate clean from contaminated areas, and also laboratories from large kitchens. The distribution of → bacteriophages by air plays an important role in dairies. UV light may be used to reduce the MO load in air, but it has only little significance in practice.

route of contamination
Contamination route. It is often also described as the "infection route", although → infection is characteristic of a different process. Simplified for the education in personnel hygiene, reference is made to the "3 F's" (faeces, flies, fingers) and 5 M's (MO, man, machine, material, milieu). These mark the complexity of the possibilities (Figure 16).

S

Sabouraud-medium

Nutrient medium for the cultivation or (agar) plate count of yeasts and moulds. Peptone, glucose and/or maltose (agar) are the main ingredients. Especially used for the estimation of the surface microbial population of packaging materials, as well as the examination for antimycotic agents in packaging materials: e.g., Merck no. 7662 (Cerny 1986).

saccharification

Degradation of starch and other → poly-saccharides as preparatory step in the production of ethanol with the aid of *Saccharomyces cerevisiae*. The mash is treated with enzyme preparations and the moistened material is "fermented" with *A. oryzae* or *A. niger*. In the production of ethanol with *Zymomonas mobilis*, this step is not required since the bacteria are amylolytic (Kunz 1988).

Saccharomyces

Genus of the family Saccharomycetaceae (order Endomycetales; class Asco-mycetes). Represented by ca. 30 spp.; commercially most important are the beer and baker's yeasts [Windisch, S.: *Brannt-weinwirtsch.* 121 Nr. 14 (1981) 234 – 243] (Barnett et al. 1983; Priest and Campbell 1987).

Saccharomyces cerevisiae

Sp. represented by ca. 30 commercially important strains, e.g., the wine, "cham-pagne", "spirits" and beer yeasts. The "sugar mould" was discovered in 1836 and cultivated in pure culture by Hansen in 1883. Diploid cells produce 1 – 4 spores on acetate agar (e.g., Merck no. 5265); → ascus. Technically utilised yeasts are mostly polyploid and produce no spores (Dittrich 1987; Glaubitz and Koch 1983).

Saccharomyces diastaticus

Causative agent of beer spoilage. Pro-duces enzymes that may cleave even high molecular sugars, in contrast to *S. cerevi-siae* and *S. uvarum*. This may result in "over" fermentation, where even dextrins in the wort (→ beer) are fermented to etha-nol. → Diabetics beer.

Saccharomyces rouxii → *Zygosac-charomyces rouxii.*

Saccharomyces uvarum

Bottom fermenting yeast, formerly desig-nated as *S. carlsbergensis*. Relatively large cells with polar budding; mostly polyploid; produces no ascospores [Win-disch, S.: *Branntweinwirtsch.* 121 No.14 (1981) 234 – 243].

saccharose → sucrose.

safety factor

The ratio of highest non-effective dose of a toxin, as determined in animal tests (→ NOEL), to the probable non-effective dose for man. This factor is mostly 1/100 of the NOEL dose, and takes into con-sideration insecurities in the transfer of data from experimental animal to man, as well as the higher sensitivity of infants, old-aged and weakened persons. See Figure 1, under ADI value.

safratoxins → *Stachybotrys.*

sake

Japanese rice wine with 14 – 20% eth-anol. Rice is saccharified with *A. oryzae* and fermented with *Sacch. cerevisiae* (Beuchat 1987; Reiss 1986).

salicylic acid

HO-C$_6$H$_4$-COOH, 2-hydroxybenzoic acid. Colourless needle-shaped crystals. Naturally present in black currants, cherries, bramble berries, carrots, etc., in conc. of 0.1 – 0.6 mg/kg; higher amounts in *Polygala senega* (Radix Senegae), *Gaultheria procumbens* (wintergreen), *Cornus sanguinea*. – LD$_{50}$ (rabbits) p.o. 1.3 g/kg BW. – Has been used in household for preventing surface mould growth on jams. Effectivity only satisfactory below pH 3.0. – Because of pharmacological action and the tendency towards cumulation with clinical symptoms including nausea and vomiting, and also because of slight antimicrobial activity, salicylic acid is not approved any more as preservative in most countries (Classen 1987).

Salmonella

Genus of the family Enterobacteriaceae, called after the American bacteriologist Salmon. Gram – , as a rule peritrichously flagellated, motile rods. The genus is represented by one sp. with ca. 2200 serotypes, to which also→ *S. typhi* and *S. typhimurium* belong; every serotype is however described as separate sp. – Growth: 10 – 45°C; a$_w$ >0.94; pH 4 – 8; O$_2$ requirement very low! Survival: dry faeces, dirt, feeds, foods, often for many years, and (e.g., in ice cream) even deep frozen.→ In brine with 20% salt several months. In the intestinal tract of flies (*Musca domestica*), 6 d. In dead flies up to 60 d. – Destruction: pasteurisation; radurisation at 3 – 4 kGy; sensitive to propionic acid, lactic acid, chlorine and other common→ disinfectants. Salting and smoking (as used for foods) are practically not effective. – Reservoirs and vectors: in addition to farm animals, also pets such as dogs, cats, hamsters, birds, reptiles, fish and wild animals. The farm animals' stocks are in practice infected by commercial feed products. These animals are usually symptomless, but are→ excretors for longer periods and in this way contaminate the environment. Incidences of excretors are: 1 – 30% for dogs,

1 – 15% for cats, 25 – 70% for turtles, up to 60% for doves and 12 – 20% for horses. It is estimated that poultry meat is responsible for up to 80% of→ salmonella cases; cross infection from animal to animal, contamination of carcases during slaughtering, evisceration, chilling (→ spin chiller) and freezing. In the kitchen→ cross contamination. Most human infections are caused in private households (Sinell 1985). – Detection: pre-enrichment in buffered peptone water (e.g., Merck no. 7228). Enrichment in tetrathionate broth (e.g., Merck no. 5285; Oxoid CM 29) and/or selenite enrichment broth (e.g., Merck no. 7717; Oxoid CM 395); plated onto brilliant green lactose sucrose agar (e.g., Merck no. 7232; Oxoid CM 263), and subculturing of typical colonies on crystal-violet-neutral-red-bile agar (VRB) (e.g., BBL 11807; Merck no. 1406; Oxoid CM 107). Compare Figure 17 (Baumgart 1990);→ official methods.

Salmonella typhi and Salmonella paratyphi

Causative agents of typhoid fever (''typhus abdominalis''; French: fievre typhoid; German: thyphus; Spanish: fiebre tifoidea) and paratyphoid fever. Acute infectious disease, reportable (communicable) to health authorities. Transmission either by direct contact between humans, or indirect via faecally contaminated food, water, utensils, hands, flies, etc. Main reservoir: man; special risk caused by asymptomatic carriers; localisation in gall bladder; excretion possible for several years. – Distribution: universal but especially in tropical countries. Infective dose may be <1000 cells per meal. Growth:→ *Salmonella*. Survival: faeces of diseased persons in soil >6 months; surface of fruit, glass, porcelain, etc., at 22°C, 10 – 20 d; in ice cream for several years; in drinking water up to 1 month; in ground pepper up to 3 weeks (Mitscherlich and Marth 1984).

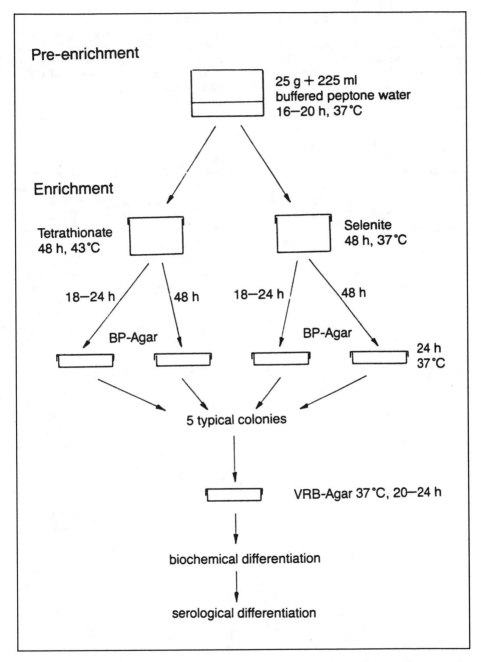

Pre-enrichment

25 g + 225 ml
buffered peptone water
16—20 h, 37°C

Enrichment

Tetrathionate
48 h, 43°C

Selenite
48 h, 37°C

18—24 h   48 h   18—24 h   48 h

BP-Agar   BP-Agar

24 h
37°C

5 typical colonies

VRB-Agar 37°C, 20—24 h

biochemical differentiation

serological differentiation

Figure 17 Detection of salmonellae, with pre-enrichment and enrichment (source: Baumgart, J.:
Mikrobiologische Untersuchungen von Lebensmitteln. Hamburg: Behr's Verlag, 1990).

– Destruction: pasteurisation; radurisation for 4 – 5 kGy. – Incubation time: 14 d. Problem arises that incriminated food cannot be withdrawn at that stage. From the third week after ingestion the organism can be detected in the stool. Early chemotherapy with antibiotics (chloramphenicol, ampicillin) mostly successful; complications rare. → Excretors (carriers) uncommon (Sinell 1985).

salmonellosis

Enteritis infectiosa. Communicable disease (e.g., according to German "Bundesseuchengesetz" §3). Caused in man by all known serotypes of → *Salmonella*. Minimal infective dose has been estimated at 10^6 to 10^7 cells per meal; recent investigations suggest numbers of 10^3 to 10^4 for some serotypes (Sinell 1985). – Incubation time: 6 – 40 h. Duration: 1 – 6 d. Symptoms: myalgia, headache, nausea, vomiting, abdominal pain, fever, nonbloody diarrhoeal stools. Initially the symptoms reminiscent of influenza. Lethality: ca. 1% of the cases, mainly infants and old aged. < Prevalence mainly in late summer and early autumn. Number of reported cases in Germany: 48,000 in 1980, 30,566 in 1985, and ca. 90,000 in 1990 (compare Figure 18). Highest frequency in children <5 a; boys more frequent than girls. Prognosis: after 1 week commonly symptomless. Many patients may however remain carriers (→ excretors) either permanently or for longer periods of time, according to German Law on Contagious Diseases (BSG) §17 and §18. Such persons are not allowed to handle or process food, e.g., in food processing plants, canteens, etc. (Hahn and Muermann 1987). – Causes: infections of man to man rare, if so, mainly in "closed" environments. Commonly from animal to man; typical → zoonosis. Poultry meat, meat and meat products, delicatessen salads, potato salad, ice cream, fish and fish products (in declining order of

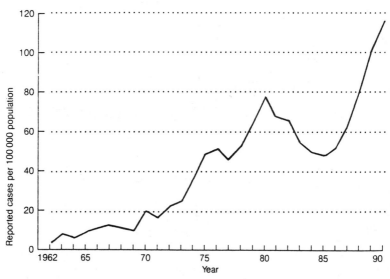

Figure 18 Reported cases of salmonellosis per 100,000 population, for Germany during the period 1962 to 1990 (adapted from Grossklaus et al. 1991).

importance). Most important is probably the risk of cross contamination in private households (ca. 80% of the cases).

salpetre → nitrate.

salt → table salt.

salt vegetables
Vegetable types (cucumbers, green beans, cauliflower, mushrooms, etc.), preserved at 2 – 4°C under air exclusion for 1 – 2 months in a 15 – 25% brine solution. Mainly semi-processed products. The a_w drops to 0.80 – 0.85, by which the lactic fermentation is inhibited. Spoilage: mainly by → surface (pellicle) yeasts on the → brine (Müller 1983a).

sampling
Samples for the microbiological examination are to be taken by trained personnel, marked, packaged and are to be accompanied by a specific report or exact indications of origin and sampling conditions. Number of samples and amount of each single sample are to be determined beforehand. Till examination samples of fresh foods and non-frozen commodities are to be stored at 0 – 5°C, but should not be frozen (→ *Cl. perfringens*). Deep frozen products should not exceed a temp. of – 18°C. Dehydrated may be stored at room temperature under exclusion of air. – In the laboratory sample fractions of specific weight or volume are drawn aseptically; these are macerated/homogenised either singly or after combination, followed by the preparation of a → dilution series. This procedure, including plating, should be performed within one hour after the sample is received by the laboratory (Baumgart 1990). – The quality of sampling will decisively determine the interpretation and reliability of the subsequent microbiological, chemical and toxicological examination. Sample plans (including collection, handling and preparation) are specified for some foods in the Official Examination Procedures (e.g. Germany: LMBG §35; also referred to in USDA and FDA specifications; see also ICMSF 1984).

sampling plan
Microbiological quality in general cannot be guaranteed by end product control only. Sampling plans are devised and adapted for in-plant process and product control. Criteria (microbial numbers) are defined for acceptability, non-acceptability and provisional acceptability of different processing steps, including the raw materials and packaging materials. The number of samples and the points of sampling should be predetermined, but the stage of sampling variable. A separate sampling plan should monitor the effectivity of cleaning (sanitation) and disinfection procedures (Krämer 1987; Sinell 1985).

Saprolegnia
Genus of Oomycetes living on fresh water animals and detritus. *S. parasitica* and other spp. may cause considerable damage in aquaculture ("fish moulds") (Müller and Loeffler 1982).

saprophytes
Saprotroph; saprobe. Opposite: → parasites. MO that decompose (e.g., with extracellular enzymes) non-living organic matter for their own nutritional requirements. Represented by most salt and fresh water and soil MO that are not dependent on photosynthesis. Also includes the spoilage organisms of foods. Produce inorganic compounds (mineralisation) and play an important role in the cycle of matter, e.g., CO_2, NO_3, H_2S, PO_4, etc.

Sarcina
Gram + cocci of the Streptococcaceae (Bergey 1986). Obligately anaerobic, occurring in cubical packets; non-motile, as-

porogenous. Produce CO_2, H_2, and butyric or acetic acid from CH. Pigments yellow to orange, insoluble. Occurrence: in air, milk and GI tract of man and animals. Salt tolerant. May cause bitter taste in milk by proteolysis.

sauerkraut

Produced by "pickling" of cabbage by lactic acid fermentation. Use of sauerkraut in D ca. 1.8 kg per person p. a (in the USA ca. 0.8 to 1.0 kg). The larger part is still produced by spontaneous fermentation, i.e., without the addition of→starter cultures. – Following wilting, washing, shredding, salting (to 1.8 – 2.2% NaCl) and tightly packing into tanks, several types of "natural" contaminating MO develop during the first phase. This is characterised by the production of formic, succinic and lactic acids, ethanol and CO_2, accompanied by foaming. Oxygen depletion in the second phase promotes the increase of facultatively anaerobic MO, especially *Lb. brevis* and *Leuconostoc mesenteroides*. The lactic acid content rises to ca. 1%, causing a lowering of the pH. Slow development of this phase (and the safe pH level) may enable the growth of *Clostridium butyricum* responsible for sensory spoilage by butyric acid production. In the third phase the metabolic activity of *Lb. plantarum* is mainly responsible for the high level of acidity finally reached. In addition, salt and acid-tolerant strains of *Pediococcus damnosus* and *Enterococcus faecalis* may also be involved. In the fourth phase, the longest, *Lb. brevis* dominates, and contributes to the final aroma formation (Müller 1983; Rehm 1985). – The large time fluctuations involved in sauerkraut fermentation (6 – 8 weeks) suggest the possible need for the use of starter cultures for its industrial production. – In the commercial product normally both stereoisomers of→lactic acid are present in the ratio 1:1.

sausage

→Fermented sausages; dry- or semi-dry sausages; → "Bruhwurst", → cooked sausages. Deboned meat, preportioned and cut to smaller sizes, is evenly mixed together with fat, salt, spices→pickle salt, water or ice, blood plasma and offal in a bowl cutter. The mixture is highly contaminated and a starter culture is added for preparation of fermented sausages and sometimes also for cooked sausages. The mixture is filled into sausage casings and, depending on the type, the sausages are either ripened, heated or smoked according to the type of production procedure. The susceptibility to spoilage differs for the particular types according to the a_w, pH, and pretreatment. The most frequent failures in commerce and at home are insufficient cooling and extended periods of storage resulting in spoilage.

sausage poisoning→botulism.

Scenedesmus

S. acutus and *S. obliquus* are unicellular green algae (→algae), that are cultivated in tropical countries for the production of→single cell protein. For this purpose channel systems have been developed, into which CO_2 gas is continuously supplied. One such system is presently in operation in Peru (Becker 1982). The yield in→biomass amounts to ca. 20 $g/m^2/d$, and exceeds that of all agricultural products by far.

Chemical composition in g/100 g DM:

crude protein	45 – 65
total nitrogen	8.3
lipids	10 – 20
carbohydrates	5 – 13
minerals	8 – 12
RNA	4.4
DNA	1.6

The biological value of the protein fraction

selection

amounts to 80 – 90% of that of casein. The amount of purine is, however, too high; yet, extraction methods used in the production of bacterial proteins, are available. Addition of purine non-reduced green algae to pastas (e.g., noodles) for improvement of the nutritional physiological quality has not found general consumer acceptance because of the green colour. Another problem is related to the lipid content that may be the cause of rancidity.

Schizosaccharomyces

Genus of the fungi (family Saccharomycetaceae); yeastlike. Reproduction by binary fission (and not by budding), or by true hyphae that fragment into arthrospores. Ferments glucose and also→ dextrins, but not galactose or lactose. Common in warm regions on dried fruit and cereals; also on jams, molasses, fruit juice concentrates and in grape juice. Forms a→skin on liquids. Osmotolerant, acid tolerant and heat tolerant; often used in the tropics as distillery yeast (fermentation temp. 38 – 42 °C). S. pombe is used in Africa for the manufacture of sorghum beer (pombe) (Dittrich 1987).

Sclerotinia

Fungal genus of the Ascomycetes (order Heliotiales). Plant parasitic spp. S. fructigena (Monilia fructigena) causing brown rot or ring rot on kernel fruit, and S. laxa (Monilia laxa) on stonefruit. S. sclerotiorum causes damage during storage of root vegetables (Beuchat 1987). May be found on celery as brick-red mould, and induces the production of→phytoalexins (5-methoxypsoralen, 8-methoxypsoralen and 4-, 5-, 8-trimethylpsoralen). The human skin is sensitised to UV light by these phototoxins. Damage during storage also on carrots, cucumbers and artichokes; the production of furocumarins has only been described for carrots [Ceska et al.: Phytochemistry 25 (1986) 81 – 83].

sclerotium

Resting form or overwintering structure, produced by several fungi. Macroscopically visible, plectenchymatous body of mostly dark coloured thick-walled→hyphae; species specific in form. Examples: Sclerotinia sclerotiorum on rapeseed or Claviceps purpurea on cereals. Under favourable conditions the sclerotia on the soil will "germinate" and produce stromata (ergot) or conidiophores (A. flavus), that promote the distribution via air and water and the conservation of the sp. In this way, however, the infection risk is greatly increased for agricultural plants (Müller and Loeffler 1982).

Scopulariopsis

Anamorph of Microascus. S. brevicaulis is the most common sp. Growth: opt. 24 – 30 °C; min. 5 °C; max. 37 °C. Min. a_w 0.85; opt. 0.92 – 0.94. pH: opt. 9 – 10. Occurrence: decaying plant material, cereals, meat, salami, cheese, eggs, etc., in the final spoilage phase. – Grows on materials containing arsenic, and produces gaseous hydrogen arsenate with a smell reminiscent of garlic (Reiss 1986).

SCP→single cell protein.

scurf→apple scurf.

secondary metabolites

Metabolites produced especially by→ fungi, but also from other MO; synthesised subsequent to the exponential growth phase (trophophase), and mostly excreted. Most important compounds of this category: → antibiotics, → toxins, → pigments and aromatic substances; produced during (e.g.) processing, transport and storage. Opposite:→primary metabolites.

selection

The manipulated promotion of the growth of certain strains or groups in a mixed microbial population, affected by changing the environmental conditions (temperature, pH, → selective media, etc.). In mixed microbial populations, typical of foods, desired spp. or groups (e.g., → thermophiles) may thus be separated from the accompanying MO. The selection within a → pure culture for obtaining MO with specific desired characteristics (uncommon of the sp.), e.g., → phage resistance of lactic starter cultures, or nitrate reduction of *Staph. carnosus* at low temperatures, is dependent on → spontaneous mutations within the population. Although such selections are based on genotypic → adaptation, they are not similar to → genetically modified MO. The mutants have been present in the original population, and would have increased selectively under specific changes of the environment. The fact that MO may change their genetic properties "spontaneously" and abruptly (i.e., mutation) has only been recognised and experimentally proved during recent decades. – This aspect should have decisive consequences for the approval of new → starter cultures or → enzyme preparations for use in food technology (DFG 1987).

selective medium

Selective growth substrate (sometimes falsely termed → differential medium). Nutrient agar or broth which, on account of its composition, will promote the growth/proliferation of one type of MO within a mixed population. Two principles are applied basically, either singly or in combination: addition of inhibitory substances that do not (or only slightly inhibit) the desired group, and/or a combination of nutrients that enables only the multiplication of the desired group. – Selective media are commercially available, mainly as powdered or granulated ready-to-use mixtures (e.g., BBL, Difco, Merck, Oxoid).

selenite broth

Enrichment broth acc. to Leifson or selenite-brilliant green-mannitol enrichment broth acc. to Stokel and Osborne for the detection of → *Salmonella* (48 h at 37 °C). Selenite inhibits the growth of coliforms and enterococci; selenite + brilliant green + taurocholate inhibit practically all accompanying MO (e.g., BBL 11608; Merck no. 7717 or 7718; Oxoid CM 395).

semi-soft cheeses

→ Edamer, → Gouda, → Tilsiter and Wilster are manufactured with mesophilic → starter cultures. Semi-hard/semi-soft cheese (→ butter cheese) is manufactured under slightly modified conditions. For mould cheeses (Roquefort, Stilton, Danablue, Bavaria Blue, Gorgonzola) a culture of → *P. roqueforti* is added to the cheese milk; in some cases *P. caseicolum* is inoculated onto the cheese surface. Microbiological defects: → early blowing and → late blowing (ICMSF 1980; Teuber 1987).

sensoric

Examination of food by man as "measuring instrument". Taste, flavour, colour and consistency are judged, and the sensory observations ("senses") are comparatively evaluated in terms of quality and quantity by different persons (Fricker 1984). In addition to processing attributes (cooking, grilling, smoking, etc.), microbiological activities are of decisive importance for the sensory (organoleptic) quality of a food. It is generally accepted that sensorically detectable changes are registered only when microbial numbers exceed 10^6/g, ml or cm^2. Example: a moist microbial layer on raw meat, observable by touching and accompanied by colour change and a damp smell, is caused by 2 – 3 layers of densely packed MO (Gram + bacteria and/or yeasts), and corresponds to ca. 10^6 cfu/cm^2. In Figure 19 the taste and odour imparting metabolites

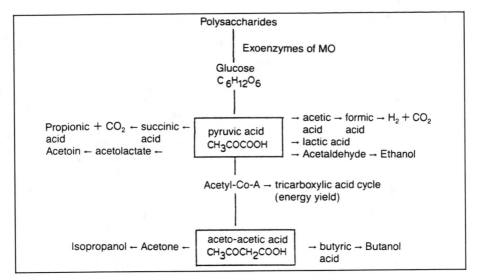

Figure 19 Some metabolic paths which may affect sensory changes in a food, in relation to Enterobacteriaceae, lactic acid bacteria, and/or clostridia (drastically simplified).

of 3 genera are summarised. – Sensory changes by MO may be evaluated in different ways by different ethnic groups; e.g., cheese is valued and accepted in Europe and North America, but is considered a spoiled product in India; the same applies to alcoholic beverages in Islamic countries.

separator slime
Clarifier slime. When milk is centrifuged to skim off the cream and for purification, solids are deposited on the walls of the tank and build up a slimy thick, yellow brown layer of several cm. It consists of dirt, cellular material (debris) of the milk and proteins. It must be removed for hygienic reasons since MO are present in large numbers and may include pathogens (*Handbuch d. LM-chemie III*, 1968). The residues from→ "bactofuges" used to remove endospores from cheese milk are generally sterilised by ultra high temperature treatment (Teuber 1987).

serotypes
Serovars. A serologically (antigenically) distinct variety within a sp. or genus, e.g., *Cl. perfringens, Salmonella, Shigella, Yersinia enterocolitica,* etc. Methods for differentiating the serotypes are still relatively complicated, and are performed mainly in specialised (clinical) laboratories.

Serratia
Genus of the→ Enterobacteriaceae. Gram –, facultatively anaerobic, peritrichously flagellated rods (Bergey 1984). Strongly proteolytic; catalase +. – *Serratia marcescens* ("*Bacterium prodigiosum*") is the most important sp. It produces a red, non-diffusible pigment on CH rich substrates; this pigment is fat-soluble and may stain plastic utensils. Growth: 21 – 37°C, min. a_w 0.945. Survival: home dust 14 d; on the skin up to 8 d; textiles up to 60 d; honey 14 d. Destruction: pasteurisation. Grows well in milk, on refrigerated meat, in the gut of flies and cockroaches, drinking water, distilled water or demineralised

water. Occurrence: surface waters, especially in autumn, drinking water, soil; milk, bread and several foods. May cause → opportunistic infections on weakened persons (Mitscherlich and Marth 1984).

shake culture

(1) Method for the rapid aerobic production of → biomass, → enzymes, → secondary metabolites under laboratory conditions. The nutrient solution is continuously shaken in shaking flasks (with 3 or 4 indents), so as to allow optimal oxygen supply. – (2) Bacterial culture in nutrient agar which has been shaken prior to solidification so as to allow even distribution of the cells. This "top-layer" method allows observations on gas production and O_2 requirement after incubation. In combination with → differential media this method also enables the determination of other characteristics, e.g., → acid production, → sulphite reduction, etc. – (3) Procedure used for the isolation of anaerobic bacteria, by gentle shaking of the inoculum in a special culture tube (e.g., Veillon tube).

Shigella

Sh. sonnei, Sh. dysenteriae, Sh. boydii and *Sh. flexneri* are causative agents of → dysentery (bacterial dysentery). Gram – , immotile rods, aerobic or facultatively anaerobic. – Min. 6 – 8 °C, opt. 37 °C. Occurrence: gut of humans and some primates. Survival: faeces on soil 5 – 10 d; on fruit and vegetables up to 10 d; in flies and ants up to 12 d; fresh water 2 – 3 d. Destruction: pasteurisation; sensitive to acids (Mitscherlich and Marth 1984).

shigellosis → dysentery.

shii-take → cultivated mushrooms.

short-time heating

Method to pasteurise milk for drinking purposes in a continuous flow heat exchanger; 15 – 40 s at a temperature of 71 – 74 °C.

silage

Animal feed made by spontaneous lactic fermentation of chopped grass, maize plants, etc., stored under anaerobic conditions. The bacteria present on the plant material (mainly LAB) metabolise the available plant sugars primarily to lactic acid, by which the pH drops and preservation is attained. Good quality silage contains ca. 130 mol of lactic acid/kg after 30 d, as well as 80 mol acetic acid/kg and 30 mol of succinic acid/kg. Butyric acid and propionic acid should not be present in significant amounts, and are indications of fermentation failures allowing the growth of clostridia. The acidification also discourages the growth of *Listeria* spp. – Because of the relatively high number of *Cl. endospores* normally present, milk of cows fed with silage should not be used for the manufacture of → hard cheese; → late blowing.

silting

Firm/solid, gelatinous meat products (with or without vegetables such as carrots, cucumbers, etc.) solidified with the aid of gelatine or product specific proteins. High spoilage potential because of high → a_w. Acidification will reduce the spoilage potential (Sinell 1985).

silver

Ag. Heavy (noble) metal. Microbistatic in low conc.; antimicrobial moiety is the silver ion. → Katadyn filters. Silver nitrate ($AgNO_3$; "Hollenstein") is microbicidal or microbistatic, depending on conc. and also virucidal (Wällhausser 1988). Approved for the preparation of drinking water and beverages; especially made use of for bottled or canned drinking water, on ships, for camping, etc. – Certified colour additive E 174.

single cell protein
SCP. Commercially available as food or feed yeast. Under the impression of the protein deficiency ("protein gap") in world nutrition, biotechnical processes have been developed for the rapid and effective production of microbial→biomass, using low cost nutrients (mineral oil fractions, methanol, ammonia, etc.). Example: $CH_3OH + NH_3 + O_2 + minerals→$microbial biomass with ca. 50% protein in DM. *Methylomonas clara* (bacterial protein); *Fusarium graminearum* (mycoprotein); → *Spirulina*; → *Scenedesmus*. These products, however, contain >6% of nucleic acids (→ DNA) in DM; for human nutrition this conc. should constitute <2% of the DM. Elaborate processes are necessary to achieve this substantial reduction, which is important so as to protect the health of the consumer on consumption of relatively large amounts (gout). Only in a situation of continued uncontrolled growth of the world population, SCP may regain importance; presently the "protein gap" seems to be closed by soya production.

single strain culture
→ Starter culture, containing only one MO sp., e.g., *P. nalgiovensis*. Opposite: multistrain culture, e.g., for cultured creamery butter.

skatol
β-Methyl-indole. Unpleasant smelling compound. Produced (in addition to indole) by Gram – bacteria by the oxidative deamination of tryptophane during protein degradation in meat and meat products (Krämer 1987).

skin microorganisms
The human skin is typically colonised by surface microbes that may be transferred by contact with foods. Typical skin microbes: *Staphylococcus* spp., *Enterococcus, Sarcina, Corynebacterium*, coliforms, *Micrococcus* spp., etc.

Skyr
Islandic lactic fermented milk product, manufactured with *Lactococcus* spp.

slant → slope.

slime components
Macromolecular swelling compounds produced by some MO (→ capsule). Undesired in beverages because of increase in viscosity and complicating filtration. Desired in Swedish "Long milk". Polysaccharides of different composition produced by *Str., Leuconostoc, Enterobacter, Sarcina*, etc., and polypeptides by *B.* (ropiness).

slime layer
Multiple layers of bacteria and yeasts on the surface of meat or cooked sausages; indication of onset of spoilage. At chill temperatures (0 – 2 °C) the spoilage association is determined by the type of product and packaging; e.g., under aerobic conditions mainly *Pseudomonas-Acinetobacter-Moraxella* association; under vacuum packaging mainly *Lactobacillus sake/Lactobacillus curvatus* and *Leuconostoc* spp. (*Lc. carnosum, Lc. gelidum*). At room temp. *B. Micrococcus*, Enterobacteriaceae and yeasts. The slimy layer often smells dampmusty to putrid (Sinell 1985). – Similar layers are commonly observed on synthetic surfaces (pipes, water containers of coffee machines, etc.); smell mostly neutral.

slope
Agar slope (slant). A solid nutrient agar medium (ca. 7 ml) which has been allowed to set in a diagonally oriented test-tube ("culture tube"). In this manner the surface in contact with the air is enlarged. Used for cultivation and keeping of aerobic MO (→ culture collection).

slow ripening

Part of the production process of dry and semi-dry (fermented) sausages requiring climatised cold storage for 4 – 8 weeks. The desired aroma is obtained by promoting the development of desired organisms with the aid of temperature, rH and fresh air supply control (Krämer 1987; Sinell 1985).

smoke

A mixture of solids (ash, soot), droplets (water, tar) and numerous gaseous compounds obtained by smouldering or glowing chips or sawdust of deciduous wood. Coniferous wood is seldomly used. 1 m^3 smoke contains up to 3 g of components. Several ingredients are microbicidal or microbistatic, e.g., formaldehyde and other aldehydes, phenols, cresols, guajocols, thymol, p-hydroxybenzaldehyde, formic acid, acetic acid, propionic acid, etc. [Baltes, W. and Söchting, I.: *Z. Lebensm. Unters. Forsch.* 169 (1979) 9 – 16]. It also contains carcinogenic hydrocarbons, e.g., 3,4-benzpyrene acting as indicator substance because of its favourable analytical properties. It is estimated that ca. 10,000 chemical substances are present in smoke of which ca. 500 contribute to the aroma (Classen et al. 1987). – Smoke is allowed as food additive in the sense of the "ZZulV" and meat regulations of 21.1.82 for the treatment of meat and meat products. The finished product should not contain more than 1 μg of 3,4-benzpyrene/kg (Sinell 1985).

smoke aroma preparations

Also called liquid smoke or smoke essence (concentrate). It is allowed in some countries instead of the conventional→ smoking for the production of sausages. An advantage is that these products are free from carcinogens, typically present in natural smoke, and which constitute a risk factor (→ smoke). The condensate is a brown liquid with varying water content. It is directly added to the mixtures of the different cooked sausages, salami and other fermented dry sausages, and then well mixed. At present the use is still not allowed in Germany, however, smoke aroma flavours bound to salt or glucose are considered as foodstuffs and their use is therefore approved. – The possible microbicidal or microbistatic effect is negligible at the concentrations typically used. The inhibitory effect of drying on microbial growth, otherwise obtained by the smoking process, should be achieved by reducing the a$_w$ value to < 0.95.

smoking

Traditional procedure for preparation of meat, sausages and fish with extended storage life. It has a characteristic taste and smell. Antimicrobial effects are in the first instance due to the rapid drying of the surface, with damage and partial destruction of the MO on the surface as secondary effects. Different types can be distinguished: cold smoking also "Kaltenräucherei" at < 30 °C for salted fish, smoked salmon, fermented sausages and raw ham for several hours up to weeks; warm smoking or rapid smoking (30 – 60 °C) in combination with curing salt and a$_w$ values < 0.95 to prevent rapid multiplication of MO internally in the product; hot smoking (65 – 80 °C) numerous ocean and fresh water fishes, e.g., sprats, trout, herring, etc., and also cooked sausages. Following hot smoking < 60 °C meat products are heated subsequently for a short period to destroy the vegetative MO in the core (Krämer 1987; Sinell 1985).

SO₂ → sulphur dioxide.

sodium hypochlorite

NaOCl. Ingredient of disinfectants used in the food industry and large kitchens (Wallhäusser 1988).

sodium phenyl phenolate→ ortho-phenyl phenol.

soft cheese
Camembert, Brie, Romadour, Limburger, "Münster" cheese, etc. Produced with the aid of mesophilic cheese cultures. Also→ rennet cheese on which a protective (dry) coat or layer is not formed, since the dry mass only amounts to ca. 44%. The surface is protected by a dense layer of moulds (→ P. caseicolum) in the case of Camembert and Brie. Other types are protected by→ Brevibacterium linens. The mould culture is already present in the cheese vat milk, but the bacterial cultures are streaked onto the surface 2 – 3 h after the treatment in the salt bath and repeated up to 8 times (Teuber 1987).

soft rot
Soft discoloured spot in stored fruit and vegetables caused by extracellular→ pectinases or→ cellulases of parasites and→ opportunistic parasites, e.g., Erwinia carotovora in the case of root vegetables and P. expansum in stonefruit (Müller 1983a; Pitt and Hocking 1985).

soil culture
Approved method for storage of→ pure cultures of fungi. Dried garden soil + clay (1:1) is thoroughly mixed and 6 cm filled into a test tube, 8 ml of water added, and autoclaved two times each for 1/2 h. Inoculation with a drop of mould suspension; incubation for 10 d at room temperature followed by storage in refrigerator. Subculturing with moist inoculation needle, and streaking out on agar medium surface.

soil filter→ biofilter.

soil microorganisms
Bacteria and moulds inhabiting soil. By estimation 1 g of soil contains $>10^9$ viable MO, most of which are→ saprophytes. They play a key role in the mineralisation of plant and animal residues. Soil contact with food may constitute an important source of→ contamination with spoilage microbes.

solid-state fermentation
"SSF". Novel term used for traditional fermentations of solid or pulpy (semi-solid) substrates. Examples: cheese, fermented sausages (dry sausage), raw ham, sourdough, sauerkraut, pickles, cocoa, east Asian specialities.→ Fluidised bed fermenter (Kunz 1988).

sorbic acid
$CH_3 - CH = CH - CH = CH - COOH$, 2,4-hexadienoic acid. LD_{50} rat p.o. 7.4 – 10.5 g/kg BW. Long-time feeding of rats with 5% sorbic acid in feed did not cause any detectable defects over two generations.→ ADI value 0 to 25 mg/kg BW (FAO/WHO). Metabolised in the body as caproic acid.→ Preservative, pH-dependent as other org. acids (pH <6). Mainly fungistatic. Growth of toxinogenic moulds prevented with 0.1 – 0.2%. Catalase +, Gram – bacteria relatively sensitive, in contrast to catalase-negative bacteria (LAB and clostridia). No effect against→ endospores (Wallhäusser 1988). EC no. E 200; sodium sorbate E 201; potassium sorbate E 202; calcium sorbate E 203. – Maximum conc.: see FDA §121.101; German Regulation "ZZulV" of 22.12.1981, Appendix 3, List B.

sorbitol
Commercial names: sionone, karione. Glucitol. Polyol corresponding to glucose, with a sweet taste and suitable for diabetics as sugar substitute. Utilised by gut bacteria and many other MO. Occurrence: berries, cherries, plums and other fruit.

sorbose
Ketohexose. L-sorbose formed by bac-

terial (*Acetobacter* and *Gluconobacter* spp., especially *Gl. suboxydans*) oxidation of D-sorbitol. Important intermediate in the manufacture of vitamin C (ascorbic acid). Sweet tasting sugar; not fermented by *Sacch. cerevisiae*.

Sordaria fimicola

Fungal sp. (order Sordariales); belongs to Eurotiales. Universally found in the faeces of herbivores. Grows extremely fast in soil and on nutrient media (at 18°C ca. 2 cm/d). Found on cereals, peanuts and vegetables. No information on → mycotoxins (Domsch et al. 1980).

sour → sourdough.

sour milk cheese

Cheese produced from quarg or acid-rennet-quarg, covered with proteolytic surface microbes responsible for ripening from the outside to the inside. The surface is inoculated with → *Brevibacterium linens* or → *P. caseicolum*. Examples: "Harzer" or "Mainzer" cheeses, Romadur, "Olmützer", Herbal cheese, etc. (Teuber 1987).

sourdough

A dough prepared from rye or mixtures of rye and wheat flours, and containing metabolically active MO (LAB and yeasts). Most typical is the heterofermentative *Lactobacillus sanfrancisco*. By selection of appropriate temperatures, holding times and consistency ("dough yield"), etc., the rate of microbial growth and metabolic activity can be controlled. Dough from a previous batch is used as inoculum. Commercial dough "starter" is available typically in semi-moist state, but also either as dehydrated or lyophilised culture, containing $10^6 - 10^{10}$ LAB/g and $10^3 - 10^7$ yeasts/g. – Sourdough is mainly used for the manufacture of rye bread or mixed-flour breads containing some quantity of rye flour. Entrapment of CO_2 bubbles in the elastic flour gluten complex promotes the lightening of the texture of the dough; the metabolites of the heterofermentative → LAB contribute to the soury-aromatic flavour typical of sourdough breads [Bode and Seibel: *Getreide, Mehl u. Brot* 36 (1982) 1, 11 – 12] (Spicher and Stephan 1987). – The ratio of yeasts to LAB changes with the respective stages of sourdough preparation (succession). Depending on the cultural methods or direct counting method (→ microbial count) used, results may differ (Table 37). The different steps from the initial phase to the finished dough are depicted schematically in Figure 20. – Considerable time input is required

Table 37 Share of bacteria and yeasts in the population of sourdough (Spicher and Stephan 1987, modified).

Sourdough Stage	Viable Cell Counting Yeasts: Bacteria (%)	Average Population Microscopic Counting ($\times 10^6$/g)	
		Yeasts	Bacteria
Preliminary	60:40	18.4	1050.0
Fresh sourdough	60:40	17.5	1121.0
Basic "sour"	40:60	30.4	1693.7
Finished sourdough	70:30	20.9	799.9
Dough	80:20	15.6	696.8

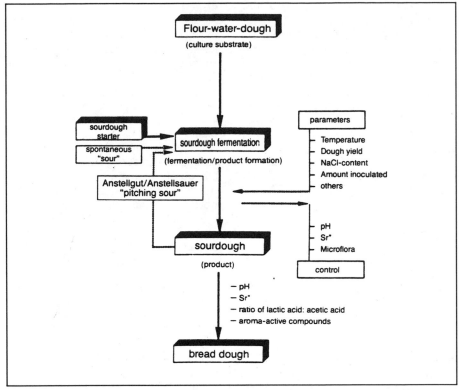

Figure 20 Flow scheme for the preparation of rye dough with the use of sourdough (acc. to Spicher, G. and Stephan, H.: *Handbuch Sauerteig. Biologie, Biochemie, Technologie*. Hamburg: Behr's Verlag, 1987, p. 17).

for the manufacture of sourdough bread. Problems may arise as a result of the regulatory standardisation of the length of a modern working day.

sourdough culture
Sourdough produced from a→pure culture. It is described in the "Codex alimentarius austriacus" of 1966 as follows: fermentation or acidification (acid production) can either occur spontaneously or introduced by the addition of cultures containing sourdough yeasts and acid producers. According to the origin it is called natural sourdough or culture sourdough (Spicher and Stephan 1987).

sourdough spoilage
A shift in the microbial balance of sourdough or the presence of ("foreign") microbes may result in defective fermentations. This is associated with atypical aroma and taste of the bread, and results from abnormally high amounts of butyric acid, acetic acid and propionic acid. Description of undesired (deleterious) microorganisms in literature is insufficient; these cannot be classified by presently available phenotypical systems (Spicher and Stephan 1987).

souring inhibitory test
Method for the examination of foods for

the possible presence of inhibitory substances (→antibiotics; →disinfectants; →mycotoxins) for acidifying bacteria. It is possible, e.g., to inoculate milk with a yoghurt culture and incubate at 42 – 43 °C; the time till coagulation is compared with that of a sample free from inhibitory substances. → Enzyme preparations are incubated at 68 °C with *Bacillus stearothermophilus* in a nutrient solution containing bromocresol purple as pH and brilliant black as Eh indicator, till colour change is observed. – Both tests may be performed under non-axenic conditions.

soy cheese

In China: "Sufu". Prepared from soy milk with different Mucoraceae; ripening in 12% salt brine for 2 months. – In Japan: "Natto". Manufactured with *B. subtilis* from soy milk (Beuchat 1987; Krämer 1987; Steinkraus 1983).

soy sauce

Shoyu. Condiment prepared by fermentation of soy beans and roasted wheat with *Aspergillus oryzae* at 30 °C for several d (= "koji"). The "koji" is suspended in 17 – 19% salt brine to form "moromi", which is then fermented for at least 6 months. During this period proteins are hydrolysed by enzymes from the koji organism to low MW substrates which are readily available for the dominating LAB population (*Lactobacillus delbrueckii, Pediococcus* spp.) and yeasts (*Saccharomyces* spp., *Zygosaccharomyces rouxii*). After fermentation and "maturing" the product is separated from the solid residue by filtering and pressing (Beuchat 1987; Krämer 1987; Steinkraus 1983).

sp.

Abbreviation for species; – spp. plural of sp., i.e., more than one sp.

sparkling wine

Similar to "Champagne" from a specific region in France. Wine with 8 – 12% ethanol is fermented following the addition of sugar and a yeast culture. The fermentation is performed either in closed tanks or in thick-walled bottles ("traditional process"). The CO_2 pressure gradually increases and the fermentation is completed within 1 – 2 months. *Brettanomyces* may cause sensory defects by ester formation during bottle fermentation; the quality of the wine substrate is of decisive importance, and filtration prior to bottle fermentation may be an important preventative measure (Dittrich 1987).

species

A small, more restrictive taxonomic grouping of biological entities. Genetical similarity within the grouping, i.e., genetic interchange or interbreeding are possible. Important is the fertility within the grouping or collection of populations.

Sphaerotilus natans

"Sewage fungus". Gram – , polar or subpolar tuft of flagella, long bacterial rods; aerobic to micro-aerophilic. Long chains or bunches of cells in sheathed trichomes. Especially in nutritionally rich media at the inlet of sewage works, or in wastewater systems of food processing plants. Massive growth may result in a blocking of pipelines and canals.

spices

Dried plants or parts of plants which are used in whole or ground form for flavouring, seasoning and sensory improvement of foods and foodstuffs. Spices, as raw or natural agricultural products, generally contain high microbial numbers (Table 38). Until 1986 ethylene oxide has been used for decontamination; reduction to $< 10^5$/g assured sufficient safety levels for use in processing, e.g., of meats. However, the possible formation of carcino-

Table 38 Maximum aerobic microbial numbers expected for untreated spices (in cfu/g).

Spice	Aerobic bacteria	Moulds
Pepper (black)	7×10^8	2×10^6
Pepper (white)	2×10^5	3×10^6
Paprika	3×10^7	2×10^6
Nutmeg	3×10^3	3×10^4
Coriander	2×10^6	up to 10^4
Marjoram	8×10^5	up to 10^5
Thyme	up to 10^6	2×10^5

genic ethylenechlorohydrin prompted authorities to withdraw approval of ethylene oxide for treatment of spices and dehydrated vegetables, and also for protection of stored materials. Radiation treatment offers a superior alternative; with the exception of some European countries (e.g., D, DK) its use for spices is approved in several countries, e.g., B, BR, F, H, N, NL, SU, ZA including a number of developing countries. Several of the natural components show antibacterial and/or antioxidative activity: e.g., p-hydroxybenzoic acid and its esters (→ preservatives), as well as salicylic acid and hydroxy-cinnamonic acids are found in a large number of spices [Schulz and Herrmann: *Z. Lebensmittel Unters. Forsch.* 171 (1980) 193 – 199]. An effect towards shelf life extension of the processed food can hardly be expected in view of the small concentrations used, but also because of the increase of the contamination level caused by most spices. Contamination with → food poisoning MO cannot be excluded; after addition of spices foods should not be kept for longer periods without refrigeration. – Several extracts of spices contain genotoxic substances (→ AMES test) [Lafont et al.: *Microbiol.-Aliments-Nutrition* 2 (1984) 239 – 249] (Peppler and Guarino 1984).

spin chiller
Chill water bath for washing and/or cooling of fruit (e.g., peaches), melons, sweetcorn, carrots, radishes, etc., before packaging, storage and transport. Similar systems are used for the chilling of poultry for 15 – 30 min, prior to freezing, but are hygienically questionable. The distribution and increase in numbers of salmonellae throughout the washing water, with concomitant transmission of cross-infections between carcasses, is promoted by this process (Heiss and Eichner 1984; Krämer 1987).

spiral plate count method
Semi-automatic method for determining the viable numbers (→ microbial count) in a liquid sample. The inoculum from a serial dilution or liquid sample is distributed by means of a syringe onto a rotating plate, along a spiral path starting from the centre of the plate containing the nutrient medium. Following incubation, the colonies are counted on a particular region of the plate, using a template. For routine counting of large numbers of plates, a laser colony counter can be used. Also suitable for use with the → AMES test.

Spirillum
Genus of the Spirillaceae (Bergey 1984). Gram – , asporogenous; helical, relatively long rods. Motile by bipolar tufts of flagella. Aerobic to microaerophilic; respiratory metabolism; oxidase + , catalase – . Commonly halophilic. Occurrence: surface waters, wastewater; on ocean fish in coastal waters; salted fish (Sinell 1985).

Spirulina platensis
Filamentous → cyanobacteria found in saline, alkaline lakes in the tropics with high conc. of bicarbonate/carbonate. Used as food for centuries, e.g., along Lake Chad (Africa), and cultivated in open basins in India as → single cell protein. Collection (e.g., filtration) of the masses of

growth is relatively easy because of the size of the cells (ca. 0.5 mm in length) – by contrast to → Scenedesmus (0.01 mm). The thin cell walls enable the rapid sun-drying of the → biomass, and easily extractable. The fluorine content (ca. 95 mg/kg) is relatively high.

Chemical composition (g/100 g DM):

crude protein	55 – 65
total nitrogen	11
lipids	2 – 6
carbohydrates	10 – 15
crude fibre	1 – 4
minerals	5 – 12
RNA	3.9
DNA	1.2

The biological value of the protein amounts to 80 – 90% of that of casein (Becker 1982). The → purine content exceeds 2%, and constitutes a health risk if the product is consumed over extended periods. Supplementation of carbohydrates, e.g., of pastas or rice, however seems promising.

spoilage association
Group(s) of MO typically involved in the spoilage of a particular food product.

spontaneous heating
Temperature increase in grains stored in bins, bags, heaps, silos, etc. Moisture, oxygen availability and temperature are the main factors determining probable mould growth. At water activity values > 0.72 an additive effect is observed between the respiration of MO and the grains; this again adds moisture and warmth to the grain bulk that accelerate further respiration or microbial activities. Initially, mesophilic aerobes and moulds dominate; when ca. 35°C is reached *B.* spp. and *Thermoactinomyces* spp. cause localised ("pocket") heating to 50 – 60°C. This may result in a loss of the germination potential of the grains, sensory changes, nutrient losses, and a reduction of the baking quality (Müller, G. 1983, 1983a).

spontaneous mutation
"Natural" → mutation of a particular MO; occurs only at low frequencies and in absence of any detectable mutagens. The ratio of lethal mutants to survivors (with specialised properties) differs from sp. to sp. Spontaneous mutation forms the basis of → selection, → adaptation and phylogeny (Schlegel 1985).

spontaneous sourdough
"Natural sour". LAB and yeasts which develop during a spontaneous fermentation after mixing water and flour. The majority MO present in the flour initially multiply, but the LAB soon dominate and outcompete and inhibit the other MO. Gradually the yeasts increase as well, although relatively little is still known about this group. The populations in solid sourdoughs (50% water) differ from those in the soft sourdoughs (70 – 80% water). – These "spontaneous" sourdough preparation procedures are no longer practised (Spicher and Stephan 1987).

sporangium
Sac-like structure for the asexual production of sporangiospores (Figure 21). A striking, dark coloured organ, visible with the naked eye, formed by a number of lower fungi, e.g., → Mucor, → Rhizopus, etc.

spore formers
Gram + bacteria of the family Bacillaceae (→ endospores), able to form dormant spores under limiting conditions (e.g., nutrient limitation) which are extremely resistant to adverse physical conditions (e.g., heat, radiation, desiccation) and to chemical agents. Especially the endospores of the genera *Bacillus* and *Clostridium* are of extreme importance for food processing and handling, both concerning keeping

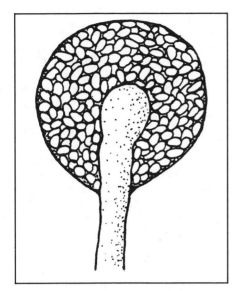

Figure 21 Sporangium with sporangiospores
of *Mucor* sp.

quality (shelf life) and the hygienic inte-
grity. They include a.o. food poisoning
bacteria such as→*B. cereus,*→*Cl. botulinum,* and→*Cl. perfringens.* Their high resistance served as decisive criterion for
the development of sanitation and disinfection methods, as well as heat processing programmes (→D value,→F value).
– Further endospore-forming genera include *Sporolactobacillus, Desulfotomaculum, Sporosarcina* and *Oscillospira.*

Sporendonema→ *Wallemia.*

spores
Small→dormant, distribution or reproductive forms of MO. In some bacterial
genera (→spore formers) one spore is
formed asexually within a vegetative cell
(→endospore). It contains a dehydrated
protoplast, and is surrounded by a thick
multilayered wall. Such endospores are
extremely resistant to all processing
measures, including radurisation, and

may survive for several years under dry
conditions. Formation and factors involved are quite similar for all bacterial
spores. Food processing should be aimed
at the prevention of conditions that induce
sporulation. – Spores of yeasts and fungi
are formed in different manners. Generally
distinguished between sexual and asexual formation, thick-walled and relatively
resistant spores, and those which only
slightly differ from the vegetative organs. – The relatively low resistance is
balanced in general by the formation of
enormous numbers. The→conidia, produced asexually externally to the phialides
are distributed mainly by wind and insects, and constitute a severe contamination risk within food processing plants
(→elevators). Often possess more than
one nucleus (→heterokaryosis); forms the
basis for the extreme→adaptability of
fungi.

Sporolactobacillus
Endospore-forming, monotypic genus of
Gram +, non-motile and catalase – , rod-
shaped bacteria, producing lactic acid
from CH. Occurs in the rhizosphere,
faeces of herbivores and in sewage. Germination of endospores and growth even
in presence of 2 mg nitrite/ml.

Sporotrichum
Imperfect fungal genus of the
Moniliaceae. *S. carnis* is commonly involved in the spoilage of fresh meat (Reiss
1986).

spray drying
Preheated, liquid or pasty foods are dispersed into fine droplets by jets or rotatory
plates, and sediment gradually against an
air stream at 200°C to the bottom. This
process enables the drying of such
products, although some MO, especially
endospores and some thermoduric MO
may survive. This fact has consequences
for the hygiene of instant products.

Energetically unfavourable process, although economical for products such as milk, infant foods, coffee-extract, etc. (Heiss and Eichner 1984).

sprout mycelium
Pseudohyphae; pseudomycelium. Chains or groups of cells formed by some yeasts under specific conditions, reminiscent of a→mycelium on microscopical examination. Distinguished from a real mycelium by the absence of septa, e.g., for Candida albicans growing on Maize Meal Agar, and intercalary growth (Müller and Loeffler 1982).

St. Anthony's fire→ ergotism.

stab culture
Culture of bacteria in the butt of an agar or gelatine slope. A straight inoculation needle or loop is stabbed vertically 3 – 4 cm into the medium. The MO partly remain at the surface and partly distributed along the path of the stab. They replicate, according to their oxygen demand, preferably either in the top or bottom section. In→gelatine the liquefaction may be determined simultaneously. – Simple method for the collection of→characteristics for the→identification (Baumgart 1990).

stabilisers
Thickening and gelatinising agents (E 400 and following). Food additives with the aid of which the desired consistency or viscosity of a food product is attained. Mainly → polysaccharides from plant or microbial origin; may be dissimilated enzymatically by bacteria and moulds. Swelling hydrocolloids which provide excellent growth conditions for MO because of their water binding capacity (Classen 1987). – In food microbiology inhibitory substances (→ preservatives) are sometimes termed "stabilisers", because of their "stabilising" effect against spoilage.

Stachybotrys
Genus of the→ Deuteromycetes, → black moulds. S. alternans is the most common sp. in connection with food spoilage. Growth: 5 – 40°C; pH 3 – 9.8; a_w 0.94 at 37°C. Non-fastidious; cultivation possible on moist filter paper. Occurrence: soil, mouldy hay and other plant products. May cause damage as cellulose degrader (paper, textiles from plant tissues). – Forms safratoxins (→trichothecenes) that may cause stachybotryotoxicosis of horses, cattle, sheep and poultry; employees in agriculture may contract the disease after contact with contaminated feeds, etc., which manifests as infections of the skin and mucous membranes (Gedek 1980; Reiss 1986).

staining procedures
Staining procedures are applied for studying morphological features (the shape: cocci, rods with different lengths and diameters, etc.; chain formation, tetrad arrangement, etc.) under the bright-field microscope, especially when no phase-contrast or interference-contrast microscope are available. For this purpose a simple staining procedure, e.g., with methylene blue solution acc. to Loeffler, is used. The→gram-staining procedure is applied for determining the gram-reaction of a bacterial cell.→Spore-staining is necessary to confirm the presence of→endospores. Acid-fast bacteria (e.g., Mycobacterium or Nocardia) are stained with carbolfuchsin solution acc. to Ziehl-Neelson. – For moulds either methylene blue or→lactophenol blue solutions are used, especially when the slides are to be kept for a while. – All staining reagents are commercially available, e.g., from Merck, Fluka, Sigma, etc. (Baumgart 1990).

standard nutrient media
Standard plate count agar is CH free and is used for the determination of viable microbial numbers (→cfu) in milk, milk

products, water and sewage (e.g., Merck no. 1621). – Standard nutrient agar or broth with glucose is suitable for the cultivation of more fastidious bacteria (e.g., Merck no. 7881 or 7882). – Standard-II-Agar or broth without glucose is suitable for the cultivation and enrichment of less fastidious bacteria, and as base for the preparation of special nutrient media (e.g., Merck no. 7883 or 7884; Oxoid CM 3, CM 1 and CM 9).

staphylococcal food poisoning

A form of acute gastroenteritis caused by the ingestion of food contaminated with the→ enterotoxins (exotoxins) of *Staph. aureus*. The organism is not "toxic" as such, but causes infection, e.g., of wounds, and may often be resistant to penicillin (common "hospitalism" pathogen of the first years after the second World War). After an incubation period of 2 – 6 h, following oral ingestion of 0.5 – 1 μg of toxin/adult, symptoms such as nausea, abdominal pain and diarrhoea without fever appear. It may last from 8 h to 3 d. Cure is normally complete, although weakness may persist some time. Complications may occur in infants and old-aged, weakened persons. Because of the short incubation time and severe symptoms, it may cause uneasy and complex situations within travelling groups (flights, train journeys) and closed communities. – Products regularly involved: sliced meats, sausages, sliced cheese, inferior quality soft cheese, low-acid delicatessen foods and salads containing poultry meat, meat, bologna-sausages, cheese and potato, as well as hors d'oeuvres with aspic, poultry meat fillings, pastries, tongue, roastbeef, home-made mayonnaise, confectioneries with cream filling, etc. (Sinell 1985).

Staphylococcus

Genus of the family Micrococcaceae. Gram + , non-motile, asporogenous, cata-

lase-positive cocci. Grows relatively easily on a large number of common nutrient media, and may produce typical haemolysis on blood agar. Colonies round, smooth, shiny, white to orange. – *Staph. epidermidis* is typically associated with the skin and mucous membranes of warm-blooded animals and man (Bergey 1984; Werk, R.: *Staphylokokken*. Stuttgart: Wiss. Verl.-ges. 1990).

Staphylococcus aureus

Most common food poisoning MO. Coccus-shaped cells single, in pairs or irregular bunches; aerobic to facultatively anaerobic; colonies light to dark yellow. Causes inflammation. Min. 10°C, opt. 37°C, max. 45°C; pH > 4.5; a_w > 0.86; inhibited by 17.5% NaCl and killed by 20% NaCl. Toxin production > 7°C. Survival in ice cream, cheese, meat products, etc., at – 18° to + 20°C more than 150 d. Some isolates produce heat stabile enterotoxins; this characteristic is typically accompanied by properties such as pos. coagulase reaction, lecithinase production, hemolysis, pos.→ tellurite reaction and resistance against penicillin (Mitscherlich and Marth 1984). – Transfer to foods by hands, flies, cockroaches, contaminated utensils, rinsing cloths, towels, cutting boards or chopping blocks, etc. (Krämer 1987; Sinell 1985).

Staphylococcus carnosus

Starter culture used for the manufacture of fermented (dry/semi-dry) sausages. – Cocci, 1 μm, single or in pairs; colonies round, entire, slightly convex, white, opalescent; older colonies brownish in the centre. – Facultatively anaerobic, catalase + , coagulase – , nitrite + , acetoin + , non-hemolytic. Temp. 15 – 45°C; growth at 15% NaCl [Schleifer and Fischer: *Int. J. Syst. Bacteriol.* 32 (1982) 153 – 156; Bergey 1986]. – As starter culture in mixture with LAB, used for fermented sausages. *Staph. carnosus* reduces nitrate to nitrite and by means of catalase activity removes H_2O_2 formed by peroxide producing bacteria. In this way rancidity is prevented in the product. By maintaining the NO_2 conc. the→ reddening of the haemoglobin is promoted at pH < 5.5. These bacteria, however, do not reproduce during ripening (DFG 1987).

Staphylococcus xylosus
Coagulase − , hemolysis − , $NO_2(+)$; growth at 10% NaCl + , at 15%(+). – The sp. plays a similar role as *Staph. carnosus* during the ripening, and may occasionally be added to starter cultures (Bergey 1986; DFG 1987).

starch
Amylum. Reserve CH of all plants and main component of many basic foodstuffs such as cereals, rice and potatoes. Hydrolysed by extracellular→ amylases of many MO. Fermentative valuable yeasts are not able to utilise starch.

starter cultures
A→ pure culture or controlled mixed strain culture of selected, defined and viable MO. Used as inoculum of at least 10^6 cells/g of food with the aim at initiating a fermentation for improving the appearance, smell, taste and keeping quality of the product. Starter cultures are commercially available as suspensions, or lyophilised powder, and are applied by the manufacturer either directly, or are subcultured under the strictest safety precautions against contamination, both for increasing the inoculum volume, and for obtaining cultures for further inoculum propagation. – Primarily→ LAB and/or→ yeasts or mixtures of them (→ sourdough; → kefir) are used.→ Immobilised MO are strictly spoken not starter cultures, and can be considered rather comparable to→ enzyme preparations. – Starter cultures are used for the manufacture of ca. 35% of our total daily diet, by fermenta-

tion. Examples are: bread and bakery products,→ sauerkraut, dill cucumbers, → yoghurt and fermented milk products, →cheese, →fermented sausages, → beer,→ wine and spirits (DFG 1987).

steamer
Vessel for steam treatment (ca. 100°C) of utensils or materials at atmospheric pressure. Typically used for→ tyndallisation (fractionated sterilisation on each of three consecutive days); contact time ca. 20 min for each treatment. Comparable to household type steam vessel (pot) for canning of different foods.

Stemphylium
Genus of the Fungi Imperfecti, especially referred to in connection with cellulose decomposition;→ wall fungus. The conidia may cause asthma in sensitive persons upon inhalation (Reiss 1986).→ Black moulds.

stereoisomers→ lactic acid.

sterigmatocystin
→ Mycotoxin, that may be produced by different→ *Aspergillus* spp. during growth on foods (cereals, fruit, fruit products, cheese). Chemically quite similar to the→ aflatoxins, and is also carcinogenic and hepatotoxic (Reiss 1981). Fluoresces brick-red on thin-layer chromatography plates under short-wave UV light, and may easily be confused with chlorophyll.

sterilisation
The killing, destruction, elimination or irreversible inactivation of all MO and their→ endospores in or on a product or utensils. Applied in practice by dry or moist heat, filter sterilisation, radappertisation (→ radiation treatment), etc. In the laboratory moist heat sterilisation is applied in an autoclave at 121°C for 15 – 30 min. In this sense the process of sterilisation is not identical to heat treatments

resulting in "commercial" → sterility, e.g., in the manufacture of canned foods; → F value. Radiation treatment in this sense is applied mainly for disposable medical equipment.

sterility

Freedom from viable organisms or such capable of replication, including their spores and dormant forms. In practice this condition can be attained by heating or → radappertisation. The success of the process is strongly dependent on the composition of the product. Control or monitoring of the effectivity is difficult in practice, especially when a large quantity of units is treated. In practice "over-treatment" is accepted to some extent, so as to achieve a reasonable safety margin in the interest of the consumer and for protection against spoilage. – In the laboratory sterility is generally achieved by autoclaving at 121 °C for up to 30 min. In this manner (a.o.) false positive results are prevented, e.g., in the determination of the → plate count. – "Commercial sterility" (Germ.: "Handelsübliche Sterili-tät") is a term used in food technology. It refers to the condition of a product, e.g., a canned commodity, in which no MO are able to replicate, as long as the condition (pH, a_w, O_2 content, etc.) remain un-changed. Example: canned fruits or fruit juices with a pH value < 4.5 may contain endospores of *Clostridium* and/or *Bacillus* that will not germinate as long as the criti-cal pH is not exceeded.

sterility control

Difficult task for the microbiological labora-tory if it does not concern membrane filter-able products. In practice statistically determined numbers of the sterilised goods are incubated at elevated tempera-ture. Tropical conserved foods are incu-bated/stored at 55 °C so as to enable the germination of *Bacillus stearothermo-philus*. – Sterility control of packaging ma-terials, e.g, for → aseptic systems, is per-formed by means of the → swab method. Bottles or glasses are coated on the inside with nutrient agar. Sterilised air, e.g., for → fermentors, is monitored with membrane filters in an adjacent flow cur-rent, through gas washing bottles with nutrient broth, or by means of aerial count-ing devices. → Air examination.

stock protection

It is inconceivable without pest control. Rodents and birds decimate the crops and contribute to contamination with their excretions. Arthropods cause damage and invasion sites for the entry of MO or small pores creating an ideal microclimate for fungi. → Post harvest protection. The precautionary steps against microbial spoilage and the production of myco-toxins are: continuous pest control espe-cially at main collection and storage locations. Of special importance is the control of the a_w value in the stored products (< 0.72) and the use of bags and containers resistant to rodents and insects (Heinz 1983).

stomacher

Device for preparing the first step of a → serial dilution. Well suited for cooked and prepared foods of which a measured amount is mechanically agitated together with a specific volume of diluent at 400 – 500 cycles per min for 2 min. The stomacher is reliable and applicable to automised systems of plate counting (Baumgart 1990).

stonefruit

Apples, pears and quinces must be har-vested in the preclimacteric stage when stored for long periods of time. The spe-cific type determines the temperature at which it should be stored and the compo-sition of the air in the CA storage (Nicolaisen-Scupin 1985). Cold storage and CA storage cannot fully inhibit the growth of *P. expansum* (→patulin) and *Botrytis cinerea*. Precautionary tree care is the most important step in the prevention of not only *Pezicula multicorticis* and *Pez. alba*, but also *Colletotrichum gloeosporioides*. *Alternaria alternata* develops on cold sensitive types when the storage temperature is too low. *Stemphylium botryosum* and *Cladosporium herbarium* preferentially attack the fruit growing on the southern side of the trees (Sommer, N. F.: in Kader 1985).

storage conditions
The general adjustment of temperature, humidity, light and the composition of the atmosphere in such a way that the chemical, physical and sensory characteristics of the stored goods be retained as well as possible, and the growth of the MO be inhibited as much as possible or even prevented.

storage moulds
Refers to part of the fungi present on the surface of grains→in addition to→field fungi, which develop during storage. Opt. 25 – 30 °C, min. 0 – 5 °C, max. 40 – 45 °C, min. a_w value 0.8 – 0.90. They are partially responsible for the self-heating (self combustion) effect of grain, e.g., *P. cyclopium* or *P. funiculosum*. Typical representatives of this group are: *A. amstelodami, A. candidus, A. chevalieri, A. ochraceus, A. repens, A. restrictus, A. ruber, P. chrysogenum, P. verrucosum* var. *verrucosum*. *A. flavus* is of importance in the southern states of the USA and tropical countries (Reiss 1986).

storage rots
Diseases of stored fruit, vegetables and potatoes caused by different MO. Appear more frequently amongst riper (maturer) or older products.→Opportunistic parasites.→Vegetables.

strain
No systematic unit, but represents a→pure culture or microbial isolate of which the origin and particular physiological characteristics are known; the cells of such a culture are accepted to be genetically similar. May differ in one or more minor properties from another representative of the same taxon (sp.). May be cultivated for longer or shorter time periods in the laboratory and may serve as (e.g.) reference culture in the→identification of newly isolated strains.

strain collection
MO kept, e.g., on→agar slants, in→stab cultures, or in vacuum sealed ampoules as lyophilised cultures, or in→soil cultures for fungi, etc. Method for retaining isolates and their typical characteristics, e.g., as reference cultures. Of importance in the development of new fermented products and/or for modification of processes or specific steps (see Table 31, under ecology). – Every subcultivation should be on fresh medium and the culture stored in the refrigerator only after full development, so as to minimise selective changes. Incubation following inoculation should be within the temperature range in which the culture is to be used, or in which the damage (spoilage) occurred. Fungi should not be incubated in the dark, but in a day-night-rhythm;→culture collection.

strawberries
Microbiological damage mainly by *Botrytis cinerea*; development may set in on ageing blossoms followed by growth into the maturing fruit. Cross contamination from infected to healthy berries. Precaution by hand-sorting of infected berries prior to packaging. Mechanical damage should be avoided; rapid refrigeration to + 1 °C will increase shelf life. Keeping quality during transport will also benefit from dry ice by which the CO_2 conc. is adjusted to 10 – 15% (Sommer,

N. F.: in Kader 1985). Radurisation of up to 1 – 2 kGy has proved effective for shelf life increase of up to 7 days.

streak culture

Culture of bacteria, yeasts and sometimes also moulds, inoculated on the surface of a solid (agar) growth medium, e.g., to obtain→pure cultures, or on agar slants for the keeping of strains (Baumgart 1986).

streaking

A sample, e.g., of a starter culture, is inoculated in such a way onto the surface of a solid medium that individual colonies will appear during incubation. For this purpose either a Drigalski spatula (→surface plating) or loop may be used. By drawing the loop, carrying an inoculum, on the surface of an agar medium, and flaming in between, a dilution effect of the cells is achieved; a single viable cell will give rise to a single colony. Streaking may be applied as simple method for testing disinfectants, e.g., of rinsing water in combination with the→disc diffusion test.

Streptobacterium

"Subgenus" of Lactobacillus, originally suggested by Orla-Jensen; represented by facultatively heterofermentative lactobacilli, including spp. which produce L(+) lactic acid (*L. casei*), DL lactic acid (*L. curvatus, L. plantarum, L. sake*) and D(–) lactic acid (*L. coryniformis, L. pentosus*) (Bergey 1986).

Streptococcus

Genus of Section 12, Gram+ cocci (Bergey 1986) which has previously been subdivided in→"Lancefield groups" A to O, based on their cell wall C substances and other cell wall (group specific polysaccharide) components. Presently a number of new genera have been recognised as separate taxa, in addition to the genus *Streptococcus*, represented by (a.o.) *S. pneumoniae, S. pyogenes* and *S. thermo-*

philus. Other genera are→*Enterococcus, Vagococcus* and→*Lactococcus*. The streptococci occur quite typically in pairs (diplococci) and chains; they are nonmotile, anaerobic to micro-aerophilic, catalase – , nitrite – , and produce L(+) lactic acid by homofermentation (glycolysis). They are grouped into pyogenic (causing infection and septicaemia) hemolytic streptococci (*S. pyogenes*), anaerobic streptococci and others (e.g., *S. thermophilus*). Occurrence: milk and milk products; some are components of starter cultures; on plants. Some spp. are hemolytic, some pathogenic for domestic animals and/or man. Some colonise the human mouth (e.g., *S. mutans, S. salivarius, S. sanguis*) and may cause→caries.

Streptococcus agalactiae

Causal agent of bovine→mastitis. Main representative of→Lancefield group B. Sensitive to penicillin. Survival: 10 – 20 weeks in straw, on clothes or in spoiled milk.

Streptococcus cremoris

Correct name: *Lactococcus lactis* ssp. *cremoris* (→synonyms). Strains used as starter often produce→diacetyl. No growth at 4% NaCl, 40 °C or pH 9.2.

Streptococcus lactis

Correct name: *Lactococcus lactis* ssp. *lactis*. Typical component of cheese and cultured butter starter cultures. Produces 0.7 – 1% of lactic acid, depending on strain. Growth at 4% NaCl. Some strains, previously grouped as *Str. diacetilactis*, produce→diacetyl from citric acid in presence of lactose in the milk. Some produce the antibiotic→nisin (→bacteriocins). The variety *taette* produces slime and is used in Scandinavia for the manufacture of "Longmilk", Longfil, Taette or Viili. In other countries occasionally add mixed to yoghurt for increasing the→consistency (Teuber 1987).

Streptococcus salivarius ssp. thermophilus

Streptococcus thermophilus. Extremely sensitive to antibiotics, especially penicillin in milk (→ mastitis; → waiting times). Opt. 45 °C, max. 52 °C. Thermoduric. Sensitive to NaCl. Develops small, pin-point colonies only in pour plate methods used for plate counting. – Symbiosis with *Lactobacillus delbrueckii* ssp. *bulgaricus* (*Lb. bulgaricus*) in starter culture for yoghurt and also present in starter cultures for Emmental (Swiss) cheese.

Streptomyces

Large genus of the → Actinomycetales (Bergey 1986). Aerobic, forming an aerial mycelium (with exospores) and a non-fragmenting substrate mycelium. Gram +, nitrite +, liquifies gelatine. Colonies firm, dry, with rough surface, coloured, and a diffusible pigment formed by several spp. *S. scabies* causes potato scurf; *S. griseus* produces streptomycin and *S. rimosus* → tetracycline. *S. griseus* is suggested to contribute to aroma production in → fermented sausages.

sublethal damage

Damage to MO by heat, cold, dehydration, irradiation, chemicals, preservation, etc., may result in an inhibition of replication or growth, without killing the cell. Such organisms may regenerate under favourable environmental conditions after some time, e.g., by repairing damage to the cell. → Resuscitation. One consequence may be the underestimation of a hygienic risk due to false negative plate counts.

submerged culture

Submerged fermentations. Culture of MO surrounded by a free aqueous (often agitated) nutrient substrate. Typical → fermenter process; also in shake culture, wastewater treatment; in static fluids for the manufacture of fermented foods (milk products, beer, wine, etc.). Applicable to aerobic and anaerobic processes. Opposite: → surface culture (Kunz 1988).

subspecies

Abbreviation: ssp. By moulds also termed variety (var.). Systematic unit below a sp., e.g., *Lactococcus lactis* ssp. *cremoris*.

substitutes

Alternatives to conventional substances/additives with a similar function as the original in a technical process. Example: → rennet substitute, → chymosin.

substrate

Nutrient substrate. Nutrient medium. General designation for solid or liquid substrates, in or on which MO grow and from which they obtain nutrients. It refers to synthetic, semi-synthetic and natural media; the latter may also refer to foods and food products which serve as substrates for the growth of MO.

substrate mycelium

Mycelium of fungi or → Actinomycetes growing on or in a substrate into which it excretes exoenzymes, and from which small molecules are assimilated. Opposite: aerial mycelium above the substrate giving the optical impression of cotton wool, etc.

succession

Change in microbial community with time. Caused either by biotic factors, related to activities of the present populations (= autogenic succession), or by abiotic factors, e.g., by external factors of cyclic nature (= allogenic succession). In food processing: e.g., by subsequent brining (→ sauerkraut) or temperature changes during cheese ripening, the desired population shifts or successions are achieved.

succinates

Salts of → succinic acid.

succinic acid

$HOOC-CH_2-CH_2-COOH$; colourless crystals. Salts: succinates. Intermediary product in → tri-carboxylic acid (Krebs) cycle in the cell. By-product of the alcoholic fermentation; reacting partly with sulfurous acid during → sulfuring of wine. Succinic acid produced by several moulds. In the intestines and rumen *Bacteroides* ferments CH to succinic acid and → acetic acid. In nature found in amber, ureum and other body fluids, in unripe berry fruit, beet juice and rhubarb. – Preliminary EC no. 363.

sucrose

Saccharose; cane sugar; beet sugar; crystal sugar. A disaccharide of glucose and fructose; utilised as C source by a large number of MO. → Sugar.

Sufu

Chinese cheese type manufactured from soybean milk. Soybean curd (Tofu) is incubated for one week with *Actinomucor elegans* and other Mucorales, and then stored in 12% brine for several weeks (Reiss 1986; Steinkraus 1983).

sugar

Common name for → sucrose (saccharose) (candy sugar, sugar beet sugar). Chemically heterogeneous group of CH ($C_nH_{2n}O_n$) acting as sources of energy in all living organisms and with a fundamental function as reserve material. The following monosaccharides can be distinguished: pentoses (5-carbon atoms), e.g., arabinose, xylose, rhamnose, etc.; hexoses (6-carbon atoms), e.g., glucose, galactose, fructose, etc.; heptoses (7-carbon atoms), e.g., mannoheptulose, sedoheptulose and other disaccharides including sucrose, lactose, trehalose, etc. Trisaccharides include gentianose, raffinose, etc., are widely distributed in nature. Polysaccharides (cellulose, starch, glycogen, etc.) are found in plant cell walls or serve as reserve materials. Sugar is the most important source of carbon and energy for MO. – Since single species cannot metabolise all kinds of sugars or only use them in a specific order when available in a mixture, methods for → differentiation have been developed on this basis, e.g., "sugar spectrum" of fermentation or utilisation. Also for the degradation of polysaccharides particular species can only excrete a few → exoenzymes. – Isoglucose (a mixture of 53% glucose, 42% fructose and the rest oligosaccharides) is obtained from glucose produced from starch with the aid of glucoamylase which is then further hydrolysed with → invertase. Such syrups are sweeter than sucrose and are used to a large extent in the industry. – In high concentrations sugars reduce the → water activity and inhibit most MO. It is not possible to reduce the → a_w level with sucrose to the same extent as with a similar concentration of → invert sugars, isoglucose or glucose syrup. See Table 5 (a_w value).

sugar substitutes

Sweet-tasting saccharides independent of insulin. They can be used in a diabetic diet. Sorbitol, xylitol, isomalt, etc., present in foodstuffs, are, contrary to glucose and sucrose, only metabolised after an adaptation phase by a mixed bacterial population typical of the oral cavity and GI tract. This delay may be related to the characteristic enzymes and/or selective effects. Sorbitol occurs in nature as sugar alcohol. Isomalt (Palatinit®) is produced from sucrose by enzymatic modification into isomaltulose in the first step with → immobilised cellular systems (*Protaminobacter rubrum, Serratia plymuthica, Erwinia rhapontici*), and then hydrated in an aqueous solution in the final step [Schiweck, H.: *Lebensmittelchem. Gerichtl. Chemie* 41 (1987) 49 – 52].

sulphate reducing MO

Sulphate reduction is a common metabolic activity among MO and plants. Sulphide necessary for the synthesis of sulphur-containing amino acids is derived from $SO_4^=$. Sulphate respiration is limited to obligatory anaerobic bacteria (*Desulfovibrio, Desulfotomaculum*). These "desulfuricants" are indirectly responsible for corrosion of iron in the soil (water and sewage pipelines) or in oxygen-free deep-sea waters (bore towers). See Table 42, page 276.

sulphide → hydrogen sulphide.

sulphite → sulphur dioxide.

sulphite reducing sporeforming bacteria

Cl. amylolyticum, Cl. nigrificans, Cl. perfringens, Cl. putrefaciens, Cl. sporogenes, etc. Important group for assessing the microbiological quality of → drinking water.

sulphite reduction

Reduction of $SO_3^=$ to H_2S during the fermentation of sulphurised must. Formation of larger amounts may result in sensory defects in wine (Dittrich 1987).

sulphur dioxide

SO_2; pungent gas; dissolved in water forms the unstable sulphurous acid:SO_2 + H_2O H_2O_3. Salts: sulphites. Sulphurous acid can be produced by yeasts from intrinsic sulphate ($SO_4^=$) in fermenting must. – SO_2 has antimicrobial properties, and it is used for fumigation, as food → preservative (restricted), and in wine making for inhibiting the "wild" yeasts and bacteria (the desired wine yeasts are relatively resistant). In combination with heat, a stronger antimicrobial effect is exerted, e.g., the effective elimination of thermoresistant ascospores of *Byssochlamys fulva* by → pasteurisation has special advances in the fruit juice industry. – In wine yeasts may form → hydrogen sulphide from sulphate and sulphite, causing sensory defects (sulphur odour) in the product (Dittrich 1987). – "MAK" value of SO_2: 5 mg/m^3 of inhaled air. LD$_{50}$ rat p.o. ca. 2 g/kg BW, rabbit p.o. 600 – 700 mg/kg BW. ADI value 0 – 0.7 mg/kg BW. It is considered that 1 – 5% of asthmatic patients react sensitively to sulphite (Classen et al. 1987). – EC no. E 220; sodium sulphite E 221; sodium hydrogen sulphite E 222; sodium bisulphite E 223; potassium bisulphite E 224; calcium sulphite E 226; calcium hydrogen sulphite E 227. For limitations, see FDA §121.101, 121.1031, and also FDA §8.303 (for sulphurous acid), or German Regulation "ZZulV" of 22.12.1981 §4, Add. 4, list B.

sulphuring

Addition of sulphurous acid (H_2SO_3) or bisulphites (e.g., sodium or potassium metabisulphite) (→ sulphur dioxide) to wine. Acetic acid bacteria, moulds (*Botrytis*) and undesired "wild" yeasts (apiculate yeasts) are inhibited and the *Saccharomyces* strains, initially present in low numbers, selectively favoured on account of their relative resistance to sulphurous acid. Sensory defective musts can be treated intensively with sulphurous acid in order to prevent "acetic sting" (Dittrich 1987).

superoxide radicals → radicals.

surface contact methods

Microbiological examination methods for the determination of the surface contamination of foods or other commodities. It is also used to evaluate sanitation and disinfection procedures, and as a control measure for staff hygiene. For nutrient agar discs different types of "containers" are commercially available, e.g., RODAC (Replicate Organism Direct Agar operation Contact) plates. Metal containers or disposable plastic plates are either filled in the laboratory with test agar medium or can be purchased prefilled and "ready-to-use". The entire sterile agar surface of ca. 10 cm^2 is pressed against the surface to be sampled and incubated either in petri dishes or after replacing the cover of the commercial container. Another variation is represented by the agar sausage method in which a dialysis tube of 2 cm diameter is used. After surface contact a disc of ca. 1 cm thick is sliced off and incubated in a petri dish with the new surface downwards. This procedure should only be used on flat surfaces that allow easy cleaning and disinfection (Baumgart 1990; Favero et al. 1984).

surface culture
Emerged culture. Method of cultivating MO on the surface of a→culture medium for determination of the→microbial count. MO can be cultured biotechnically on liquids or solid media.→Penicillin and→citric acid were formerly produced in this manner. Camembert and Romadur contain surface cultures in contrast to→submerged cultures.

surface growth
Microbial growth on surfaces of different kinds of materials in drinking water containers and other types of surfaces in contact with drinking water. Representatives of the genera *Pseudomonas, Flavobacterium, Acinetobacter, Caulobacter*, etc., are mainly involved. Less pronounced in presence of protozoa.

surface plating
Cultural method commonly used for determination of viable bacterial and mould numbers (→cfu). An agar plate, containing a layer of ca. 15 ml of the particular agar medium, is dried to remove excess moisture, and 0.1 ml of a sample suspension from the→dilution series pipetted onto the surface. The moisture droplet is distributed evenly over the nutrient medium surface with the aid of a→Drigalski spatula; preferably with the use of a rotating disk. Parallel plates are stacked and incubated with the lid downwards (Baumgart 1990). Compare Figure 14 (pour plate culture). – Procedure a.o. suitable for studying colony morphology for preliminary differentiation, or when→differential media are used. Plates with 30 to 300 colonies are counted. For calculating the weighted arithmetic mean, plates of the next highest dilution step with 1/10 of the colony numbers are also counted. Compare: (1) German Official Examination Methods acc. to §35 LMBG; (2) American Public Health Association: *Standard Methods for the Examination of Dairy Products, 14th Ed.* Washington, DC: APHA, 1987.

surface spoilage
Meat, pickled meat and sliced cold meats tend to undergo spoilage by aerobic bacteria, most commonly of the genera *B., Ps., Enterobacter, Proteus, Alcaligenes, Vibrio*, etc. Changes in the odour, taste and colour are prominent (ICMSF 1980; Kunz 1988).

survival rate
Proportion of surviving MO after a technical processing step. It does not depend on the susceptibility of the species or its→dormant forms only, but also on the time of exposure and the medium composition (CH, fat, protein).

suspension culture → submerged culture.

swab contact method
Method for the microbiological examination of contaminated surfaces that are not easily accessible. Originating from the clinical microbiology, this method may be applied to "swab" areas on equipment, threadings and the inner surfaces in

pipelines, valves, etc. Sterile non-absorbent moistened cotton swabs are rubbed thoroughly over a designated surface area at a contact angle of ca. 30 degrees. The cotton head is separated aseptically into a vial containing sterile dilution buffer. After shaking,→ serial dilutions are prepared. Alternatively, the swab may be streaked (upon turning) onto the surface of a nutrient agar plate. Moist surfaces are sampled with a dry swab, and vice versa (Baumgart 1990; Favero et al. 1984).

sweet cream butter→ butter.

sweet precipitation
Sweet curdling. Spoilage of pasteurised milk during refrigerated storage, especially by *B. cereus*, of which the endospores have not been destroyed, or after secondary contamination with→ *Pseudomonas* spp. In H milk the curdling may result from heat resistant → proteases, formed bacteria that have been killed by the heat treatment. Bitter→ peptides may thus be formed. The defect may also occur in condensed milk. The term sweet curdling is used sometimes when referring to rennet curdling (Teuber 1987).

swell
Blown can; "blower". Swelling (blowing) of a can or flexible container (pouch) through increasing internal pressure caused by gas production by microbial or chemical activity. Often found in the tropics. Mainly caused by action of heat-resistant bacteria: outgrowth of surviving endospores of *Clostridium* spp. Results from either underprocessing in relation to high pH of product, or→ leakages of the can which allow contamination via the cooling water or during transportation [also associated with products hermetically sealed into flexible laminates (e.g., → salad mixes, sliced cheese, etc.), and caused by heavy contamination of contents and non-refrigerated storage, etc.]. – Cans/containers showing swelling should be discarded/destroyed because of potential danger of→ botulism.

symbiosis
Community of two or more organisms (symbionts) for the (temporary) benefit of both. The differentiation of other relationships (e.g., parasitism) is not clear. Examples of microbial symbioses: between algae and moulds, or cyanobacteria and moulds (lichens); bacteria in a multi-strain culture (yoghurt). Even the gut bacteria in their relationship to the macroorganism can be considered as representative of this category: they receive nutrition and contribute to the absorption of the food, supply vitamins and assimilable metabolites.

symptomless
Condition after an infection disease, that may be associated with the undetected presence of pathogens in the organism. →Salmonellosis, → dysentery, → excreters (carriers). Especially critical in domesticised animals.

Syncephalastrum
Fungal genus of the→ Zygomycetales. *S. racemosum* occurs on cereals, maize grains and nuts, especially after insect invasion. May also be found on meat products (Pitt and Hocking 1985).

synonym
Used in taxonomy. Refers to names of MO that have formerly been cited in literature under different designations which have become obsolete or invalid. – During the last 15 years a large number of scientific names of MO have been changed as a result of extensive systematic and taxonomic studies. Especially chemo-structural and metabolic characteristics have been considered in comparative patterns. These changes or modifications were

Table 39 Synonyms of bacteria important in food technology, acc. to Bergey 1984, 1986, 1989 (obsolete, invalid names are given in parentheses).

(Acetomonas)	*Gluconobacter*
(Achromobacter)	*Alcaligenes* (in part)
Acinetobacter calcoaceticus	*(Alcaligenes durans)*
Acinetobacter calcoaceticus	*(Alcaligenes tolerans)*
(Aerobacter aerogenes)	*Enterobacter aerogenes*
Alcaligenes (zum Teil)	*(Achromobacter)*
(Alcaligenes durans)	*Acinetobacter calcoaceticus*
(Alcaligenes tolerans)	*Acinetobacter calcoaceticus*
(Bacillus acidophilus)	*Lb. acidophilus*
Bacillus coagulans	*(Bacillus thermoacidurans)*
(Bacillus thermoacidurans)	*Bacillus coagulans*
(Bacterium prodigiosum)	*Serratia marcescens*
(Betabacterium breve)	*Lb. brevis*
Bifidobacterium bifidum	*(Lb. bifidus)*
(Campylobacter fetus ssp. *jejuni)*	*Campylobacter jejuni*
Campylobacter jejuni	*(Campylobacter fetus* ssp. *jejuni)*
Carnobacterium divergens	*(Lb. divergens)*
(Cl. nigricans)	*Desulfotomaculum nigricans*
Cl. perfringens	*(Cl. welchii)*
(Cl. putrificus)	*Cl. sporogenes*
Cl. sporogenes	*(Cl. putrificcus)*
(Cl. welchii)	*Cl. perfringens*
Desulfotomaculum nigricans	*(Cl. nigricans)*
Enterobacter aerogenes	*(Aerobacter aerogenes)*
Enterococcus faecalis	*(Str. faecalis)*
Enterococcus faecium	*(Str. faecium)*
Gluconobacter	*(Acetomonas)*
(Klebsiella aerogenes)	*Klebsiella pneumoniae*
Klebsiella pneumoniae	*(Klebsiella aerogenes)*
Lb. acidophilus	*(Bacillius acidophilus)*
(Lb. bifidus)	*Bifidobacterium bifidum*
Lb. brevis	*(Betabacterium breve)*
(Lb. bulgaricus)	*Lb. delbrueckii* ssp. *bulgaricus*
(Lb. casei)	*Lb. casei* ssp. *bulgaricus*
(Lb. casei)	*Lb. casei* ssp. *casei*
Lb. casei ssp. *casei*	*(Lb. casei)*
Lb. casei ssp. *rhamnosus*	*(Lb. rhamnosus)*
Lb. casei ssp. *tolerans*	*(Lb. tolerans)*
(Lb. delbrueckii)	*Lb. delbrueckii* ssp. *delbrueckii*
Lb. delbrueckii ssp. *bulgaricus*	*(Lb. bulgaricus)*
Lb. delbrueckii ssp. *delbrueckii*	*(Lb. delbrueckii)*
Lb. delbrueckii ssp. *lactis*	*(Lb. lactis)*
Lb. delbrueckii ssp. *leichmannii*	*(Lb. leichmannii)*
(Lb. divergens)	*Carnobacterium divergens*
Lb. helveticus	*(Lb. joghurti)*

(continued)

Table 39 (continued).

Lb. helveticus	(Thermobacterium helveticum)
(Lb. joghurti)	Lb. helveticus
(Lb. lactis)	Lb. delbrueckii ssp. lactis
(Lb. leichmannii)	Lb. delbrueckii ssp. leichmannii
Lb. plantarum	(Streptobacterium plantarum)
(Lb. rhamnosus)	Lb. casei ssp. rhamnosus
(Lb. tolerans)	Lb. casei ssp. tolerans
Lactococcus lactis ssp. cremoris	(Str. cremoris)
Lactococcus lactis ssp. diacetilatis	(Str. lactis ssp. diacetilactis)
Lactococcus lactis ssp. lactis	(Str. lactis)
(Leuconostoc citrovorum)	Leuconostoc lactis
(Leuconostoc cremoris)	Leuconostoc mesenteroides ssp. cremoris
(Leuconostoc dextranicum)	Leuconostoc mesenteroides ssp. dextranicum
Leuconostoc lactis	(Leuconostoc citrovorum)
(Leuconostoc mesenteroides)	Leuconostoc mesenteroides ssp. mesenteroides
Leuconostoc mesenteroides ssp. cremoris	(Leuconostoc cremoris)
Leuconostoc mesenteroides ssp. dextranicum	(Leuconostoc dextranicum)
Leuconostoc mesenteroides ssp. mesenteroides	(Leuconostoc mesenteroides)
(Pasteurella X)	Yersinia enterocolitica
(Pediococcus cerevisiae)	Pediococcus damnosus
Pediococcus damnosus	(Pediococcus cerevisiae)
(Peptococcus)	Peptostreptococcus
Peptostreptococcus	(Peptococcus)
Serratia marcescens	(Bacterium prodigiosum)
(Streptobacterium plantarum)	Lb. plantarum
(Str. cremoris)	Lactococcus lactis ssp. cremoris
(Str. faecalis)	Enterococcus faecalis
(Str. faecium)	Enterococcus faecium
(Str. lactis)	Lactococcus lactis ssp. lactis
(Str. lactis ssp. diacetilactis)	Lactococcus lactis ssp. diacetilactis
Str. salivarius ssp. thermophilus	(Str. thermophilus)
(Str. thermophilus)	Str. salivarius ssp. thermophilus
(Thermobacterium helveticum)	Lb. helveticus
Yersinia enterocolitica	(Pasteurella X)

considered necessary from the viewpoint of the biological→ systematics. However, this has caused considerable confusion in the applied disciplines (here: food microbiology) and resulted in a "co-existence" of "new" and "old" names. This also has complicated and retarded legislatory procedures concerning the approval of novel beneficial organisms (→starter cultures; → enzymes) considerably. In the applied literature – including this dictionary – different names are therefore found for the same beneficial or deleterious organism. In Tables 39 and 40, the most important synonyms of MO found in starter cultures or foods are compiled; the

Table 40 Synonyms of moulds important in food technology (obsolete, invalid names are given in parentheses).

Acremonium	(Cephalosporium)
Alternaria	(Macrosporium)
Alternaria alternata	(Alternaria tenuis)
(Alternaria tenuis)	Alternaria alternata
Alternaria tenuissima	(Helminthosporium tenuissimum)
Aureobasidium	(Pullularia)
Bipolaris zeicola	(Helminthosporium carbonum)
Candida utilis	(Torulopsis utilis)
(Cephalosporium)	Acremonium
Cladosporium	(Hormodendrum)
Claviceps purpurea	(Secale cornutum)
Colletotrichum musae	(Gloeosporium musarum)
Deuteromycetes	(Fungi imperfecti)
(Endomycopsis)	Saccharomycopsis
Fungi	(Mycota)
(Fungi imperfecti)	Deuteromycetes
Geotrichum candidum	(Oidium lactis)
(Gloeosporium musarum)	Colletotrichum musae
(Helminthosporium carbonum)	Bipolaris zeicola
(Helminthosporium tenuissimum)	Alternaria tenuissima
(Hormodendrum)	Cladosporium
(Macrosporium)	Alternaria
Monascus	(Xeromyces)
(Mycota)	Fungi
(Oidium lactis)	Geotrichum candidum
Penicillium chrysogenum	(Penicillium notatum)
(Penicillium notatum)	Penicillium chrysogenum
Penicillium patulum	(Penicillium urticae)
(Penicillium urticae)	Penicillium patulum
(Pullularia)	Aureobasidium
Rhizomucor miehei	(Rhizopus miehei)
Rhizomucor pusillus	(Rhizopus pusillus)
(Rhizopus arrhizus)	Rhizopus oryzae
Rhizopus microsporus var. oligosporus	(Rhizopus oligosporus)
(Rhizopus miehei)	Rhizomucor miehei
(Rhizopus nigricans)	Rhizopus stolonifer
(Rhizopus oligosporus)	Rhizopus microsporus var. oligosporus
Rhizopus oryzae	(Rhizopus arrhizus)
(Rhizopus pusillus)	Rhizomucor pusillus
Rhizopus stolonifer	(Rhizopus nigricans)
(Saccharomyces carlsbergensis)	Saccharomyces uvarum
(Saccharomyces rouxii)	Zygosaccharomyces rouxii
Saccharomyces uvarum	(Saccharomyces carlsbergensis)
Saccharomycopsis	(Endomycopsis)
(Secale cornutum)	Claviceps purpurea

(continued)

Table 40 (continued).

Stachybotrys alternans	(Stachybotrys atra)
Stachybotrys alternans	(Stachybotrys chartarum)
(Stachybotrys atra)	Stachybotrys alternans
(Stachybotrys chartarum)	Stachybotrys alternans
(Xeromyces)	Monascus
Zygosaccharomyces rouxii	(Saccharomyces rouxii)

Table 41 Systematic groups of bacteria acc. to Bergey 1984 and 1986; of direct or indirect relevance to food microbiology acc. to Kunz 1988, modified.

Section	Genera
1. Spirochaetales	Spirochaeta, Treponema, Borellia, Leptospira
2. Aerobic/microaerophilic mobile, helical/vibroid Gram –	Spirillum, Campylobacter, Bdellovibrio
3. Nonmobile, Gram –, curved bacteria	
4. Gram –, aerobic rods and cocci	Pseudomonas, Xanthomonas, Zoogloea, Azotobacter, Azotomonas, Rhizobium, Agrobacterium, Methylococcus, Methylomonas, Halobacterium, Halococcus, Acetobacter, Gluconobacter, Legionella, Neisseria, Moraxella, Acinetobacter, Flavobacterium, Alcaligenes, Brucella, Bordetella
5. Facultatively anaerobic, Gram – rods	Escherichia, Shigella, Salmonella, Citrobacter, Klebsiella, Enterobacter, Erwinia, Serratia, Hafnia, Edwardsiella, Proteus, Providencia, Morganella, Yersinia, Vibrio, Aeromonas, Plesiomonas, Zymomonas, Chromobacterium
6. Anaerobic, Gram –, straight, curved and helical rods	Bacteroides
7. Sulfate- or Sulfur-reducing bacteria	
8. Anaerobe Gram – cocci	Veillonella
9. Rickettsias and Chlamydias	Chlamydia
10. Mycoplasmas	
11. Endosymbionts	
12. Gram + cocci	Micrococcus, Staphylococcus, Streptococcus, Leuconostoc, Pediococcus, Aerococcus, Peptococcus, Peptostreptococcus, Ruminococcus, Sarcina
13. Endospore-forming Gram + rods and cocci	Bacillus, Sporolactobacillus, Clostridium, Desulfotomaculum, Sporosarcina
14. Regular, nonsporing Gram + rods	Lactobacillus, Listeria, Erysipeltrix, Brochothrix

Table 41 (continued).

Section	Genera
15. Irregular nonsporing Gram + rods	*Corynebacterium, Arthrobacter, Brevibacterium, Caseobacter, Microbacterium, Aureobacterium, Cellulomonas, Propionibacterium, Acetobacterium, Actinomyces, Bifidobacterium*
16. Mycobacteria	*Mycobacterium*
17. Nocardioforms	*Nocardia*

old, invalid or obsolete names are given in parentheses. Since the taxonomy continuously produces new names, completeness cannot be claimed.

synthetics
Organic polymers not found in nature. MO have the ability to adapt and colonise these surfaces. The MO excrete enzymes and aggressive metabolites, e.g., organic acids, NH_4 and pigments in the presence of sufficient water. The surface can thus be damaged and the physical and chemical characteristics of the materials may change. Application of microbiological test specifications should serve as support for the choice of materials for packaging, or isolation of refrigerators or other utensils [Frank: *Forum Mikrobiologie* 8 (1985) 339 – 345].

systematics
Syn.: taxonomy. The principle of classification or grouping of organisms, on the basis of their natural relationship, as proposed by Carl von Linné in 1738, who also introduced the binary→nomenclature. The most important units are the species that are grouped together in a genus. "Similar" genera are grouped into families (familia), these into orders (ordo), and the orders into classes (classis). The basis of the biological systematics is the genetic relationship. Since the majority of bacteria, moulds and yeasts reproduce only asexually, such a principle is unpracticable. It has therefore been substituted by a system that takes into account specific similarities in composition, metabolism, in cell structure, in morphology, etc. [Kandler, O.: *Zbl. Bakt. Hyg. I. Orig.* C 3 (1982) 149 – 160]. From the food microbiological viewpoint the biological system (Table 41) only has limited application. Reference is therefore made to the→division of MO, in which the approach has been limited consciously to the physiological groups of significance in practice (Table 20).

T

table salt

NaCl (sodium chloride). Salt used in cooking, ordinary salt. LD_{50} rats p.o. 3.75 g/kg BW. The lethal dosage for human beings is 200 g. The recommended daily allowance (RDA) is 6 g. The preservative effect depends on the lowering effect of the a_w value in principle. 1.7 wt.% NaCl in water at 25°C corresponds to an a_w value of 0.99; 14.1% to a_w 0.90 and 26.5% (saturated) to a_w 0.75. Cheese (according to type) 1 – 5% in the water phase; mildly salted fish 8 – 10%; heavily salted fish ("salzgar") 20 – 24% in tissue water. → Brine is constituted of 99.6% table salt and 0.4% sodium nitrate. Physiological salt solution (Ringer's) used as a → diluent contains 0.85 – 0.9% NaCl; brine used in the preservation of vegetables 15 – 25%; *Halococcus* spp. and *Halobacterium* spp. are able to reproduce under these conditions. MO able to grow in a medium with >7% NaCl are marked as salt tolerant, e.g., *Brevibacterium linens, Serratia* spp., *Micrococcus* spp., *Staph. aureus* (17.5% inhibitory, 20% lethal). Marine bacteria, e.g., *Staph. carnosus* and other halophiles cannot reproduce when the NaCl <1%. → Cell division.

table water

Bacteria always present in water are *Pseudomonas, Flavobacterium, Cytophaga,* etc. The following organisms should be controlled: *E. coli,* coliforms, enterococci, endospore-forming aerobic bacteria, cfu's (total population), as well as *Pseudomonas aeruginosa*. Methods by Baumgart (1990), and the → official methods.

Talaromyces

Teleomorph to several *P.* ssp. *T. flavus* is found in fruit juices from warmer regions. Ascospores are relatively heat resistant and may grow out to cause spoilage of fruit juices.

tallowiness

Changes in the consistency and melting point of fats. Caused by microbial hydrolytic decomposition and subsequent oxidative → abiotic spoilage.

tank fermentation → sparkling wine.

tannins

→ Gallates. Easily soluble in water, as are similar compounds such as catechins and ellangens. Astringent taste. Mild antibiotic activity. Components of many plants, especially in tea, red wine and bark.

tartaric acid

HOOC-CHOH-CHOH-COOH. The salts are tartrates. Natural substances present in numerous fruits. Moderate antibiotic activities (→ raisins). The potassium salt crystallises in wine during prolonged storage. – Antioxidant. L(+) tartaric acid E 334, is not approved as additive in foodstuffs in Germany. Sodium tartrate E 335; potassium tartrate E 336; sodium potassium tartrate E 337.

tartrates

Salts of → tartaric acid.

taxonomy

Science of the nomenclature of organisms and their systematic arrangement into biological systems. → Systematics; → synonyms.

TDT curve

Thermal death time curve. Determined by the MO species and even strain. The time required to reduce the total count with one decimal factor (decimal reduction time → D value) for a specific species at a con-

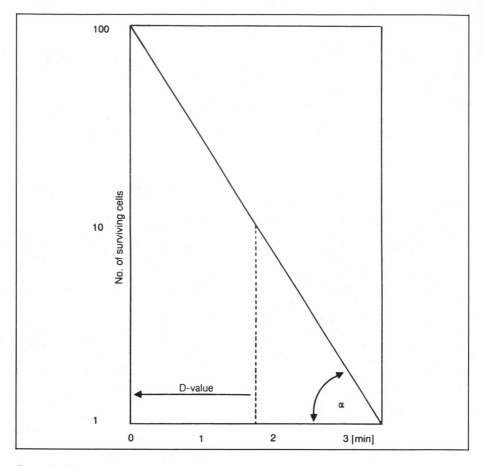

Figure 22   Thermal death time curve of a bacterial strain.

stant temperature can be read from the curve. The slope of the graph (see Figure 22) is $1/D$ = tan α. The data are presented semi-logarithmically, and the curve is steeper for more heat sensitive MO, i.e., with a smaller D value. The pH value and the composition of the medium are additional factors with an effect on the slope of the curve (Heiss and Eichner 1984).

tea

Withered leaves of the teabush *Camellia* *sinensis* are pressed and allowed to ferment in thin layers at 35 – 40 °C for several hours. Microbial enzymes play a minor role, but the intrinsic plant enzymes are involved to a great extent (Krämer 1987). All sorts of tea are highly contaminated with MO from the field and when dried possess up to $10^6 – 10^7$ cfu/g as in the case of spices. Only endospores and some of the thermoduric MO may survive when the beverage is prepared with hot to boiling water.

tea mould culture

A symbiotic culture between *Acetobacter xylinum* and specific yeasts, e.g., *Saccharomyces ludwigii* and *Schizosaccharomyces pombe*. The culture grows in sweetened black tea or other beverages. The yeasts ferment the sugar to ethanol and some → lactic acid and → gluconic acid are produced. The acetic acid bacteria oxidise a proportion of the ethanol to acetic acid. The organisms grow in a thin surface skin which becomes jelly-like and finally leathery. The excreted slime keeps the mass together and contains cellulose. – The refreshing beverage is claimed to stimulate digestion. The culture can be obtained from "Interpilz", Dr. Meixner GmbH, Stuttgart. A similar culture, named "Kombucha" acc. to Dr. med. Sklenar is also commercially available [Reiss, J.: *Deutsche Lebensm.-Rund-schau* 83 (1987) 286 – 290].

technical enzymes → enzyme preparations.

technological necessity

The technological necessity as well as the → "unobjectionability" from a health point of view (absence of any health risk) must be proven before official permission (approval) is given for the use of → food additives and → genetically modified MO or products manufactured by such MO, e.g., enzyme preparations. It is only legally accepted if the substance considerably contributes to the quality or safety of the foodstuffs and if the technological, sensory, dietetic or nutritious effects cannot be obtained by existing measures or procedures. The highest amount allowed is also directed by these criteria.

technology

General term referring to the total of all technological procedures and processing and handling measures of foods. Involved in particular physical processes such as the treatment with high and low temperatures, desiccation or dehydration and radiation treatment aimed at increasing the shelf life by destruction or temporary inactivation of MO.

teleomorph

The perfect form of a given fungus with nucleus phase exchange and sexual reproduction, e.g., → Ascomycetes, → Basidiomycetes. Opposite: → anamorph; → Deuteromycetes.

tellurite reaction

The addition of potassium tellurite to *Staphylococcus* selective agar acc. to Baird-Parker or Vogel-Johnson. Tellurite is reduced to tellurium by *Staph. aureus* causing a black colouration of the colony. The media are commercially available.

temperature

An extremely important environmental factor (→ extrinsic factor) influencing reproduction, sporulation, growth, metabolic activity (→ Q_{10} value) and toxin production of MO. Damage by undesired MO groups can be minimised or the → specific growth rate optimised by selecting and maintaining a specific temperature, e.g., in biotechnology or cheese ripening, etc. – Commonly used terms in connection with temperature (in °C) are:

deep freezer rooms	-30 to -18
deep freezer	-12 to -5
refrigerator	$+3$ to $+8$
	(optimum $+5$)
room temperature	$+17$ to $+30$
	(average $+21$)
body temperature	$+37$
serving temperature	$+40$ to $+60$

temperature optimum

The most favourable temperature for reproduction, germination of spores, growth, metabolism and production of toxins. It is species specific. MO can be

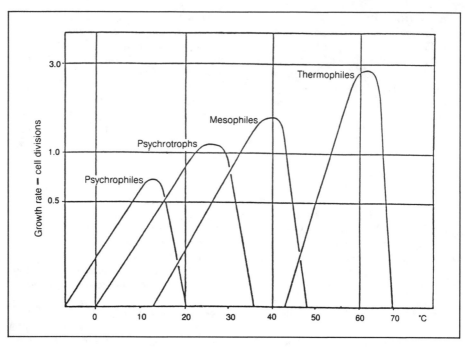

Figure 23 Number of cell divisions per h, typical of the particular temperature dependent groups (source: Zschaler, R.: "Verderb und seine Ursachen". Behr's Seminare, March 1988, 8 – 9, Hamburg).

arranged into 4 different groups (see Figure 23). Three of these groups are of practical importance. Strictly psychrophilic organisms are of little importance in the food industry; they live in deep water with a constant temperature of 4 °C. Psychrotrophic (psychrotolerant) and mesophilic organisms are the most important groups causing spoilage, and generally occur in the environment and are responsible for recycling organic residues on the mainland. Thermophilic organisms may cause severe damage or losses in several food processing operations, e.g., in the sugar industry.

temperature ranges
Given for relevant spp. or groups under the particular "key words" as minimum (min.), maximum (max.) or optimum (opt.) temperatures for reproduction or growth, with respect to its bearing on the shelf life or differentiation between organisms. The following generalised temperatures can be given for: yeasts and moulds – 2 to 50 °C; bacteria causing spoilage + 2 to 46 °C; food poisoning bacteria + 5 to 48 °C; heat resistant bacteria + 12 to 65 °C. Smaller temperature ranges are usually linked to the different species of the above mentioned groups.

tenacity
Resistance; "toughness". Property referring to the capacity of MO to survive in different adverse conditions, e.g., high temperatures, osmotic pressure, disinfectants.

tenderiser
→Proteases mixed with table salt and used to sprinkle on meat before heat processing; a prickle roller, etc., may also be applied to enable penetration of the tenderiser enzymes into the meat tissues. MO are also transported into the meat and in part not killed with insufficient heat treatment. For the same purpose meat is treated in the tropics with papaya juice, or it is laid between 2 pieces of pineapple. This preparation procedure is not allowed in restaurants or canteens.

thallus
A vegetative form of an organism in the lower "plant" kingdom which is not differentiated into roots, stems and leaves, e.g., algae, fungi and lichens or thallophytes.

Thamnidium elegans
A monotypical genus of Mucorales with characteristic dichotomically branched sporangiophores. Psychrophilic fungus occurring in soil, on grains, bread, chill stored meat, etc. Growth conditions opt. 18°C; min. −7°C: max. 27°C; min. a_w 0.92−0.94; opt. 0.98. Considerable amounts of→ethylene are produced during growth which accelerates the ripening of stored fruit and vegetables (Domsch et al. 1980; Pitt and Hocking 1985).

thawing rate→freezing rate.

thermal death point
The temperature (in °C or K) at which all MO are killed within 10 min.

thermal death time
Time (in min) required for destruction of all MO at a given temperature (°C or K).→F value.

thermisation
Heat processing using the lowest possible temperature with the aim at destroying the largest possible number of MO within a technologically justifiable time period. May also be directed at desired selection.

thermodures
MO which can survive relatively high temperatures for a specific period of time. The term in general refers to those organisms which survive→pasteurisation, e.g., *Microbacterium lacticum, Micrococcus varians, Lb. delbrueckii* ssp. *thermophilus* and endospores of bacteria, e.g., ascospores of *Byssochlamys* spp. See Table 29, under pasteurisation of milk.

thermophiles
MO preferring higher temperatures with opt. 60°C; min. rarely below 40°C; max. 70°C. Mostly bacteria, rarely fungi. Causes spoilage in the sugar industry. See Table 20 under classification of MO; Figure 23.

thermophores
Portable, solid or flexible containers, well insulated against heat or cold loss and used in the transport of ready-to-eat foods, or refrigerated or frozen commodities. The temperature range of 15−65°C is critical as the growth or reproduction of MO causing spoilage or food poisoning can either be accelerated or inhibited.

thermostat
Instrument for maintaining the selected temperature over a long period of time. In the laboratory the term refers to an incubator. The temperature should not fluctuate more than 0.5°C.

thermotolerants
MO still able to grow/reproduce at higher temperatures, although slowly. Their optimal temperature is lower than that of →thermotolerant organisms.

thiabendazole
Approved→preservative for the surface

(peels) treatment of→ bananas. The minimal inhibitory conc. (→ MIC) for moulds is 1 – 100 ppm. LD_{50} for mice, rats, hamsters is 3 – 4 g/kg BW p.o. There is no evidence of genetic toxicity. ADI value: 0 – 0.3 mg/kg BW (Classen et al. 1987).

thickener→ stabilizer.

threshold dose
Threshold value. Amount of a foreign substance (toxin) of number of MO (pathogens), exceeding of which may have deleterious effects on the organism. This value is intermediate between ineffectivity (→ NOEL) and effectivity, and is to be calculated. Similar relationships are found for damage to MO by inhibitory substances (→ MIC).

Tilsiter→ semi-hard cheese.

tobacco
The leaves of *Nicotiana tabacum* are dried in air and moistened with CH containing water, after harvesting. It is then bundled and fermented by "spontaneous" MO. The product specific enzymes are activated, followed by bacteria and fungi. Temperatures up to 70°C may be reached in the core of the bale and contribute to improvement of the structure and aroma production. The following organisms are to be found in shredded tobacco (in order of decreasing frequency): *P.* spp., *A.* spp., *Alternaria* spp., *Cladosporium* spp., *Rhizopus* spp. and a number of other fungi. Experimental work on the use of starter cultures to accelerate the process and decrease the nicotine content, has not reached a level where it could be applied in practice (DFG 1987; Krämer 1987).

tolerable values
"Tolerance values". Official regulatory hygienic and microbiological standards for foodstuffs, consumables and com-

modities in CH. Apart from→ limits ("Grenzwerte", Germ.), the exceeding of which renders a product unsuitable for consumption or distribution, the tolerable values refer to the numbers of MO that are by experience generally not exceeded under→ GMP conditions of production, storage and distribution: e.g., pasteurised milk, soured milk 10/ml enterobacteria; soft cheese 10^6/g enterobacteria; hard cheese, ready-to-eat soups, eggs or egg conserves 10/g of *E. coli*; drinking water no *E. coli* detectable in 100 ml. – Compare to guide lines in Germany.

top fermenting yeasts
Strains of→ *Saccharomyces cerevisiae* remaining in suspension instead of forming a sediment during fermentation.→ Brewer's yeast. They are used in the production of "Weissbier" (ale beer),→ "Kölsch",→ baker's yeast and fodder yeast.

"Topfen"
Southern German name for quarg;→ fresh cheese, cottage cheese.

Torula
An obsolete name for genera today named *Candida, Torulopsis* and *Rhodotorula* (Lodder: *The Yeasts.* Amsterdam, 1970).

Torulopsis
An imperfect yeast genus related to *Candida. T. stellata* is osmotolerant and occurs in grape must. Fermentation potential is limited and restricted to glucose, fructose and saccharose. Some of these species form part of the→ kefir MO and involved in the production of→ East Asian specialities or in the utilisation of whey.

total microbial population
Total microbial (bacterial) numbers. The total number of (aerobic) colony-forming units (cfu) of bacteria and/or moulds grow-

ing on a nutrient rich laboratory medium; aerobic→plate count. Important quality criterion for drinking water and raw milk; controversies for foods since no direct relation has been established between a high microbial population and potential health risk. On the other hand, low microbial numbers are no guarantee for hygienic safe foods. Sensory changes may be observed generally when microbial populations exceed $10^6 - 10^7$/g or ml. – Decisive factors for judging the safety of a food are the type of product and its previous history (heating, preservatives, etc.), and the composition of the microbial population (→indicator organisms; → pathogens, etc.), as well as the possible presence of toxins (Classen et al. 1987; Krämer 1987; Sinell 1985; Speck 1984).

toxic infection
Clinical manifestation after the ingestion of a large number of salmonellae.→Salmonellosis. The bacteria multiply in the intestine (→infection) and the lipopolysaccharides in the cell walls are toxic to the intestinal epithelium.

toxicology
Study of poisoning and its treatment. It gives information on the possible outcome in the human body, whether it is excreted slowly, metabolised, detoxified or concentrated in organs, the symptoms caused in the organism (fever, diarrhoea) and whether the behaviour of the living entity changes, e.g., photophobia, aggression. Food toxicology is mainly concerned with the effect of oral ingestion, but also resulting from dermal contact, inhalation of dusts or aerosols that may be of importance in the production and processing procedures in, e.g., mills and bakeries. – Toxicological investigations are performed with the use of animals, cell cultures, MO and/or enzymes. The results form the base for the → safety evaluation of foodstuffs in view of their approval for con-

sumption by man or animals (Classen et al. 1987; Fullgraff 1989).

toxin
Poison. → Bacterial toxins, → mycotoxins and → phycotoxins are produced by MO in foodstuffs and may cause food poisoning. These are natural metabolites with different kinds of chemical structures. A given toxin can be characterised according to the target of its effect, or according to the general type of damage caused after its oral ingestion. Examples are: hepatotoxins (toxic to the liver), e.g., sterigmatocystin; nephrotoxins (toxic to the kidneys), e.g., ochratoxin A; neurotoxin (toxic to the nerves), e.g., botulinum toxin; emetics (cause vomiting), e.g., vomitoxin; mutagens (cause mutations), e.g., fusarin C; teratogens (cause malformations), e.g., ochratoxin A; carcinogens (cause cancer), e.g., aflatoxins. In addition, substances causing inflammation of the skin upon contact are known, e.g., trichothecenes or phytoalexins.

Toxoplasma gondii
Protozoa with a complex life cycle which a.o. entails the involvement of different hosts. The parasite may be associated with more or less all mammals and many birds. The infection typically occurs perorally, and for man it may result after ingestion of food contaminated, e.g., with cat faeces or raw meat contaminated with tissue cysts. Diverse clinical symptoms occur, depending on the organ that has been affected (central nervous system, liver, heart, muscles). Toxoplasmoses are mostly clinically inapparent and may often be asymptomatic. Transplacental transfer from mother to fetus is well known and causes severe malformations or still births.

training of personnel
Education and training of employees in a food production plant is one of the most important responsibilities of the manage-

ment. Training of employees to put into practice the different aspects of GMP, including hygienic food processing with the aim at end product safety assurance. It also concerns the handling and application of cleaning and disinfection procedures. By definition training refers to the "imparting and acquiring of skills to perform specific tasks", and it may concern both management, supervision and operators (Shapton and Shapton 1991).

transmission→ carry over effect.

tricarboxylic acid cycle
Citric acid cycle. Krebs cycle; forms part of the metabolism of all living cells. It produces intermediates for fat and protein synthesis and it yields energy in the form of→ATP (Schlegel 1985).

Trichoderma
A genus of Deuteromycetes. Bright green colonies due to pigmented conidia in mucous drops. *T. hazianum* produces smooth conidia. Opt. 30°C; min. 5°C; max. 36°C; min. a_w 0.91 at 25°C. *T. Lignorum* and *T. viridae* (rough conidia) catabolise wood, paper and other cellulose products. Also occurs on maize, rice, wheat and other cereal types. It also causes spoilage of citrus fruit. Growth conditions: pH 2.5–9.5. All species produce→trichothecenes. *T. reeseri* is used for the production of→cellulase (Pitt and Hocking 1985).

Trichosporon
Imperfect yeast genus belonging to the Ustomycetes. *T. beigelii* produces buds, arthrospores, blastospores and a mycelium. Occurs on fresh crabs, poultry, cooked sausages and bread on which it may cause spoilage.

trichothecenes
A large group of mycotoxins (presently >60). Mainly produced by *Fusarium*

spp., but also *Trichoderma, Trichothecium, Stachybotrys, Cephalosporium* and *Acremonium* spp. Toxin production is promoted by cool, humid weather with chilly nights. The most important toxins are deoxynivalenol (DON, vomitoxin), nivalenol, T-2-toxin, diacetoxyscirpenol and safratoxin. Mainly found in cereals and animal feed. It causes severe illness in pets (compare: Gedek 1980). Medical historians believe that Mozart might have died from mycotoxicosis.

Trichothecium
Monotypical genus of Deuteromycetes with bicellular conidia embedded in slime. *T. roseum* is coloured pink and causes store rot in stonefruit; pink rot in gherkins and is also to be found on bread. Growth: opt. 25°C; min. 15°C; max. 35°C. Germination of the conidia from 5°C; min. a_w 0.86–0.88; opt. 0.96. Produces trichothecenes. It is thought to be the cause of endemic oesophagus cancer in the Transkei (ZA) due to contaminated maize (Reiss 1986). Identification key by Baumgart (1990) (Pitt and Hocking 1985).

"Trockenbeerenauslese"
Noble high quality wines made from must in which the grapes are overgrown to greater or lesser extent with *Botrytis cinerea*→"noble rot". The must contains >30% sugar as the grapes have partially been dehydrated, resulting in such a high osmotic pressure that "spontaneous" yeast development is rarely found. The must also contains large numbers of →acetic acid bacteria which may inhibit the desired fermentation by acetic acid production and may utilise ethanol in addition. The addition of pure cultures of osmotolerant yeasts is therefore recommended (Dittrich 1987).

trophophase
Stage of development in fungi, after germination of the spores or the conidia. It is

characterised by rapid assimilation of nutrients, the formation of substrate mycelia and growth of the colony. This stage of development is followed by the→idiophase, localised in the centre of older colonies while the growing edge stays in the trophophase.

turbidity measurement

MO growing in clear nutrient solutions cause turbidity through the scattering of light at the interface between the water and cellular surfaces. Turbidity resulting from 10^6/ml bacteria can just be detected with the naked eye. It can be measured with a photometer by determining the difference between the incident and measured (shade) light. More reproducible results can be obtained with square cuvettes than with test tubes. Reproducible results can also be obtained by measuring the scattered light perpendicular to the incident light (Tyndall effect) with a turbidometer. It is always important that the specimen is well mixed before measuring (Baumgart 1990). – In continuous processes such as in→bioreactors (turbidostats) the measurement of turbidity can be used as control parameter for maintaining a given cellular density (Kunz 1988).

tyndallisation

Fractional sterilisation. Named after the British physicist John Tyndall. It is intended to kill the vegetative cells of MO by heating the material up to 80 – 100°C for a period of up to ca. 1 h on each of three successive days.→Endospores are heat shocked by which in more rapid germination is induced than at a constant temperature. The→vegetative cells are killed during subsequent heating. It is domestically used for non-acidic vegetables, e.g., green beans, peas, etc., in order to prevent→"flat sour" spoilage.

typhus (typhoid fever)→*Salmonella typhi* and *S. paratyphi.*

U

udder infection→mastitis.

UHT method→ultra high temperature.

UHT milk
Technologically "sterile" milk; does not contain any viable MO. Sterilised by continuous heat treatment (appertisation) at "ultra high" temperatures (UHT) and short contact times ("UHTST"), either by direct (steam injection) or indirect (heat exchanging) heating at 135 – 150°C for a few seconds (→UHT processes). Filled and packaged in an aseptic system. The raw material (milk) should contain only low numbers of MO, since microbial→ proteases and→proteases produced before treatment may not be inactivated completely by the thermal process; may cause sensory defects during marketing and distribution for which at least 6 weeks should be allowed. →Milk hygienic guidelines prescribe examination of UHT milk with the→limulus test that reacts on→lipopolysaccharides of Gram – bacteria. Sufficient exo-enzymes may be produced by 10^4 of these bacteria/ml to cause deleterious effects leading to spoilage and shelf life reduction (Teuber 1987).

ultra high temperature
Sterilisation at a high temperature for a short period. Uperisation; ultra high temperature or UHT. Used for milk (→H milk), sensitive soups, e.g., cream of asparagus soup, aroma-sensitive fruit juices, and puddings to be→sterilised. It is only practical when aseptic packaging of the product is guaranteed (→aseptic plan). It is done at a temperature of 130 – 150°C in continuous flow and the period of heating is very short (2 – 10 s). A distinction is made between direct heating with injected vapour or indirect heat supply by stirring. The quality of the product is affected only slightly since the D value for chemical changes (→Q_{10} value) is much higher than the D value of MO and → endospores. The disadvantage is that several enzymes, being chemical compounds (e.g., lipases, proteases and peroxidases), are not inactivated and may cause spoilage on prolonged storage (→abiotic spoilage). The raw material should contain the lowest possible numbers of MO so that only small amounts of extracellular enzymes can take part in the process (Heiss and Eichner 1984; Krämer 1987).

under-processing
Insufficient processing. The goal of killing all MO or specific groups in a technical process is not reached, when either the contact time has been too short or the temperature too low. It represents a far reaching technical error with severe consequences, since the surviving MO may reproduce without competition. Pasteurisation equipment in general is automatically secured against under-processing by recirculation and repeating of the operation. In this manner it can be guaranteed that all pathogens are killed by the process.→Pasteurisation.

uperisation→ultra high temperature.

urea
H_2N-CO-NH_2 (carbamide) and uric acid from→DNA or→RNA are used in→culture media for the production of→starter cultures and/or→enzyme preparations, as sole or part of N source. Some doubt about possible health risk since (a.o.) carcinogenic ethylcarbamate (urethane) (N_2N-COOC_2H_5) may be formed during growth or fermentation (DFG 1987).

urethane→uric acid.

Ustilaginales

Parasitic fungi belonging to the Basidio-mycetes and causing damage mainly to monocotyledons. *Ustilago zeae* causes smut in maize, *U. tritici* in wheat and *Tilletia tritici* common blunt in wheat. → Ear moulds.

utensils

Referred to in different food regulations, and include instruments, machines, working and cutting surfaces, containers, brushes, etc., that are to be cleaned and → disinfected easily. Wood is a less acceptable working material on or into which nutrient rich liquids and suspended MO may establish, and may escape the cleaning and disinfection procedures. In addition, the necessary heat treatment is deleterious to the material. Glass, china (porcelain), stainless steel and smooth synthetic surfaces are preferred.

UV decontamination

Approved method to reduce the bacterial load in water and on the surfaces of fruit and vegetable products and hard cheese during storage. UV light can also be used to sterilise packaging material used in → aseptic techniques. Air decontamination does not constitute any health risks, but direct contact with UV rays of employees and products or stored foodstuffs should be prevented by precautionary measures. It is useful to use UV radiation continuously in the inoculation area of a laboratory, but care should be taken that air is circulated moderately in order to bring the organisms in the vicinity of the light source (Sinell 1985). – The most effective wavelength ranges from 200 to 300 nm with a max. of 260 nm, being the absorption maximum of nucleic acids. Thymine, cytosine and uracil in → DNA are extremely sensitive. Following treatment with UV or ionising radiation, a repair mechanism may become operative in MO cells when transferred to a favourable environment.

V

vacuum drying
Rapid and careful drying method, allowing the use of lower temperatures concomitantly with an increase in negative pressure. Most commonly used in→ freeze drying. Almost all MO survive this method of water extraction. Protective gas packaging is recommended so as to avoid oxidation damage in view of the enlarged surface created with this method (Heiss and Eichner 1984).

vacuum packaging
Airtight plastic bags, evacuated and sealed after filling with highly perishable products, e.g., meat, sausages, fish, etc. Aerobic bacteria quickly utilise the residual oxygen so that ideal conditions for the multiplication of→ anaerobes are created, unless chilled at the same time (4 °C). An acute danger of→ botulism exists if not chilled;→ fresh water fish.

variability
A basic survival mechanism of all living organisms, common for MO and the prerequisite for→ adaptation and→ selection. In the case of bacteria it is promoted by the exchange of→ plasmids and in fungi by→ heterokaryosis.

variety
Abbreviation: var. A geographic or ecologically deviating species or type, belonging to a specific species and extremely well adapted (through selection) to the habitat or host, e.g., *Fusarium avenaceum* var. *fabae*.

vat milk
Milk with adjusted fat contents, often pasteurised, in open or closed containers, e.g., baths, tanks, trucks, cheese tanks and used for the production of sour milk products or cheese. The milk is warmed to specific temperatures required for the product or starter, inoculated (final concentration: 10^6/ml bacteria) and rennet and/or rennet substitute is added;→ cheese processing suitability.

vector
Vehicle. Way of transport for MO to hosts or habitats. Wind, water, insects, rodents, pets, man, utensils.

vegetable juice
Juices prepared from minced vegetable types, with or without added lytic enzymes. Commercial distribution after pasteurisation or mild lactic fermentation and pasteurisation. Enzyme preparations of *B. polymyxa* with→ pectinase, polygalacturonidase and pectate-lyase activities used for hydrolysis of tissues. Lactic fermented vegetable juice from white cabbage, carrots, beetroot, celery or tomatoes. Spontaneously fermented products contain both stereoisomers of lactic acid, as in sauerkraut. For production of juices with only $L(+)$ lactic acid→ starter cultures of *Lb. bavaricus, Lb. casei* and *Lactococcus lactis* ssp. *lactis* are used. – Effective thermisation and prevention of recontamination render vegetable juices microbiologically stabile. Main spoilage association: *B. subtilis, B. coagulans, B. thermodurans* and *Cl. pasteurianum* (Kunz 1988).→ Nitrate content of raw materials should be determined before processing.

vegetables
All vegetables are heavily contaminated at the surface even prior to harvest. The type of epidermis and the presence of chemical protective substances (phytoalexins, → phytoncides), will determine the storage potential of the commodity, e.g., storageable (cabbage types, carrots, onions),

short-time storage potential (cucumbers, paprika, tomatoes, asparagus), and no storage potential (spinach, lettuce). Storage rots are enhanced by mechanical damage (insects, birds, harvesting, transport, etc.). Strict control before and during storage necessary, as well as selection and removal of infected pieces that may cause cross contamination (Table 42). Sufficient aeration and refrigeration should be continuously assured during storage; heat generation by living plant tissues (see Table 36 for respiration energy), and→ethylene production.→CA storage effective for shelf life extension of some vegetable types, since, in addition

to the metabolism being retarded, the growth and metabolic activity of harmful MO are inhibited or prevented (Bünemann and Hansen 1973; Müller 1983a; Nicolaisen-Scupin 1985).

vegetative cells
Cells able to divide and reproduce the same (identical) form or→dormant forms which again produce vegetative cells when germinating. In the vegetative state more susceptible to heat, radiation, drying, preservatives, disinfectants and other noxious elements than the dormant forms. With a few exceptions vegetative cells constitute the dominant group in

Table 42 Storage rots of vegetables (acc. to Frank, H. K.: in Heinze, K. 1983, modified).

Vegetable Type	Parasite	Comment
Cabbage types	Botrytis cinerea Rhizoctonia solani Rhizopus stolonifer Sclerotinia sclerotiorum	grey mould
China cabbage	Botrytis cinerea	in CA store
Carrots	Erwinia carotovora Xanthomonas sp. Botrytis cinerea Ceratocystis fimbriata Rhizoctonia carotae Fusarium spp.* Sclerotinia sclerotiorum	wet rot wet rot grey rot phytoalexin formation red to white phytoalexin formation
Celery	Erwinia carotovora Botrytis cinerea Sclerotinia sclerotiorum	wet rot grey mould phytoalexin formation
Horseradish	Botrytis cinerea Penicillium hirsutum Sclerotinia sclerotiorum	grey mould green mould
Onions	Aspergillus flavus* A. alliaceus, A. niger, A. glaucus Botrytis allii Fusarium moniliforme* Colletotrichum circinans	wet rot grey rot at onion stem "dirt" stains

foodstuffs, with either a positive or negative influence on quality and characteristics. Bacterial→starter cultures consist only of vegetative forms.

Venturia→ apple scab.

Verticillium
Genus of the imperfect fungi. Teleomorphs: *Nectria, Cordyceps* and *Torrubiella*. Saprophytic species cause damage in all foodstuffs. Some species are plant pathogens, a few are host specific and cause withering diseases in cultivated and wild plants.→ Kojic acid is produced by *V. dahliae* on suitable substrates. – *V. psalliotae* is described as the cause of brown spots on cultured mushrooms. Opt. 23 – 24 °C; min. 8 – 10 °C; max. 35 °C. No infection of mushrooms if the temperature is maintained <22 °C. pH 3.9 – 7.4 (Domsch et al. 1980).

Vibrio cholerae
Gram – , comma-shaped to curved rods, motile due to a polar flagellum, aerobic, nitrite + . The most common biotype *eltor* is α-haemolytic. Reproduction: 8 – 42 °C, opt. 37 °C. Survives: Coca Cola 30 °C for 4 h; smoked fish 30 °C for 2 – 4 d; on the surface of several fruits, 30 °C for 1 – 2 d; honey 25 °C for 12 h; ice cream 4 °C for 14 – 18 d; meat and meat products (prepared) 30 °C for 2 – 5 d; sea water 5 – 10 °C for 60 d, 30 °C for 10 – 13 d; fresh water 21 °C for 12 – 51 d; glass, porcelain (dry) 21 °C for 4 h. Killed by pasteurisation (Mitschirlich and Marth 1984). It causes→ cholera in man and is most commonly transmitted by water or ice cream.

Vibrio parahaemolyticus
Halophilic vibrios requiring > 1% NaCl for multiplication. Habitat: coastal waters and seafoods/marine animals from such environments. It produces a heat stable, haemolytic exotoxin causing human gas-troenteritis after consumption of contaminated raw fish, crabs or mussels. Incubation time: 2 – 48 h. Infection dose in a healthy person is 10^7 living organisms. Intermittent diarrhoea, body pains and low fever remain for approx. 2 – 5 d. It occurs often during summer time in Japan where fresh fish is generally consumed (Krämer 1987; Sinell 1985).

vinegar
Contains at least 4% of→ acetic acid. May be produced from virtually any (alcoholic) fermented liquor (e.g., wine, beer, cider, honey beer, palm wine, etc.); differences in taste and flavour, with commercial designation often referring to the kind of raw material used. Aerobic (oxidative) microbiological conversion of alcohol, with a high oxygen requirement. Early vinegar production from an alcoholic liquid was carried out in shallow pans or wooden vats with large surface areas for aeration. A bacterial scum, containing *Acetobacter* spp. from the air and utensils, developed on the surface. Pure cultures, e.g., of *Ac. orleanense*, were selected increasingly for inoculation. Modern vinegar production relies to some extent on the→ Frings submerged culture "acetator", in addition to the "generator" process, developed early in the 19th century by Schuezenbach in Germany. Modern generators are packed with beech-wood shavings or chips from air-dried wood; a film of acetic acid bacteria develops gradually on the surface of these wood shavings (→immobilisation) while the alcoholic broth trickles down and air is passed in counter-current from the bottom to the top to provide oxygen and control heat generation. Most commonly used spp. for commercial production are *Ac. curvum* and *Ac. schuezenbachii*, neither of which tends to "overoxidise" (→ acetic acid bacteria) alcohol.→ Submerged and→ continuous culture methods are used mainly for→ acetic acid production. – The liquid is filtered when the

desired concentration is reached, and may additionally be flavoured and pasteurised. Main uses are for flavouring and acidification of foods (Rehm 1985).

violacein
A purple, water insoluble→pigment of *Chromobacterium violaceum*.

viral gastroenteritis
Collective term for gastroenteritis caused by viruses, mostly transmitted to man by water, raw salads and vegetables. The following are described as pathogens: Rotaviruses Group A and others, Norwalk viruses and large "related" types, (27 – 32 nm), Adeno-, Astro- and Caliciviruses. Serological methods to detect the viruses are not available for all the pathogens, but electron microscopic examination of the stools within the first 48 h after the clinical symptoms have started is successful [Brede, H. D.: *Forum Mikrobiol.* 11 (1988) 139].

viral hepatitis A
Epidemic jaundice. Transmitted in warmer regions by contaminated water and foodstuffs or utensils washed in it. Period of incubation: 15 – 20 d. The faecal-oral route of transmission is indicative of poor hygiene. – The initial symptoms are not clear, but are flu-like with fever followed by jaundice. Viruses are excreted in the faeces.

virus
Dormant viruses are typically symmetric filterable particles of macromolecular size not visible with a light microscope. They have immunogenic characteristics and genetic properties are contained in the nucleic acid. A virus consists of nucleic acid, either DNA or RNA and a protein. It does not possess enzymes for the production of energy (except for some phages which possess ATPase), and requires the cellular functions of a suitable host cell for multiplication. The obligatory intracellular multiplication is linked to the genome which means that the cells of the host are reprogrammed to produce new viruses, usually associated with harmful effects for the host (Grafe 1977). A few viruses are transmitted by food and cause illnesses. Vectors can be: raw milk, contaminated water and salads, fruit and vegetables washed with contaminated water, raw meat and especially mussels, shrimp and other marine animals from coastal waters (Sinell 1985). HIV virus causing AIDS cannot be transmitted by food.

vitamin C→ascorbic acid.

Voges-Proskauer test
The VP test is used to detect→acetoin (acetyl methylcarbinol) in liquid media and it is produced by several bacterial species or genera from glucose and lactose. Most of the time it is combined with the methyl red test for the→differentiation of→*E. coli*, *Citrobacter* and *Enterobacter*. Methyl red VP broth, e.g., Merck no. 5712 (Baumgart 1990).

Vollsauer→sourdough.

volutin
Reserve material of phosphate and (in some MO) of energy.

vomitoxin→trichothecenes.

VRB agar
Violet red bile agar; crystal violet neutral red bile agar. Selective nutrient medium. Use to detect and count→coliforms and→*E. coli* in water, ice cream, meat, etc.; may involve prior pre-enrichment for→*Salmonella*. Commercially available from BBL, Merck, Oxoid, etc.

VRBD agar
Crystal violet neutral red bile glucose agar acc. to Mossel. Selective nutrient medium

for the isolation and determination of the bacterial count of all Enterobacteriaceae in foodstuffs. The nutrient medium should not be autoclaved, but treated for 30 min in a→ steam cooker. Commercially available from BBL, Merck, Oxoid.

W

waiting time
Period between the last administration of a pharmaceutical agent (medicine) in meat animals to be slaughtered, or in the production of foodstuffs (regulations on pharmacologically active substances of 3.8.1977). Of special interest are→antibiotics administered against→mastitis in cows with mastitis. Milk may not be delivered to dairies within 5 – 17 days following administration of antibiotics i.m. or i.p. per injection, or for direct treatment of the udder, as it may cause an allergic or sensitising reaction in consumers – especially penicillin. In addition it may be responsible for economical losses due to inhibition of starter cultures for, e.g., yoghurt production (Sinell 1985).

wall mould
(Germ.: "Wandschimmel"). Moulds growing on the inner walls of store- or ripening rooms (cellars) with a high humidity (>70% rH) and with a considerable difference between the inner and outer temperatures. It may also occur in cold storage rooms or refrigerators. Ideal "growing" possibilities for the "specialists" are created by the volatile organic substances, e.g., ketones, oils of mustard, indoles, other aromatic substances and→ condensation. The paint is attacked and the colour and physical properties are changed. Species commonly isolated belong to the following genera: *Alternaria, A., Acremonium, Cladosporium, Mucor, P., Spicaria, Stemphylium* and *Trichoderma*. More or less all may cause spoilage of foodstuffs. People working in such rooms are at risk of developing→allergies due to the inhaled conidia or spores. Control includes: washing of the walls with a formalin solution (careful: inhalation protection) or (less efficient) a solution of 90 parts of water, 10 parts of

spirits and 2 parts of salicylic acid (Reiss 1986).→ Mould protection paints. → Filling materials.

Wallemia sebi
Monotypical, osmophilic genus of the → Deuteromycetes. Large numbers of conidia are produced in chains, without phialides, at the distal ends of the conidiophores. The colonies are orange brown to black brown. Growth conditions: opt. 24 – 30°C; min. 5°C; max. 37 – 40°C. Min. a_w 0.75; pH 5.5 – 7.3. NO_3 is toxic, but ammonia and urea are well metabolised. Habitat: bread, marzipan, dates, fruit, nuts, malt extract, jam, jellies, ham, salted fish (Domsch et al. 1980). The fungus is also isolated from human abscesses and organs.

Waltari wine
Alternative for sulphur dioxide treatment of wine→sulphuring ("Schwefelung"). The wine is bottled and stored in a nitrogen atmosphere free from oxygen (Classen et al. 1987). In practice difficult to apply.

water activity
The microbial spoilage of a product mainly depends on its available water content or water activity (a_w value). The a_w value is an indication of the water available for growth. With→abiotic spoilage the concept of water potential is used where $= a_w$ × 100, and is expressed as % of the aqueous saturation of the gas in which the foodstuff is found at a specific temperature. – The relationship between the water activity and the water content is established by sorption isotherms, not only in foodstuffs, but also for packaging material and other products. The water activity is inversely related to→osmotic pressure. These factors are essential for the survival

Table 43 Water contents of some foods with storage potential, based on a_w value of 0.70 (limit for microbiological "safety" acc. to Mossel 1977).

Food	Water Content in %
Nuts	4 – 9
Full milk powder	7
Cocoa	7 – 10
Soybeans	9 – 13
Dehydrated full egg	10
Skim milk powder	10
Lean dried meat	10
Oats	13
Rice, cereals	12 – 15
Dehydrated vegetables	12 – 22
Wheat flour, noodles	13 – 15
Dehydrated soups	13 – 21
Dried fruit	18 – 25

of MO and depend a.o. on the presence of substances not able to absorb water from the atmosphere, e.g., fat droplets in nuts. These relations are given in Table 43. Water activity levels <0.70 are considered as "safe levels" since they do not allow reproduction, germination of spores and growth of MO (Heiss and Eichner 1984). → Water requirements of different MO groups.

water bloom
Massive development of → blue algae on the surface of standing water by sudden eutrophication, e.g., soil or plant material washed into reservoirs or lakes after a thunderstorm. Some species produce → phycotoxin.

water requirement
Practically all MO multiply well at → water activities of 1 – 0.98, where sufficient water with dissolved nutrients is available. The energy requirement for nutrient uptake increases with the decreasing a_w

value or increasing → osmotic pressure, in order to prevent excessive loss of cell contents to the environment, and to still allow uptake of nutrients. The ability to satisfy their water requirements differs between species (Table 44). – Water requirements cannot be satisfied in frozen products (physiological drought). The concentration of dissolved substances and the osmotic pressure increase during the freezing process. The water content in dry products can be increased locally by insects through their droppings. It may satisfy the water requirements of → xerophilic fungi and through the process of respiration of increasing growth, water is released and other species may start to grow.

Weisslacker
→ Semi hard cheese with a dry mass of 47% in the full cream product (Teuber 1987).

well cooking
Heat treatment used to improve the digestibility, taste and consistency of raw foods (meat, poultry, fish, vegetables, potatoes). For → canned foods the destruction/killing of all MO is important; for acid foods all viable/vegetative forms should be destroyed; → F value. In high-temperature/short-time treatment the desired degree of "commercial sterility" may be achieved without the product being thoroughly cooked (Heiss and Eichner 1984).

wet brining
Salt bath containing 18 – 20% NaCl and used in the treatment of → rennet cheese. Whey is extracted from the cheese curd and the salt concentration increases up to 2 – 3% with the concomitant selection of the desired ripening MO (Teuber 1987).

wet curing
Curing of products in a → brine solution

containing → nitrite curing salt (12 – 20 °C) and the desired salt tolerant curing MO, e.g., *Staph., Micrococcus, Lb., Enterococcus* and *Vibrio* spp. (Krämer 1987).

wheat beer → beer.

whey
Yellow-green liquid formed as a by-product during the manufacture of cheese; separates from casein curd. Distinction can be made between rennet whey (sweet whey) from rennet cheese and sour whey from → sour milk or → cottage cheese. Rennet whey contains 93 – 95% water, up to 0.8% fat, 0.9% protein, 3.8 – 5% milk sugar (lactose), 0.5 – 0.8% minerals. In acid whey the greatest part of the lactose is converted into lactic acid. Excellent culture medium for bacteria and carrier of a large number of → bacteriophages. The most important source of phages in the industry is whey residues in or on utensils even in the dried form (Teuber 1987). Pigs can be fed whey without complications. It can be added to

Table 44 Minimum water activity necessary for the growth of MO commonly associated with foods.

a_w	Bacteria	Yeasts	Fungi
0.98	*Cl. botulinum C* *Ps.* spp.	—	—
0.97	*Cl. botulinum E*	—	—
0.96	*Flavobacterium,* *Klebsiella, Shigella*	—	—
0.95	*Alcaligenes, B.,* *Citrobacter, Cl. botulinum* *A u., B., Cl. perfringens,* *E., Proteus, Salmonella,* *Serratia, Vibrio*	—	—
0.94	*Microbacterium,* *Pediococcus*	—	—
0.93	*Lb. plantarum*	—	*Mucor, Rhizopus*
0.92	—	*Pichia, Rhodototorula*	—
0.91	*Corynebacterium,* *Str. Staph.* (anaerobic)	—	—
0.90	*Micrococcus, Pediococcus*	*Saccharomyces*	*Byssochlamys nivea*
0.88	—	*Candida, Torulopsis*	*Cladosporium*
0.87	—	*Debaryomyces*	—
0.86	*Staph.* (aerobic)	—	*Paecilomyces, Alternaria* *alternata*
0.80	—	*Saccharomyces*	*P. martensii*
0.75	halophilic bacteria	—	*A. amstelodami,* *Wallemia sebi*
0.70	—	—	*A. ruber, Chrysosporium* spp.
0.65	—	*Saccharomyces bailii*	*A. repens, A. chevalieri*
0.62	—	*Saccharomyces rouxii*	*A. echinolatus,* *Monascus ruber*
<0.60	—	osmophilic yeasts (in honey)	—

processed cheese, thickened or as a powder, to increase the lactose content.

whey powder→ milk powder.

whey utilisation
It can be used as medium to produce → single cell protein with the aid of lactose fermenting yeasts, mainly *Kluyveromyces fragilis*. Another method is the bacterial conversion of lactose into lactic acid (*Lb.* species) which then serves as growth medium for *Candida, Saccharomyces, Torulopsis, Brettanomyces*, etc. Ethanol production is possible, but is rarely practised and then mostly for the production of whey vinegar (Kunz 1988).

white beer ("Weissbier")→ beer.

white bread
Wheat bread. Bread made from wheat flour with the aid of baker's yeast as leaven. Wheat sourdough is occasionally used to improve the characteristics of the dough, e.g., an aromatic taste, improved ability to keep fresh and a longer shelf life (Spicher and Stephan 1987).

white rot
Common name for the spoilage of potatoes caused by *Fusarium coeruleum* and in the case of carrots and other root vegetables by *Sclerotinia sclerotiorum*.

white wine→ wine.

wine
Alcoholic beverage prepared from grape juice or other fruit. The fruit is pressed and a fermentation tank is filled with the fresh juice (must). The must and pulp are separated for white wines. The must of red grapes is also white, but, for red wines, it is fermented together with grape skins and stalks for a few days. During fermentation anthocyans and tannin are partially leached by ethanol from the cells, and dissolve in the must. Rosé wines can be made from either red grapes allowed to ferment for a few hours together with the skins or with a portion of the skins remaining in the fermenting liquid. In France white wine is also produced from cheaper red grapes, but it is distinguished from wine made from white grapes being named "blanc de blanc". Fermentation is caused by "natural" yeasts on the skins and stalks. Presently mainly→ pure yeast cultures are used for must inoculation.

wine grapes
Table grapes. Often attacked by *Botrytis cinerea, P. expansum, Cladosporium herbarum* or *Alternaria alternata* before harvest in a moist climate. Protected during transport by refrigeration to 0 °C and treatment for a short period of 20 min with 0.5% SO_2 (Sommer, N. F.: in Kader et al. 1985).

wine spoilage
Numerous kinds of microbial wine spoilage are known (Dittrich 1987).→ Acetic sting;→ cork taste. Wine spoilage must be distinguished from wine errors since the latter are caused by chemicophysical factors related to processing failures.

wine yeasts
Yeasts present in must and involved in its fermentation are divided into strong fermentors, e.g., *Saccharomyces cerevisiae, S. uvarum, S. roseum*, etc., and weak fermentors such as *Kloeckera apiculata, Torulopsis stellata*. The top-fermenting yeast *Metschnikowa pulcherrima* only develops at the surface (e.g., when the tank is not filled to the top) and is rarely involved in the fermentation. The weak fermenting yeasts are the most abundant in fresh must, but are quickly overgrown by the strong fermenting yeasts. The→ pure yeast cultures are mostly polyploidal. They have developed selectively and are well adapted to the "non-physiological" conditions in the fermentation tank. Spe-

cial yeast types are commercially available as starter cultures and, depending on processing requirements, are able to survive low temperatures, the addition of H_2SO_3, high osmotic pressures and O_2 depletion (Dittrich 1987).

wood

Unsatisfactory material for→ utensils that come into direct contact with food; decontamination under household conditions is problematic.

wort

Maltose containing substrate, produced from malt and fermented to→ beer. The ground malt is made into a mash by mixing with water. The starch is broken down to sugars by product specific amylases. Part of the mash is boiled, to improve starch degradation, to stop hydrolysis of the starch and to retain the desired solids content of the wort in the final product. The wort is separated from the grain hulls by filtration (in the "lauter" tun) and boiled with hops. After dilution to the desired concentration, it is inoculated with yeasts and the sugars fermented.

wort agar

Wort broth used as nutrient media to culture, isolate and cultivate fungi and yeasts, e.g., BBL 11826; Merck no. 5448; Oxoid CM 247. Used for the cultivation of the → pitching yeasts in the brewing industry. Elsewhere the nutrient medium is only used in special cases. More commonly malt agar is used since it is difficult to obtain standardised beer wort.

wort bacteria

Designation used in the beer industry when referring to bacteria of the genera *B., Cl., E.* and *Enterobacter* typically associated with wort. Most of these bacteria are killed during fermentation, but due to their metabolites produced in the wort, the sensory quality of the beer may be negatively affected (Baumgart 1990; Priest and Campbell 1987).

X

Xanthomonas
A genus of the Pseudomonadaceae. Gram −, aerobic, polarly flagellated rods producing a yellow water insoluble pigment. Can liquify gelatine, catalase + (Bergey 1984). *X. campestris* occurs on leaf and sprout vegetables and causes spotted soft rot. Often detected in water.

Xeromyces
Obsolete name for the xerotolerant genus → *Monascus*.

xerophilic moulds
More correctly xerotolerant, since growth of these spp. improves with increasing content of available water. The lowest level is ca. a_w 0.62. Especially representatives of the → *A. glaucus* group and → *Wallemia sebi* are of interest because of their detrimental role in the environment, and as "pioneer" organisms for other moulds (Reiss 1986).

xerotolerants
MO growing in habitats with a low water content, e.g., marzipan, jam, confectioneries and honey. Comprise practically only yeasts and moulds. → Water activity, → water requirements. The range of conditions for growth depends on the species but varies between a_w 0.62 − 0.70 (Table 44). Xerophiles or → osmophiles which prefer a sugar concentration above 50% have not been found yet. As selective medium either 50% glucose agar or potato agar with 60% sucrose, pH 5.2, are suitable after enrichment with relevant nutrient solutions (Baumgart 1990).

XLD agar
Xylose-lysine-deoxycholate agar according to Taylor. It is used as selective medium for the detection of *Salmonella* and *Shigella* (Baumgart 1990): e.g., BBL 11838 or Merck no. 5287.

xylitol
Pentitol formed by the reduction of xylose (wood sugar). It has a sweet taste and can be used by diabetics as sugar substitute. It is utilised by the intestinal MO and has a light laxative effect.

xylose
Wood sugar. Aldopentose. It is fermented by several bacterial and yeast spp., but not by *Saccharomyces cerevisiae*.

Y

yeast extract
German: Hefeextrakt. Prepared by→
autolysis of baker's or brewer's yeast.
Water-soluble; important ingredient for
numerous microbiological growth media
and as supplement in industrial fermentations. Rich in peptides, amino acids and
vitamins (especially B vitamins). Commercially available as powder or granulate,
e.g., Difco 0127; Merck 3753; Oxoid L20.

yeasts
German: Hefen; French: levure; Span.:
levadura. A category of fungi belonging to
different fungal taxonomic groups. Mostly
unicellular; formation of pseudomycelium
occurs, depending on cultural conditions
and sp. Asexual reproduction most typical
by budding (e.g., *Sacch. cerevisiae*) and
in some by binary fission (e.g., *Schizosaccharomyces* spp.). Most typical fermentative and non-pathogenic (saprophytic),
e.g., *Sacch. cerevisiae*, but some are non-
fermentative (*Sporotrichum*), and pathogenic (e.g., *Candida albicans*). Some
representative of typical spoilage association of some foods (e.g., *Zygosaccharomyces rouxii*). Some are used for the
biotechnical production of→single cell
protein (*Candida utilis = Torulopsis utilis*),
and of→enzyme preparations for the food
industry (Müller and Loeffler 1982; DFG
1987).→Differentiation as for bacteria is
mainly based on physiological criteria
(Baumgart 1990; Pitt and Hocking 1985).

Yersinia enterocolitica
Species belonging to→Enterobacteri-
aceae. Able to multiply in the refrigerator.
Opt. 28°C, min. 1°C, max. 44°C, pH
4.4 – 9. Causes enteritis and enterocolitis,
commonly with secondary pathological
effects, e.g., arthritis or meningitis, caused
when the bacteria invade the intestinal
wall and enter into the blood system.
Period of incubation 7 – 10 d. Transmitted
by domestic animals (→zoonosis) and
foodstuffs, e.g., milk and water (Krämer
1987; Sinell 1985). Survives in water at
4°C >540 d. Multiplication in distilled
water and drinking water has been
proven. Killed by pasteurisation (Mitscherlich and Marth 1984). Detection: enrichment at low temperature and streaking
onto a selective medium according to
Wauters or Schiemann, e.g., Merck no.
15209 or 16434 on which a suspected
species can be confirmed biochemically
(Baumgart 1990; Speck 1984).

yoghurt
A thermophilic mixed culture of *Str.
salivarius* ssp. *thermophilus* and *Lb. bulgaricus* (1:1) is employed in the production of fermented sour milk. Both organisms form a symbiotic relationship at 45 °C
where *Str.* lowers the Eh value rapidly and
produces amino acids and small amounts
of CO_2 which stimulate the growth of *Lb.*
The *Lb.* spp. are proteolytic and produce
amino acids for the *Str.* strain. After an
incubation period of 2.5 – 3 h the bacterial
count reaches ca. 10^9/g. The lactic acid
content amounts to 0.8 – 0.9% (mild) and
1 – 1.3% (strong). Aromatic substances,
e.g., acetaldehyde, acetone, ethanol,
butanone-2, diacetyl- and ethyl acetate
are produced. In addition to lactic acid,
citric, acetic, formic and succinic acids are
found. Milk used in yoghurt production
should be heat treated for 5 – 40 min at
75 – 95°C before inoculation. Milk powder can be added to increase the dry
matter content. When a pH value of 4.6 is
reached the product must be cooled
rapidly to 5 °C or the culture must be inactivated by heat treatment to prevent subse-

quent acid formation (the latter is to be declared). By adding lactase ($\rightarrow \beta$-galactosidase) to milk it is possible to increase the sweetness and omit the addition of sugar, since lactose is cleaved into glucose and galactose.

Z

z value
The increase in the processing temperature in K necessary to obtain a ten-fold ($\rightarrow$ D value) destruction rate (i.e., requiring 1/10 of the treatment time). The flatter the destruction temperature curve, the larger the z value (D value plotted against the temperature). Vegetative bacteria have z values around 5 K, endospores of *Cl. botulinum* 8.3 – 9 K, *Cl. perfringens* and *B. cereus* 9.7 K and *Cl. sporogenes* 9 – 11 K. It should be noted that the D and z values strongly depend on the product, pH, fat, protein and CH which may exert a $\rightarrow$ protective colloidal effect (Heiss and Eichner 1984).

zearalenone
F-2 toxin. A $\rightarrow$ mycotoxin produced by several $\rightarrow$ *Fusarium* spp. during growth on cereals. Oestrogenic activity. Feedstuffs contaminated with zearalenone can cause infertility in cattle and pigs and vulvovaginitis in sows. It is transformed into the more active form zearalenol in the animal body (Reiss 1981).

zoonoses
Communicable diseases which can be transmitted to man by mammals. A summary is given in Table 45 of examples where foodstuffs may act as vectors (vehicles). The first group can be transmitted by milk; the second includes enteric infections or toxi-infections; the third group concerns foodstuffs in contact with the animal and group four contaminated by animals with possible multiplication during processing. Also refer to Table 30, under pathogenic MO.

Zygomycetes
Class of the division Zygomycota. Unseptated hyphae with sexual reproduction by zygosporangia, most of the time reproduced by heterothallic individuals upon contact. Primary sporangia with numerous sporangiospores germinate from the zygospore (Figure 21, under sporangium). $\rightarrow$ *Mucor*, $\rightarrow$ *Rhizopus*, $\rightarrow$ *Thamnidium*.

Zygosaccharomyces rouxii
(*Saccharomyces rouxii*). $\rightarrow$ Osmotolerant yeast. Plays a role in the spoilage of must and fruit juice concentrates, fruit pulp for the production of fruit yoghurt, maraschino, confectionery, marzipan raw

---

Table 45   Zoonotic MO responsible for food-borne infections (acc. to Sinell 1985).

Gr. 1		Gr. 2	
Gr. 1	*Mycobacterium bovis, M. tuberculosis*	Gr. 2	*Salmonella*
	Brucella spp.		Enteropathogenic *E. coli* strains
	β-Hemolytic *Str.* spp.		Other Enterobacteriaceae
	Str. agalactiae		*Yersinia enterocolitica*
	Staph. aureus		*Vibrio parahaemolyticus*
	Coxiella sp.		*Campylobacter jejuni*
	Ticks – encephalitis virus		"Opportunists", e.g., *Proteus, Ps.*, etc.
	Mouth-and-foot disease virus		
	Vaccinia virus		
Gr. 3	*B. anthracis*	Gr. 4	*Listeria monocytogenes*
	Francisella tularensis		*Erysipelothrix* sp.
			Leptospira sp.

mass, honey, etc. Fermentation may start even after a long dormant period and the tins or glasses can swell due to the gas produced.

Zymomonas mobilis

(*Thermobacterium mobile*). Gram– rods, polar flagella, anaerobic to aerotolerant, catalase+. Opt. 30°C. Up to 10% ethanol, CO_2 and a little lactic acid is produced via the Entner-Doudoroff pathway. Habitat: in fermenting plant juices (pulque and palm wine), spontaneous fermentation of cocoa beans. Successfully implemented in a large experimental plant for the fermentation of crude glucose syrup from starch waste in a continuous fermentation process (Gottschalk et al. 1986) and on a technical scale for the production of ethanol as fuel.

References

Alberts, B., Bray, D., Lewis, J., Raff, M., Roberts, K., Watson, J. D.: *Molekularbiologie der Zelle.* Weinheim: VCH Verlagsges.m.b.H., 1986.

Ayres, John C., Mundt, J. Orvin, Sandine, William E.: *Microbiology of Foods.* San Francisco: W. H. Freeman and Company, 1980.

Bacus, J.: *Utilization of Microorganisms in Meat Processing.* Letchworth: Research Studies Press Ltd., 1984.

Barnett, J. A., Payne, R. W., Yarrow, D.: *Yeasts, Characteristics and Identification.* Cambridge: Univ. Press, 1983.

Baumgart, J., in collaboration with Firnhaber, J. and Spicher, G., Timm, F., Zschaler, Regina: *Mikrobiologische Untersuchung von Lebensmitteln. 1st Ed.* Hamburg: Behr's, 1986.

Baumgart, J.: *Mikrobiologische Untersuchung von Lebensmitteln. 2nd Ed.* Hamburg: Behr's, 1990.

Becker, E. W.: *Mikroalgen – Ergebnisse aus drei Versuchsvorhaben.* Schriftenreihe der GTZ 124. Eschborn: GTZ, 1982.

Berg, H. W., Diehl, J. F., Frank, H. K.: *Rückstände und Verunreinigungen in Lebensmitteln.* UTB 675. Darmstadt: Steinkopff, 1978.

Bergey's Manual of Determinative Bacteriology. 8th Ed. Baltimore: Williams and Wilkins Co., 1974.

Bergey's Manual of Systematic Bacteriology. Vol. 1 and Vol. 2. Baltimore: Williams and Wilkins, 1984 and 1986.

Bergquist, D., Krafta, A., Cotterill, O. and Magwire, H.: "Eggs and Egg Products". In Speck, M. L. (ed.): *Compendium of Methods for the Microbiological Examination of Foods. 2nd Ed.* Chapter 45. Washington, DC: APHA, 1984.

Betina, V. (ed.): *Mycotoxins – Production, Isolation, Separation and Purification.* Amsterdam u. a.: Elsevier, 1984.

Beuchat, L. R. (ed.): *Food and Beverage Mycology. 2nd Ed.* New York: Van Nostrand Reinhold Co. Inc., 1987.

Bötticher, W.: *Technologie der Pilzverwertung.* Stuttgart: Verlag Eugen Ulmer, 1974.

Bruchmann, E.-E.: *Angewandte Biochemie.* Stuttgart: Ulmer, 1976.

Bundesgesundheitsamt (Publisher and Editor): *Amtliche Sammlung von Untersuchungsverfahren nach §35 LMBG.* Berlin, Köln: Beuth Verlag GmbH. Continuously extended.

Bünemann, G. and Hansen, H.: *Frucht- und Gemüselagerung – Eine Anleitung für den Lagerwart.* Stuttgart: Verlag Eugen Ulmer, 1973.

Cerny, G.: *RGV-Handbuch Verpackung. Die Lebensmittelverpackung aus mikrobiologischer Sicht.* Berlin: Verlag Erich Schmidt GmbH, 1986.

CFR: *Current Good Manufacturing Practice in Manufacturing, Packing or Holding Human Food; Title 21; Code of Federal Regulations.* Chapter 1, Part 110.10. Washington, DC: US Government Printing Office, 1987.

Chichester, D. F. and Tanner, F. W., Jr.: "Antimicrobial Food Additives". In Furia, T. E. (ed.): *CRC Handbook of Food Additives. 2nd Ed., Volume 1.* Chapter 3. Cleveland, Ohio: CRC Press, Inc., 3rd Printing, 1977.

Classen, H.-G., Elias, P. S. and Hammes, W. P.: *Toxikologisch-hygienische Beurteilung von Lebensmittelinhalts- und -zusatzstoffen sowie bedenklicher Verunreinigungen.* Berlin and Hamburg: Verlag Paul Parey, 1987.

Console, M. O. and Cowman, S.: "Bottled Water". In Speck, M. L. (ed.): *Compendium of*

Methods for the Microbiological Examination of Foods. 2nd Ed. Chapter 57. Washington, DC: APHA, 1984.

Crueger, W. and Crueger, A.: *Lehrbuch der angewandten Mikrobiologie.* Wiesbaden: Akademische Verlagsgesellschaft, 1982.

D'Aoust, J. Y.: *"Salmonella".* In Doyle, M. P. (ed.): *Foodborne Bacterial Pathogens.* Chapter 9. New York: Marcel Dekker, Inc., 1989.

Demain, A. L. and Solomon, N. A.: *Manual of Industrial Microbiology and Biotechnology.* Washington, DC: ASM, 1986.

DFG, Deutsche Forschungsgemeinschaft: *Maximale Arbeitsplatzkonzentrationen und Biologische Arbeitsstofftoleranzwerte 1985.* Weinheim: VCH, 1985.

DFG, Deutsche Forschungsgemeinschaft: *Maximale Arbeitsplatzkonzentrationen und Biologische Arbeitsstofftoleranzwerte 1990.* Weinheim: VCH, 1990.

Diehl, J. F.: *Safety of Irradiated Food.* New York, Basel: Marcel Dekker, Inc., 1989.

Dittrich, H. H.: *Mikrobiologie des Weines. 2nd Ed.* Stuttgart: Ulmer, 1987.

Domsch, K. H., Gams, W., Anderson, Traute-H.: *Compendium of Soil Fungi. Vol. 1 and 2.* London, New York, Toronto, Sidney, San Francisco: Academic Press, 1980.

Dörfelt, H. (ed.): *Lexikon der Mykologie.* Stuttgart, New York: Gustav Fischer Verlag, 1989.

Doyle, M. P.: *Foodborne Bacterial Pathogens.* New York: Marcel Dekker, Inc., 1989.

Doyle, M. P. and Padhye, V. V.: *"Escherichia coli".* In Doyle, M. P. (ed.): *Foodborne Bacterial Pathogens.* New York: Marcel Dekker, Inc., 1989.

Drews, G.: *Mikrobiologisches Praktikum. 4th Ed.* Berlin, Heidelberg: Springer Verlag, 1983.

Ellis, M. P.: "Dematiaceus Hyphomycetes". Commonw.: Mycol. Inst. Kew, Surrey, 1971.

Evered, D. and Whelan, J. (ed.): *Microbial Toxins and Diarrhoeal Disease.* Ciba Foundation Symposium 112. London: Pitman, 1985.

Favero, M. S., Gabis, D. A. and Vesley, D.: "Environmental Monitoring Procedures". In Speck, M. L. (ed.): *Compendium of Methods for the Microbiological Examination of Foods. 2nd Ed.* Chapter 3. Washington, DC: APHA, 1984.

FDA: *Bacteriological Analytical Manual,* 1984.

Fehrenbach, F. J. (ed.): *Bacterial Protein Toxins. Zentralblatt für Bakteriologie, Supplements. Vol. 17.* Stuttgart, New York: Gustav Fischer Verlag, 1988.

Fokkema, N. J., Heuvel, J. van den (ed.): *Microbiology of the Phyllosphere.* Cambridge a. o.: Cambridge University Press, 1988.

Frank, H. K.: *Aflatoxine-Bildungsbedingungen, Eigenschaften und Bedeutung in der Lebensmittelwirtschaft.* Hamburg: Behr's, 1974.

Frazier, W. C.: *Food Microbiology. 2nd Ed.* New York: McGraw-Hill Book Company, 1967.

Fricker, A.: *Lebensmittel mit allen Sinnen prüfen.* Berlin, Heidelberg, New York, Tokyo: Springer Verlag, 1984.

Fülgraff, G. (ed.): *Lebensmittel-Toxikologie.* UTB 1515. Stuttgart: Ulmer, 1989.

Furia, T. E.: *Handbook of Food Additives. 2nd Ed., Vol. I.* Cleveland, Ohio: CRC Press, Inc., 1977.

Gardner, Wm. Howlett: "Acidulants in Food Processing". In Furia, T. E. (ed.): *CRC Handbook of Food Additives. 2nd Ed., Vol. I.* Chapter 5. Cleveland, Ohio: CRC Press, Inc., 1977.

GDCh-Fachgruppe: "Lebensmittelchemie und gerichtliche Chemie". *Enzympräparate – Standards für die Verwendung in Lebensmitteln.* Hamburg: B. Behr's Verlag, 1983.

Gedek, B.: *Kompendium der medizinischen Mykologie.* Berlin, Hamburg: Parey, 1980.

Gilliland, S. E., Speck, W. E. and Vedamuthu, E. R.: "Acid-producing Microorganisms". In

Speck, M. L. (ed.): *Compendium of Methods for the Microbiological Examination of Foods. 2nd Ed.* Chapter 16. Washington, DC: APHA, 1984.

Glaubitz, M., Koch, R.: *Atlas der Gärungsorganismen. 4th Ed.* Revised by G. Bärwald. Berlin, Hamburg: Publisher Paul Parey, 1983.

Gottschalk, G., Beyreuther, K., Fritz, H. J., Gronenborn, B., Hammes, W., Kula, Maria Reginia, de Meijere, A., Vogel, S., Wandrey, Ch. *Biotechnologie.* Köln: Publishing Company Schulfernsehen – vgs – 1986.

Grafe, A.: *Viren – Parasiten unseres Lebensraumes.* Berlin, Heidelberg, New York: Springer Verlag, 1977.

Haensch, G. and Haberkamp de Anton, G.: *Wörterbuch der Biologie* (englisch – deutsch – französisch – spanisch). München, Bern, Wien: BLV Verlagsges.m.b.H., 1976.

Hahn, P., Muermann, B.: *Praxis-Handbuch Lebensmittelrecht.* Hamburg: Behr's Verlag, 1987.

Hatcher, W. S., Jr., Hill, E. C., Splittstoesser, D. F. and Weihe, J. L.: "Fruit Beverages". In Speck, M. L. (ed.): *Compendium of Methods for the Microbiological Examination of Foods. 2nd Ed.* Chapter 47. Washington, DC: APHA, 1984.

Hauschild, A. H. W.: "*Clostridium botulinum*". In Doyle, M. P.: *Foodborne Bacterial Pathogens.* Chapter 4. New York: Marcel Dekker, Inc., 1989.

Hayes, W. A.: "Biology and Technology of Mushroom Culture". In Wood, B. W. B. (ed.): *Microbiology of Fermented Foods. Vol. 2.* London and New York: Elsevier Applied Science Publishers, 1985.

Heinze, K. (ed.): *Leitfaden der Schädlingsbekämpfung. Band IV: Vorrats- und Materialschädlinge (Vorratsschutz).* Stuttgart: Wiss. Verlagsges, 1983.

Heiss, R., Eichner, K.: *Haltbarmachen von Lebensmitteln.* Berlin, Heidelberg, New York, Tokyo: Springer Verlag, 1984.

Henze, J. and Hansen, H.: *Lagerräume für Obst und Gemüse.* Darmstadt: KTBL-Schrift 327, 1988.

HMSO: *The Food Act.* London: HMSO, 1984.

HMSO: *The Food Hygiene General Regulations.* Statutory Instruments, S1, 1970, No. 1172. London: HMSO, 1970.

Hobbs, W. E. and Greene, V. W.: "Cereal and Cereal Products". In Speck, M. L. (ed.): *Compendium of Methods for the Microbiological Examination of Foods. 2nd Ed.* Chapter 51. Washington, DC: APHA, 1984.

ICMSF: *Microbial Ecology of Foods. Volume 1: Factors Affecting Life and Death of Microorganisms. Volume 2: Food Commodities.* New York, London, etc.: Academic Press, 1980.

ICMSF: *Microorganisms in Foods, 2. 2nd Ed. Sampling for Microbiological Analysis: Principles and Specific Applications.* Toronto, Canada: Univ. of Toronto Press, 1984.

Johnson, H., Kruger, A.: *Das grosse Buch vom Wein.* München: Gräfe und Unzer Verlag (no year).

Joly, P.: *Le genre* Alternaria. Paris: Lechevalier, 1964.

Kader, A. A., Kasmire, R. F., Mitchell, F. G., Reid, M. S., Sommer, N. F., Thompson, J. F.: *Postharvest Technology of Horticultural Crops.* University of California: Special Publication 3311, 1985.

Kielwein, G. and Luh, H. K.: *Internationale Käsekunde.* Essen: Magnus Verlag (no year).

Knothe, H., Naumann, P., Riecken, E. O., Schönfeld, H. (ed.): *Infektionen des Gastro-Intestinaltraktes.* Basel: Editiones Roche, 1978.

Korab, H. E. and Dobbs, W.: "Soft Drinks". In Speck, M. L. (ed.): *Compendium of Methods for the Microbiological Examination of Foods. 2nd Ed.* Chapter 56. Washington, DC: APHA, 1984.

Krämer, J.: *Lebensmittel-Mikrobiologie.* UTB 1421. Stuttgart: Publisher Eugen Ulmer, 1987.

Kröll, K.: *Trocknungstechnik. 2nd Volume: Trocknerund Trocknungsverfahren.* Berlin, Heidelberg, New York: Springer 1978.

Kunz, B.: *Grundriss der Lebensmittel-Mikrobiologie.* Hamburg: Behr's Verlag, 1988.

Labbe, R.: "*Clostridium perfringens*". In Doyle, M. P. (ed.): *Foodborne Bacterial Pathogens.* New York, Basel: Marcel Dekker, Inc., 1989.

Lück, E.: *Fachwörterbuch des Lebensmittelwesens* (engl. – german). Wiesbaden: Brandstetter Verlag, 1963.

Madden, J. M., McCardell, B. A. and Morris, J. G., Jr: "*Vibrio cholerae*". In Doyle, M. P. (ed.): *Foodborne Bacterial Pathogens.* New York, Basel: Marcel Dekker, 1989.

Mehlmann, I. J.: "Coliforms, Fecal Coliforms, *Escherichia coli* and Enteropathogenic *E. coli*." In Speck, M. L. (ed.): *Compendium of Methods for the Microbiological Examination of Foods. 2nd Ed.* Chapter 25. Washington, DC: APHA, 1984.

Mitscherlich, E., Marth, E. H.: *Microbial Survival in the Environment.* Berlin, Heidelberg, New York, Tokyo: Springer Verlag, 1984.

Moore-Landecker, E.: *Fundamentals of the Fungi.* Englewood Cliffs, NJ: Prentice-Hall, Inc., 1972.

Mossel, David A. A.: *Microbiology of Foods. Occurrence, Prevention and Monitoring of Hazards and Deterioration.* Utrecht: The University of Utrecht, 1977.

Mossel, David A. A.: *Microbiology of Foods. The Ecological Essentials of Assurance and Assessment of Safety and Quality.* Utrecht: University of Utrecht, 1982.

Müller, E. and Loeffler, W.: *Mykologie. 4th Ed.* Stuttgart, New York: Thieme Verlag, 1982.

Müller, G.: *Grundlagen der Lebensmittelmikrobiologie. 5th rev. Ed.* Darmstadt: Dr. Dietrich Steinkopff Verlag, 1983.

Müller, G. in cooper. with Lietz, P. and Munch, H.-D.: *Mikrobiologie pflanzlicher Lebensmittel. 3rd fully rev. Ed.* Darmstadt: Dr. Dietrich Steinkopff Verlag, 1983a.

Müller, G. (ed.): *Wörterbücher der Biologie/Mikrobiologie.* UTB 1024. Stuttgart, New York: Gustav Fischer Verlag, 1980.

Nicolaisen-Scupin, Lieselotte. New edition and extended by Hansen, H.: *Leitfaden für Lagerung und Transport von Gemüsen und essbaren Früchten. 4th Ed.* Wolfsburg: Publisher Günter Hempel, 1985.

Obst, Ursula and Holzapfel-Pschorn, Annette: *Enzymatische Tests für die Wasseranalytik.* München, Wien: R. Oldenbourg Verlag, 1988.

Peppler, H. J.: "Spices and Gums". In Speck, M. L. (ed.): *Compendium of Methods for the Microbiological Examination of Foods. 2nd Ed.* Chapter 48. Washington, DC: APHA, 1984.

Pichhardt, K.: *Lebensmittelmikrobiologie – Grundlagen für die Praxis. 2nd Ed.* Berlin, Heidelberg: Springer Verlag, 1989.

Pitt, J.: *The Genus* Penicillium *and Its Teleomorphic States* Eupenicillium *and* Talaromyces. London: Academic Press, 1979.

Pitt, J. I., Hocking, Ailsa D.: *Fungi and Food Spoilage.* Sidney u. a.: Academic Press, 1985.

Preussmann, R. (ed.): *Das Nitrosamin-Problem (Report of DFG).* Weinheim: Publisher Chemie, 1983.

Priest, F. G. and Campbell, I.: *Brewing Microbiology.* Amsterdam: Elsevier, 1987.

Raab., P.: *Pimaricin (Natamycin).* Stuttgart: Georg Thieme Verlag, 1974.

Raper, K. B., Fennell, D. I.: *The Genus* Aspergillus. Baltimore: Williams and Wilkins, 1965.

Raper, K. B., Thom, C., Fennell, D. I.: *A Manual of the Penicillia*. Baltimore: Williams and Wilkins, 1949. Reprint 1968.

Rehm, H.-J.: *Industrielle Biotechnologie. 3rd Ed.* Berlin, Heidelberg, New York: Springer Verlag, 1985.

Reiss, J. (ed.): *Mykotoxine in Lebensmitteln*. Stuttgart, New York: Gustav Fischer Verlag, 1981.

Reiss, J.: *Schimmelpilze*. Berlin, Heidelberg, New York, Tokyo: Springer Verlag, 1986.

Reuter, H. (ed.): *Aseptic Packaging of Food*. Lancaster, PA: Technomic Publishing Company, Inc., 1989.

Reuter, H. (ed.): *Aseptisches Verpacken von Lebensmitteln*. Hamburg: Behr's Verlag, 1987.

Rhodes, Martha E. (ed.): *Food Mycology*. Boston, MA: G. K. Hall and Co., 1979.

Roberts, E. H. (ed.): *Viability of Seeds*. London: Chapham and Hall, 1972.

Rodricks, J. V., Hesseltine, C. W., Mehlmann, M. A. (ed.): *Mycotoxins in Human and Animal Health*. Park Forest South, IL: Pathotox Publishers, Inc., 1977.

Roth, L., Daunderer, M. and Kormann, K.: *Giftpflanzen, Pflanzengifte – Vorkommen, Wirkung, Therapie*. Landsberg, München: ecomed verlagsges.mbH, 1987.

Roth, L.: *Sicherheitsdaten – MAK-Werte. 7th Ed.* Landsberg, München: ecomed verlagsges.mbH, 1989.

Schachta, M. and Jorde, W.: *Allergische Erkrankungen durch Schimmelpilze*. München-Deisenhofen: Dustri-Verlag Dr. Karl Feistle, 1989.

Schauff, M.: *Verordnung über Milcherzeugnisse. In der Fassung der 5. Änderungsverordnung vom 3.12.1987*. Textedit. with explan. Hamburg: Behr's Verlag, 1988.

Schlegel, H. G.: *Allgemeine Mikrobiologie. 6th Ed.* Stuttgart: Georg Thieme Verlag, 1985.

Schmidt-Lorenz, W.: "Behaviour of Microorganisms at Low Temperatures". *Bull. de l'Inst. Internat. du Froid, Paris*: No. 1 and 2, 1967.

Schmidt-Lorenz, W.: *Sammlung von Vorschriften zur mikrobiologischen Untersuchung von Lebensmitteln*. 4 Volumes. Weinheim: Publisher Chemie, 1980 – 1983.

Seeliger, H. P. R.: *Entstehung und Verhütung von mikrobiellen Lebensmittelinfektionen und -vergiftungen. 2nd Ed.* Paderborn: Ferdinand Schöningh, 1977.

SGLH/Schweizerische Gesellschaft für Lebensmittelhygiene (ed.): *Neue Erkenntnisse über die Erreger bakterieller Lebensmittelinfektionen*. Heft 14 der Schriftenreihe, Okt., 1984.

SGLH/Schweizerische Gesellschaft für Lebensmittelhygiene (ed.): *Was müssen Vertrieb und Konsument über die hygienische Behandlung von Lebensmitteln wissen?* Heft 12 der Schriftenreihe, Okt., 1982.

Shapton, D. A. and Shapton, N. F.: *Principles and Practices for the Safe Processing of Foods*. Halley Court, Jordan Hill, Oxford: Butterworth-Heinemann, 1991.

Sinell, H.-J.: *Einführung in die Lebensmittelhygiene. 2nd Ed.* Berlin, Hamburg: Publ. Paul Parey, 1985.

Smith, A. L.: *Microbiology Laboratory Manual and Workbook. 5th Ed.* St. Louis, Toronto, London: The C. V. Mosby Company, 1981.

Speck, M. L.: *Compendium of Methods for the Microbiological Examination of Foods. 2nd Ed.* Washington, DC: APHA, 1984.

Spicher, G., Stephan, H.: *Handbuch Sauerteig – Biologie, Biochemie, Technologie. 3rd Ed.* Hamburg: Behr's Verlag, 1987.

Steinkraus, K. H. (ed.): *Handbook of Indigenous Fermented Foods*. New York, Basel: Marcel Dekker, 1983.

Steyn, P. S., Vleggaar, R. (ed.): *Mycotoxins and Phycotoxins*. Amsterdam, Oxford, New York, Tokyo: Elsevier, 1986.

Stiles, M.: "Less Recognized or Presumptive Foodborne Pathogenic Bacteria". In Doyle, M. P. (ed.): *Foodborne Bacterial Pathogens.* Chapter 16. New York, Basel: Marcel Dekker, Inc., 1989.

Teuber, M.: *Grundriss der praktischen Mikrobiologie für das Molkereifach. 2nd Ed.* Gelsenkirchen-Buer: Publisher Th. Mann, 1987.

Underkofler, L. A.: "Enzymes". In Furia, T. E. (ed.): *CRC Handbook of Food Additives. 2nd Ed, Vol. 1.* Chapter 1. Cleveland, OH: CRC Press, Inc., 3rd Printing, 1977.

Vries de, G. A.: "Contribution to the Knowledge of the Genus *Cladosporium* Link ex." *Fr. Bibl. Mycol., Bd. III.* Lehre: Cramer, 1967.

Wallhäusser, K. H.: *Praxis der Sterilisation, Desinfektion – Konservierung – Keimidentifizierung – Betriebshygiene. 4th Ed.* Stuttgart, New York: Georg Thieme, 1988.

Weichmann, J. (ed.): *Postharvest Physiology of Vegetables.* New York, Basel: Marcel Dekker, Inc., 1987.

WHO Technical Report Series No. 598: *Microbial Aspects of Food Hygiene.* Report of a WHO Expert Committee w. t. partic. of FAO Geneva, 1976.